IET TRANSPORTATION SERIES 11

Navigation and Control of Autonomous Marine Vehicles

Other volumes in this series:

Volume 1 **Clean Mobility and Intelligent Transport Systems** M. Fiorini and J.-C. Lin (Editors)
Volume 2 **Energy Systems for Electric and Hybrid Vehicles** K.T. Chau (Editor)
Volume 5 **Sliding Mode Control of Vehicle Dynamics** A. Ferrara (Editor)
Volume 6 **Low Carbon Mobility for Future Cities: Principles and applications** H. Dia (Editor)
Volume 7 **Evaluation of Intelligent Road Transportation Systems: Methods and results** M. Lu (Editor)
Volume 8 **Road Pricing: Technologies, economics and acceptability** J. Walker (Editor)
Volume 9 **Autonomous Decentralized Systems and Their Applications in Transport and Infrastructure** K. Mori (Editor)
Volume 12 **EMC and Functional Safety of Automotive Electronics** K. Borgeest
Volume 38 **The Electric Car** M.H. Westbrook
Volume 45 **Propulsion Systems for Hybrid Vehicles** J. Miller
Volume 79 **Vehicle-to-Grid: Linking electric vehicles to the smart grid** J. Lu and J. Hossain (Editors)

Navigation and Control of Autonomous Marine Vehicles

Edited by
Sanjay Sharma and Bidyadhar Subudhi

The Institution of Engineering and Technology

Published by The Institution of Engineering and Technology, London, United Kingdom

The Institution of Engineering and Technology is registered as a Charity in England & Wales (no. 211014) and Scotland (no. SC038698).

First published 2019

The Institution of Engineering and Technology
Michael Faraday House
Six Hills Way, Stevenage
Herts, SG1 2AY, United Kingdom

www.theiet.org

British Library Cataloguing in Publication Data
A catalogue record for this product is available from the British Library

ISBN 978-1-78561-338-8 (hardback)
ISBN 978-1-78561-339-5 (PDF)

Typeset in India by MPS Limited

Contents

Preface

Recent years have witnessed a rapid development of control, communication, sensor and computing technologies and their successful applications in many real-world problems including control and navigation of unmanned vehicles. The aforesaid technological advances have led to development of autonomous vehicles that operate autonomously which can drive of their own without necessitating human intervention or supervision. With the advance of technologies, increasing attention is being given to the research on unmanned vehicles operating in different environmental domains including land (unmanned ground vehicle (UGV)), aerospace (unmanned aerial vehicle (UAV)) and sea (Autonomous underwater vehicle (AUV) and unmanned surface vehicle (USV)). It is however interesting to see that despite the equal importance for the development of all types of vehicles, technologies developed for marine vehicles, especially for USV, are less mature than that of other vehicles, which is in part due to the reasons of complexity of the environment i.e. marine vehicles operate at the interface of two fluid systems. Ocean covers about two-thirds of the earth. It has a great effect on the future existence of all human beings. Marine robotic vehicles including unmanned surface vehicles (USVs) and unmanned underwater vehicles (UUVs) can help us better understand marine and other environmental issues, protect the ocean resources of the earth from pollution, and efficiently utilize them for human welfare. However, there are lot issues involved for navigation, control and communication of these vehicles. This book covers different aspects of recent advances on navigation and control of marine vehicles.

Chapter 1 provides an insight into autonomous marine vehicles with particular focus on USVs by discussing the design of an autonomous navigation system. It provides a general overview of the development of USV and the motion planning (or path planning) strategies demonstrating how a practical USV is able to autonomously navigate itself on a lake. The goal of path planning of marine vehicles is to find a collision free path from start to goal point in an uncertain environment comprising of static and moving obstacles. There are umpteen approaches adopted in literature towards path planning of marine robots. These approaches are adopted from the domain of ground and aerial robots. **Chapter 2** on Efficient Optimal Path Planning of Unmanned Surface Vehicles presents an overview towards application of grid-based path planners in the area of optimal path planning of unmanned surface vehicles (USVs). Collision Avoidance of Maritime Vessels in **Chapter 3** discusses about an artificial potential field based online collision avoidance system for manned and unmanned maritime vehicles which is capable of reacting to static and dynamic obstacles in the vicinity. The standard marine 'rules of the road' are integrated into the collision avoidance framework. A risk assessment module is also introduced which is based on the standard closest point of approach method. A decision maker then selects appropriate rules based on relative heading and positions of the vessels. For the detection part, an integrated vision and laser-based system has been developed to provide sensing functionality for multiple obstacles.

Autonomous craft are preferred over manned vessels in scientific operations as they are more suited for long enduring and tedious missions in dangerous or hazardous environments. The technique presented is fairly generic and is applicable to a general class of marine vehicles ranging from a small/medium-sized craft to a large freighter or an oil tanker. The small/medium-sized vehicles have been widely employed in surveillance and scientific missions. For surveillance missions, deploying autonomous vehicles will maximise the coverage and reduce the number of personnel involved in the operation area. Simulation results are provided to include the three fundamental collision encounter scenarios i.e. overtaking, head-on and crossing. It is also shown that Dubin circles could be successfully employed to take the dynamics of the craft into account which provides a general method independent of the craft size. **Chapter 4** on Sliding mode control for path planning guidance of marine vehicles proposes implementation of the guidance commands in underwater scenarios where information about the target point is unavailable after some time. In such cases, the open loop control method is the only way to intercept the targets. A natural way of its implementation is sample and hold technique which does not yield satisfactory performance with its usual sampling method. This chapter proposes different sampling techniques to implement the guidance law in open loop. These methods are based on sampling the guidance command generated using closed loop feedback with last known estimates about the target. To ensure robustness, a sliding mode control based impact angle guidance law is used. Performance of the guidance scheme using proposed sampling methods are evaluated through extensive numerical simulations for the different engagement scenarios and shown to work well. Operating a flight style autonomous underwater vehicle in close proximity to a terrain is very often completely reliant on the vehicle sensors for terrain detection. These pose challenges on the manoeuvrability of such vehicles, which are also required to be energy efficient. **Chapter 5** on experimentally based Analysis of low altitude terrain following by Autonomous Underwater Vehicles illustrates recent experimentally based results on the performance of such vehicles. These results use the fully understood environment of a lake and are for the critical tasks of repeatability, obstacle detection and the actuation strategy used. **Chapter 6** on Nonlinear H^{∞} Control of Autonomous Underwater Vehicles is about designing of nonlinear state and output feedback H^{∞} control algorithms for an Autonomous Underwater Vehicle (AUV) for both vertical and horizontal planes which will lead to design a 3D path following control. A three-degree-of-freedom (3-DOF) nonlinear model of an AUV has been considered in both horizontal and vertical planes for developing the diving and steering control law respectively. From the results obtained, it is concluded that the proposed robust control algorithms exhibit a good tracking performance ensuring internal stability and significant disturbance attenuation. **Chapter 7** describes a real-time energy optimal trajectory planning and tracking method for an uninhabited surface vehicle (USV) in dynamically changing environment. The problem is first discretized by Legendre pseudospectral method to transform the trajectory planning problem into a nonlinear programming problem with less discrete points and higher accuracy compared with traditional discretization methods. Then a novel strategy is developed to re-plan the trajectory in real-time during the entire manoeuvring. The re-planning strategy uses the values obtained in the last re-planning as its initial guess values and upper and lower bound, which can reduce the computational complexity dramatically. Simulation results show that this novel approach could save a considerable amount of equivalent energy compared with an existing traditional method. **Chapter 8** on Cooperative Path-Following Control with Logic-Based Communications: Theory and Practice introduces an event-driven, logic-based communication system for decentralized control of a network of nonlinear systems (agents) with the objective of driving their outputs along predefined paths at desired speeds, while holding a desired formation pattern compatible with the paths. An extended cooperative path following framework is adopted where communications among agents take place at discrete time instants, instead of continuously. The set-up derived is used to solve the problem of cooperative path-following control of multiple underactuated autonomous marine vehicles. The results of experimental field tests with a group of marine vehicles are presented and discussed.

Marine cyber physical system motivated by recent fast paced developments cover quite a broad range of control, communication and computing technologies that is changing how current marine vehicles are designed and operated. Special emphasis is given to design of coordinated and cooperative navigation, guidance, and fault tolerant and reconfigurable systems for various autonomous marine vehicles (AMVs). Formation is one of the effective ways to achieve cooperative control. **Chapter 9** presents a review on various formation control and coordination strategies of multiple AMVs. Advantages and disadvantages of the control strategies for multiple agents cooperative motion together with the network and communication issues in formation are also discussed. **Chapter 10** on Hydro-acoustic Communications and Networking in Contemporary Underwater Robotics: Instruments and Case Studies describes the results of an easy-to-achieve implementation of an underwater acoustic sensor network capable of operating in different modes for the exchange of underwater acoustic data among mobile and/or stationary network nodes that may include autonomous surface vehicles (ASVs), autonomous underwater vehicles (AUVs), remotely operated vehicles (ROVs), and benthic stations. The data exchanged are crucial in a number of applications that include cooperative multiple vehicle navigation and control, mission status assessment, and environmental sensing. A key novelty of the implementation described is a cost effective and compact Under-Water Acoustic (UWA) modems incorporating the open-source, open-architecture software. Apart from providing details on the application of the EviNS Framework, the chapter offers a detailed analysis of the performance of an ad-hoc underwater sensor network operating in a shallow water area, i.e. in a hydro-acoustic environment of large practical interest.

Historically, the development and adoption of autonomous surface vehicles (ASVs) has not drawn the same attention as air, land or even sub surface systems. However, in the last ten years it has been one of the fastest and most exciting areas of robotics and autonomy to observe and partake in. The topic is made even more exciting by the fact that there is still so much to do providing benefits to safety, efficiency and productivity. **Chapter 11** focusses on the commercial applications of Autonomous Surface Vessel (ASV) technology. These are wide in terms of complexity, value, water depth, distance from shore and frequency of operation. Many of the companies and research organisations involved with ASV development have very significant overlaps between technologies, equipment and techniques.

The Maritime Environment and associated industries and operations are vast and there is always a new opportunity for marine robotic systems to make their impact. Thus it is not possible to cover all applications of the marine robotic technology in this book. Further, the pace of development is so fast that we could discuss some of the newest applications in the time frame that would be of interest to a large spectrum of readers of this research monograph.

Sanjay Sharma
Bidyadhar Subudhi
Editors

Chapter 1

Modelling and control of autonomous marine vehicles

Yuanchang Liu[1] *and Richard Bucknall*[1]

Abstract

With the advance of artificial intelligence technologies, increasing deployment of autonomous marine vehicles is taking place especially in the application for executing demanding tasks. This chapter provides an insight into autonomous marine vehicles with a special focus on unmanned surface vehicles (USVs) and the design of an autonomous navigation system. After providing a general review on the development of USVs in both civil and military applications, methods employed for modelling and controlling USVs are specifically introduced and discussed. Particular interest has been cast on the motion planning of USVs with a series of simulation results presented to showcase the effectiveness of the autonomous navigation system in a practical maritime environment.

1.1 Introduction

With the rapid advance of technologies, increasing attention has been given to the research of unmanned vehicles operating in different environmental domains including land (unmanned ground vehicle (UGV)), aerospace (unmanned aerial vehicle (UAV)), sea (Autonomous underwater vehicle (AUV) and unmanned surface vehicle (USV)). It is, interesting to see that despite the equal importance for the development of all types of autonomous vehicles, technologies developed for marine vehicles, especially for USV, are less mature than that of other vehicles, which is partially due to the reasons of complexity of the marine environment and the needs of public and private sector. This chapter provides an insight into autonomous marine vehicles with particular focus on the USV by discussing the design of an autonomous navigation system. It will provide a general overview of the development of USV and introduce the modelling and control methods. In terms of the control strategies, the motion planning (or path planning) algorithms will be specifically discussed together

[1]Department of Mechanical Engineering, University College London, UK

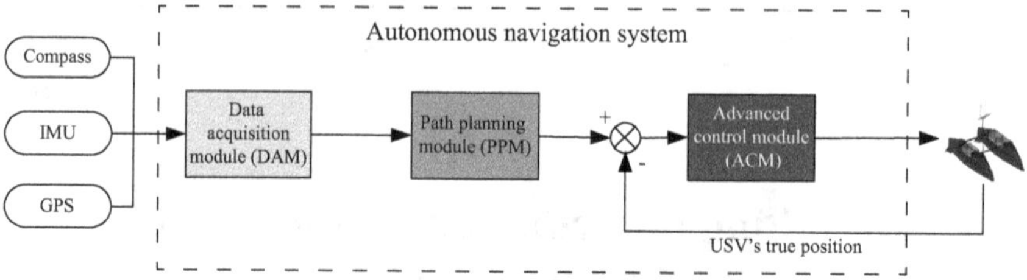

Figure 1.1 The autonomous navigation system for a USV

with simulation results demonstrating how a practical USV is able to navigate itself autonomously on an open sea area.

1.1.1 USV prototypes and core systems

USVs were first proposed and developed for military operations during World War II and mainly used as gunnery and missile targets; whereas, the educational and civilian deployment of USVs began to emerge in the 1970s [1]. From the 1990s, a number of important research programmes, such as the OWL MK II and the Roboski [2], greatly promoted the development of USVs in terms of the mission capability and vehicle autonomy, which has further increased the ability and application of USVs to include ocean exploration missions, coastal line patrol and environment monitoring tasks.

To intelligently and robustly guide and control a USV, the autonomous navigation system is vitally important (a generic system structure is shown in Figure 1.1). Within the system, the data acquisition module (DAM) monitors and obtains real-time traffic information from various marine devices to determine the USV's as well as other vessels' positions, courses and velocities. Using the information fed from the navigational sensors, the path planning module (PPM) is then responsible for generating safe trajectory in accordance with the specific mission requirements. The advanced control module (ACM) system finally calculates appropriate control commands so that a trajectory can be followed robustly and accurately.

Figure 1.2 provides an overview of some key USV prototypes as well as their corresponding navigation systems. Most of the early development of USVs focused on the navigation and control modules with the main aim of letting the vehicle follow a predefined trajectory accurately. Only simple missions could be executed because the system lacked path planning as well as collision avoidance capabilities. Since 2005, with the advance of the hardware technologies, high-level decision making systems using artificial intelligence have been implemented into the NGC system, which improves the autonomy of the most USV platforms. Obstacle detection and avoidance capabilities have been successfully integrated making the USV capable of undertaking complex missions. The following subsections review representative prototypes of USVs covering the areas ranging from military to educational and civilian applications.

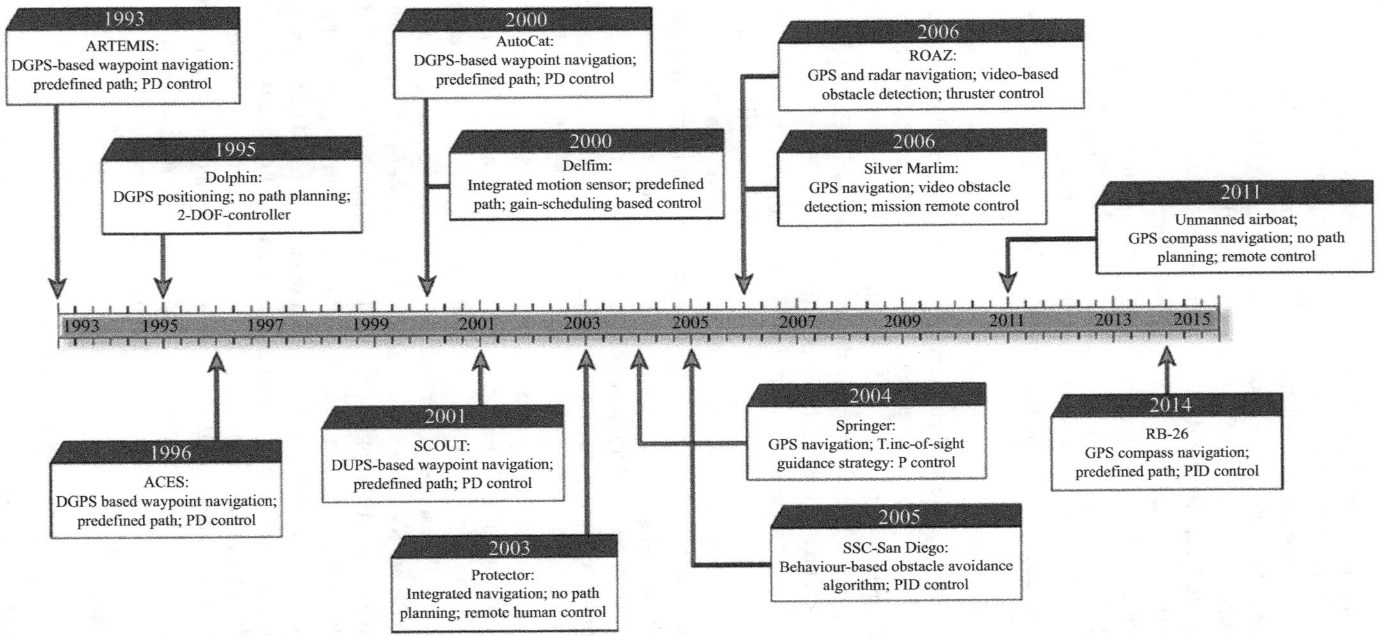

Figure 1.2 Summary of key USV prototypes and their navigation systems

1.1.1.1 USVs for education and civil applications

For nonmilitary applications, USVs have primarily been developed by several leading universities and research institutions across the USA and Europe. For example, a group of USVs have been developed at Massachusetts Institute of Technology (MIT) consisting of the fishing trawler-like vehicle *ARTEMIS*, the catamarans Autonomous Coastal Exploration System (*ACES*), *AutoCat* [3] and the kayak *SCOUT* [4]. These USVs demonstrated the feasibility of the automatic heading control using the proportional-derivative (PD) actions and the waypoint based navigation using the Differential Global Positioning System (DGPS). The applications of the developed USVs include the hydrographic data collection and the robust localisation for undersea vehicles using the distributed acoustic navigation algorithms.

The autonomous catamaran *Delfim* (shown in Figure 1.3(a)) was originally developed by the DSOR lab of Lisbon (IST-ISR) as a communication relay to support AUV missions funded by the EU project ASIMOV [5] and was then improved and converted to a stand-alone unit for collecting bathymetric mapping and marine data. The *Delfim* was equipped with an attitude reference unit, a Doppler log and a DGPS [5], and the data from these sensors were fused and integrated to provide accurate navigational information. The gain-scheduling control theory was used to develop a path following controller to track a predefined reference path [6].

The autonomous catamaran *Springer* (shown in Figure 1.3(b)) was developed by the University of Plymouth (UK). *Springer* was designed primarily for undertaking pollutant tracking, environmental and hydrographical surveys in shallow water [7]. It also serves as a testbed platform for other academic and scientific institutions. The line-of-sight (LOS) guidance strategy utilised in the vehicle for navigation is integrated with the control system to perform various missions.

A radio controlled agricultural USV named *RB-26* (shown in Figure 1.3(c)) was developed by a Japanese company named Hokuto Yamar. The vehicle is mainly deployed in paddy fields with the purpose of autonomous weeding and fertilisation [8]. A GPS compass was installed to the USV to obtain position and heading information with precision of 1 m and 0.5°, respectively, and the USV was controlled using the PID techniques to follow a predefined path to investigate the paddy. It was demonstrated that the *RB-26* USV can automatically track the predefined path; however, the tracking performance can be impaired by oscillations generated by the wind.

Several other notable USVs developed for education and civil applications include: the Measuring *Dolphin* designed and developed by Majohr and Buch [9] at the University of Rostock in Germany, which featured high accuracy positioning and track guidance for shallow water operation; the autonomous catamaran *Charlie* (shown in Figure 1.3(d)) developed by CNR-ISSIA Genova (Italy) for the collection of sea surface biological micro-layer [10]; and the commercial USV developed by Twichell *et al.* [11] to generate a three-dimensional (3D) map of an oyster habitat and to map the shallow coast.

1.1.1.2 USVs for military applications

For security reasons, there is not a wealth of available literature on military USVs. However, some information on existing platforms can be found in the review papers

Figure 1.3 *USV platforms. (a) Delfim USV, (b) Springer USV, (c) RB-26 USV and (d) Charlie USV*

with brief introductions. For example, the *SSC San Diego* USV based upon the Bombardier *SeaDoo Challenger 2000* [12] was developed as a testbed for military purposes in the USA. The waypoint-based navigation system of *SSC San Diego* was achieved using two-level processes including the receiving and parsing the path message from an Operator Control Unit (OCU), and executing the Proportional-Integral-Derivative (PID) action. An obstacle avoidance algorithm has also been integrated for close range collision avoidance.

Other military based USV prototypes have been extensively reviewed in Caccia *et al.* [13]. Among them, the Israeli *Silver Marlin* USV has received a great deal of attention. Apart from its advanced autonomy and obstacle detection capability, i.e., using an obstacle avoidance sensor to create an overall picture of both moving and static obstacles in its surroundings areas, the vehicle has been used and operated as a part of multiple platform task force that included UAVs, combat ships and ground forces. Such a concept of deploying multiple vehicles cooperatively to conduct complex missions is becoming increasingly important in recent years. Compared with the single platform, the benefits of using multiple vehicles include the wide mission area, improved system robustness and increased fault-tolerant resilience.

1.1.2 The control strategies of USV

In the research community of unmanned vehicles (including USV, AUV, UAV and mobile robots), there are three widely adopted control strategies as follows:

- **Remote control strategy:** vehicles installed with autopilot are able to robustly execute the control command. However, there is a lack of decision making or planning system, meaning the vehicle is guided by human operators;

- **Semi-autonomous control strategy:** a higher level control strategy compared with the remote control. The vehicles are not only equipped with autopilot but also with decision making. However, only simple missions can be autonomously accomplished by the vehicles, and there is a need by bringing in the human operator especially during challenging situations.
- **Fully-autonomous control strategy:** the highest control strategy. The vehicle is able to fully autonomously execute all types of missions under most conditions without any intervention from human operators.

Figure 1.4 has compared these three different strategies from the autonomy level and system reliability/robustness perspectives. Even though the remote control strategy has the lowest autonomy level, because of the involvement of the human operator, such an approach can provide the highest system reliability especially for complex missions. On the contrary, the fully-autonomous strategy sacrifices the system reliability to achieve the high autonomy. In addition, in order to satisfactorily accomplish the mission using the full-autonomous control, there is a high demand in the fields of artificial intelligence technologies, advanced sensor suites, efficient data transmission channels and intelligent decision making algorithms.

The present research trend in the unmanned vehicle community is that the UAV and mobile robots are the pioneer in terms of the autonomy level (as marked by the red block in Figure 1.4); whereas the marine community still adopts the semi-autonomous

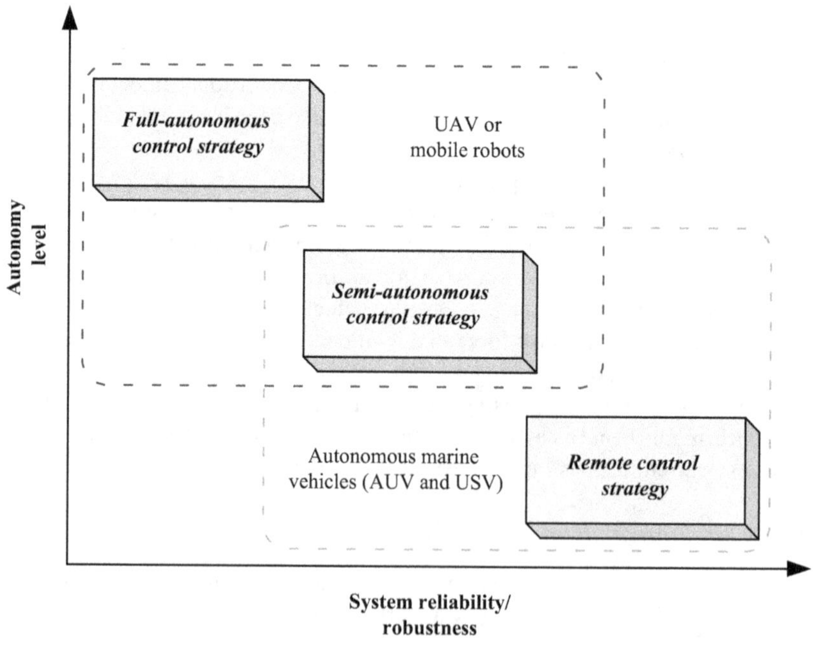

Figure 1.4 The autonomy level and the system reliability of remote, semi-autonomous and full-autonomous control strategies

or remote control strategies for most vehicles. The likely reason is that the popularity in the UAV or mobile robots for various civilian applications has largely promoted their associated research. Also, the easy access to the platforms of these two kinds of vehicle has made the verification of any new technologies much easier. Finally, the diverse nature of the maritime environment has imposed higher requirements on the development of autonomous marine vehicles, on not only the vehicle design, but its associated navigation systems.

1.2 Mathematical modelling of autonomous marine vehicles

The mathematical modelling of the motion of marine vehicles is important in order to accurately analyse the system and robustly control the vehicle. In general, a marine vehicle moves in 6 degrees of freedom (DOF) with 6 independent coordinates determining the position and orientation, respectively. As graphically shown in Figure 1.5, the position information is expressed in terms of the surge, the sway and the heave motion, which are expressed along the x, y and z axes with their time derivatives representing the translational motion. Similarly, the orientation motions are defined as the roll, the pitch and the yaw, respectively, and their time derivatives reveal the corresponding rotational motion.

The 6 DOF motion is a general form for all kinds of the marine vehicles including AUV and USV, and it can be further simplified according to the type of the vehicle or the vehicle's propulsion system. As listed in Table 1.1, for an AUV as it can freely move in a 3D environment, the 6 DOF can be used with each of these six motions been controlled. In contrast, the movement of surface vessel is constrained to the sea surface meaning its motion can be reduced to 3 DOF by neglecting the heave,

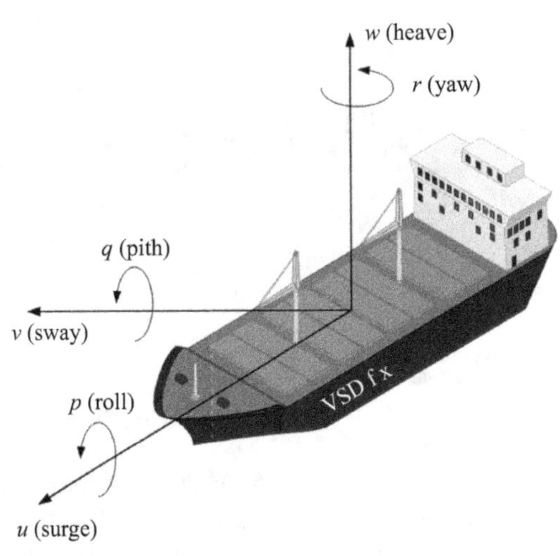

Figure 1.5 DOF motion of a surface vessel

Table 1.1 Motion types for different marine vehicle platforms

Vehicle type	Operating environment	DOF	Controlled motions
Underwater vehicles	3D environment	6 DOF	Sway, surge, heave, yaw, pitch and roll
Surface vehicles	2D environment	3 DOF	Surge, sway and yaw
Small scale USV	2D environment	3 DOF	Surge and yaw

pitch and roll motions. It is worth noting that for a small scale USV because of the capability of the installed propulsion system, even though the motion of the vehicle still belongs to 3 DOF, only the surge and yaw motion can be controlled making the vehicle under-actuated.

1.2.1 Kinematic motion of marine vehicle

When deriving the kinematic motion equation of a marine vehicle, without considering the forces acting on the vehicles that causes the motion, only the geometrical aspects of motion which are of interest are considered. However, due to the complexity of the motion, different coordinate frames to express the position and velocity of the vehicle are defined [14]. The commonly used are as follows:

- **Inertial coordinated frame** (denoted as <*e*>): It is a nonaccelerating reference frame where Newton's law of motion is applied with its origin located at the centre of the earth. Normally, the position as well as the orientation information is defined in <*e*> as

$$\boldsymbol{P}^e = \begin{bmatrix} x \\ y \\ z \end{bmatrix} \in \mathbb{R}^3, \tag{1.1}$$

$$\boldsymbol{\Theta}^e = \begin{bmatrix} \phi \\ \theta \\ \psi \end{bmatrix} \in \mathcal{S}^3, \tag{1.2}$$

  where $\boldsymbol{P}$ describes the position of the vehicle in <*e*>, while Θ reveals the orientation information.
- **Body coordinate frame** (denoted as <*b*>): This is a moving coordinate frame with its origin fixed to the vehicle. The *x*-axis is defined as being along with the transverse direction (sway motion), the *y*-axis points towards the forward direction (surge motion) and the *z*-axis is towards the vertical direction (heave

motion). The body frame is normally used in the implementation of the equation of motion with vehicle's linear and angular velocities being defined in <*b*> as

$$\boldsymbol{v}^b = \begin{bmatrix} u \\ v \\ w \end{bmatrix} \in \mathbb{R}^3, \tag{1.3}$$

$$\boldsymbol{\omega}^b = \begin{bmatrix} p \\ q \\ r \end{bmatrix} \in \mathcal{S}^3. \tag{1.4}$$

Based upon the position and velocities defined in earth and body frame respectively, the 6 DOF kinematic equation can be written as

$$\dot{\eta} = J(\eta)\boldsymbol{v}, \tag{1.5}$$

where η and $\boldsymbol{v}$ are

$$\eta = \begin{bmatrix} \boldsymbol{P}^e \\ \boldsymbol{\Theta}^e \end{bmatrix} \in \mathbb{R}^3 \times \mathcal{S}^3, \quad \boldsymbol{v} = \begin{bmatrix} \boldsymbol{v}^b \\ \boldsymbol{\omega}^b \end{bmatrix} \in \mathbb{R}^6. \tag{1.6}$$

$J(\eta)$ is the transformation matrix that is able to convert the velocity information expressed in the body frame to the earth frame. $J(\eta)$ can be written as

$$J(\eta) = \begin{bmatrix} \boldsymbol{R}_b^e(\boldsymbol{\Theta}) & 0_{3\times3} \\ 0_{3\times3} & \boldsymbol{T}_\Theta(\boldsymbol{\Theta}) \end{bmatrix} \tag{1.7}$$

with $\boldsymbol{R}_b^e(\boldsymbol{\Theta})$ for linear velocity transformation and $\boldsymbol{T}_\Theta(\boldsymbol{\Theta})$ for angular velocity transformation as

$$\boldsymbol{R}_b^e(\boldsymbol{\Theta}) = \begin{bmatrix} \cos(\psi)\cos(\theta) & \begin{matrix} -\sin(\psi)\cos(\phi) \\ +\cos(\psi)\sin(\theta)\sin(\phi) \end{matrix} & \begin{matrix} \sin(\psi)\sin(\phi) \\ +\cos(\psi)\cos(\phi)\sin(\theta) \end{matrix} \\ \sin(\psi)\cos(\theta) & \begin{matrix} \cos(\psi)\cos(\phi) \\ +\sin(\phi)\sin(\theta)\sin(\psi) \end{matrix} & \begin{matrix} -\cos(\psi)\sin(\phi) \\ +\sin(\theta)\sin(\psi)\cos(\phi) \end{matrix} \\ -\sin(\theta) & \cos(\theta)\sin(\phi) & \cos(\theta)\cos(\phi) \end{bmatrix} \tag{1.8}$$

$$\boldsymbol{T}_\Theta(\boldsymbol{\Theta}) = \begin{bmatrix} 1 & \sin(\phi)\tan(\theta) & \cos(\phi)\tan(\theta) \\ 0 & \cos(\phi) & -\sin(\phi) \\ 0 & \sin(\phi)/\cos(\theta) & \cos(\phi)/\cos(\theta) \end{bmatrix}. \tag{1.9}$$

For the surface vehicle, the 3 DOF kinematic motion equation can be simplified by neglecting the heave, roll and pitch motions, which yields

$$\boldsymbol{\nu} = \begin{bmatrix} u \\ v \\ r \end{bmatrix} \tag{1.10}$$

$$\boldsymbol{\eta} = \begin{bmatrix} x \\ y \\ \psi \end{bmatrix} \tag{1.11}$$

and the transformation matrix can now be written as

$$J(\psi) = \begin{bmatrix} \cos(\psi) & -\sin(\psi) & 0 \\ \sin(\psi) & \cos(\psi) & 0 \\ 0 & 0 & 1 \end{bmatrix}. \tag{1.12}$$

1.2.2 Dynamic motion of marine vehicle

The conventional way to express the 6 DOF non-linear dynamic motion of a marine vehicle is to use the equation as

$$M\dot{\nu} + C(\nu)\nu + D(\nu)\nu + g(\eta) = \tau + w, \tag{1.13}$$

where M is the symmetric positive-definite system inertia matrix, $C(\nu)$ is the Coriolis-centripetal matrix, $D(\nu)$ is the damping matrix and $g(\eta)$ is the vector of gravitational/buoyancy forces and moments [14]. On the left-hand side of the equation, all the forces that have effect on the motion of vehicle are listed including the vector of control input (τ) and the environmental disturbances caused by wind, waves and currents (w).

The system inertial matrix M is the sum of rigid body inertia matrix (M_{RB}) and the added mass matrix (M_A) as

$$M = M_{\mathrm{RB}} + M_A = \begin{bmatrix} m\boldsymbol{I}_{3\times 3} & -m\boldsymbol{S}(\boldsymbol{r}_g^b) \\ m\boldsymbol{S}(\boldsymbol{r}_g^b) & \boldsymbol{I}_0 \end{bmatrix} + \begin{bmatrix} \boldsymbol{A}_{11} & \boldsymbol{A}_{12} \\ \boldsymbol{A}_{21} & \boldsymbol{A}_{22} \end{bmatrix}, \tag{1.14}$$

where m is the mass of the vehicle, $\boldsymbol{I}_0 \in \mathbb{R}^3$ is inertial matrix, $\boldsymbol{r}_g^b$ is the coordinate of the centre of gravity defined as $[x_g \;\; y_g \;\; z_g]^T$. $\boldsymbol{S}(\boldsymbol{r}_g^b)$ is a skew-symmetric matrix having the form of $\boldsymbol{S}(\boldsymbol{r}_g^b) = \begin{bmatrix} 0 & -z_g & y_g \\ z_g & 0 & -x_g \\ -y_g & x_g & 0 \end{bmatrix}$. Using the concept of fluid kinetic energy, the added mass matrix (M_A) can now be further expanded as

$$M_A = \begin{bmatrix} X_{\dot{u}} & X_{\dot{v}} & X_{\dot{\omega}} & X_{\dot{p}} & X_{\dot{q}} & X_{\dot{r}} \\ Y_{\dot{u}} & Y_{\dot{v}} & Y_{\dot{\omega}} & Y_{\dot{p}} & Y_{\dot{q}} & Y_{\dot{r}} \\ Z_{\dot{u}} & Z_{\dot{u}} & Z_{\dot{\omega}} & Z_{\dot{p}} & Z_{\dot{q}} & Z_{\dot{r}} \\ K_{\dot{u}} & K_{\dot{v}} & K_{\dot{\omega}} & K_{\dot{p}} & K_{\dot{q}} & K_{\dot{r}} \\ M_{\dot{u}} & M_{\dot{v}} & M_{\dot{\omega}} & M_{\dot{p}} & M_{\dot{q}} & M_{\dot{r}} \\ N_{\dot{u}} & N_{\dot{v}} & N_{\dot{\omega}} & N_{\dot{p}} & N_{\dot{q}} & N_{\dot{r}} \end{bmatrix} \tag{1.15}$$

$C(v)$ is calculated based upon the Coriolis and centrifugal terms as

$$C(v) = C_{\text{RB}}(v) + C_A(v) \tag{1.16}$$

with

$$C_{\text{RB}}(v) = \begin{bmatrix} \mathbf{0}_{3\times 3} & -m\boldsymbol{S}(\boldsymbol{v}_1) - m\boldsymbol{S}(\boldsymbol{v}_2)\boldsymbol{S}(\boldsymbol{r}_g^b) \\ -m\boldsymbol{S}(\boldsymbol{v}_1) + m\boldsymbol{S}(\boldsymbol{r}_g^b)\boldsymbol{S}(\boldsymbol{v}_2) & -S(\boldsymbol{I}_0\boldsymbol{v}_2) \end{bmatrix}, \tag{1.17}$$

$$C_A(v) = \begin{bmatrix} 0 & 0 & 0 & 0 & -a_3 & a_2 \\ 0 & 0 & 0 & a_3 & 0 & -a_1 \\ 0 & 0 & 0 & -a_2 & a_1 & 0 \\ 0 & -a_3 & a_2 & 0 & -b_3 & b_2 \\ a_3 & 0 & -a_1 & b_3 & 0 & -b_1 \\ -a_2 & a_1 & 0 & -b_2 & b_1 & 0 \end{bmatrix}, \tag{1.18}$$

where

$$\begin{aligned} a_1 &= X_{\dot{u}}u + X_{\dot{v}}v + X_{\dot{w}}w + X_{\dot{p}}p + X_{\dot{q}}q + X_{\dot{r}}r, \\ a_2 &= X_{\dot{v}}u + Y_{\dot{v}}v + Y_{\dot{w}}w + Y_{\dot{p}}p + Y_{\dot{q}}q + Y_{\dot{r}}r, \\ a_3 &= X_{\dot{w}}u + Y_{\dot{w}}v + Z_{\dot{w}}w + Z_{\dot{p}}p + Z_{\dot{q}}q + Z_{\dot{r}}r, \\ b_1 &= X_{\dot{p}}u + Y_{\dot{p}}v + Z_{\dot{p}}w + K_{\dot{p}}p + K_{\dot{q}}q + K_{\dot{r}}r, \\ b_2 &= X_{\dot{q}}u + Y_{\dot{q}}v + Z_{\dot{q}}w + K_{\dot{q}}p + M_{\dot{q}}q + M_{\dot{r}}r, \\ b_3 &= X_{\dot{r}}u + Y_{\dot{r}}v + Z_{\dot{r}}w + K_{\dot{r}}p + M_{\dot{r}}q + N_{\dot{r}}r. \end{aligned}$$

$\boldsymbol{D}(\boldsymbol{v})$ is the damping matrix including the linear damping term ($\boldsymbol{D}$) and a non-linear damping term ($\boldsymbol{D}_n(\boldsymbol{v})$) as

$$\begin{aligned} \boldsymbol{D}(\boldsymbol{v}) &= \boldsymbol{D} + \boldsymbol{D}_n(\boldsymbol{v}) \\ &= -\text{diag}(X_u + Y_v + Z_w + K_p + M_q + N_r) \\ &\quad - \text{diag}(X_{|u|u}|u| + Y_{|v|v}|v| + Z_{|w|w}|w| + K_{|p|p}|p| + M_{|q|q}|q|) \\ &\quad + N_{|r|r}|r|), \end{aligned} \tag{1.19}$$

where X_u, Y_v, Z_w, K_p, M_q, N_r are the linear damping coefficients, $X_{|u|u}|u|$, $Y_{|v|v}|v|$, $Z_{|w|w}|w|$, $K_{|p|p}|p|$, $M_{|q|q}|q|$, $N_{|r|r}|r|$ are the quadratic damping coefficients.

Finally, the vector of gravitational/buoyancy forces and moments ($g(\eta)$) can be calculated as

$$g(\eta) = \begin{bmatrix} (W - B)\sin\theta \\ -(W - B)\sin\phi\cos\theta \\ -(W - B)\cos\phi\cos\theta \\ -(y_g W - y_b B)\cos\phi\cos\theta + (z_g W - z_b B)\sin\phi\cos\theta \\ (x_g W - x_b B)\cos\phi\cos\theta + (z_g W - z_b B)\sin\theta \\ -(x_g W - x_b B)\sin\phi\cos\theta + (y_g W - y_b B)\sin\theta \end{bmatrix} \tag{1.20}$$

where W is the weight, B is the buoyancy with the centre of the buoyancy located at (x_g, y_g, z_g).

1.3 Intelligent path planning and control of autonomous marine vehicles

By referring to Figure 1.1, among the three modules (DPM, PPM and ACM), the PPM plays the most important role in the autonomous navigation system as a number of decisions are required to be made in the PPM such that the vehicle is able to autonomously and safely navigate. Typical decisions to be made include: (1) how the vehicle can find a reference to reach the target point in the most efficient way; (2) how the vehicle should adjust its trajectory to accommodate the emergency situation such as an unexpected moving ship; and (3) how the vehicle can, to the most extent, minimise the effect from the severe weather or environmental conditions. All of these issues must be addressed in the PPM by implementing planning algorithms, and since the decisions need to be made within a short period of time the algorithm should be of high computational efficiency.

In general, the planning algorithms can be categorised into two groups based on the properties that are involved, namely

- **Path planning**: focusing on a collision-free or safe path from start to goal points, disregarding the dynamic properties, i.e., velocity and acceleration.
- **Motion planning**: superset of path planning with additional dynamic properties taken into consideration (also known as trajectory planning).

Path planning typically refers to the computation of the robot position and orientation of geometric specifications only, whereas motion planning involves evaluation of linear and angular velocities, taking robot dynamics into account. Therefore, path planning is a subset of motion planning, and the trajectory provided by the path planning algorithm is normally considered as the reference guidance route to cover a large travelling area; whereas the motion planning algorithm is more often used for small sized vessels navigating in a confined area, where the motion of the vehicle is constrained. However, in this chapter, these two different algorithms are not been specifically distinguished so both of them is referred to as ***path planning***.

Another approach to categorise the planning algorithms is according to structure, i.e., the on-line and off-line planning. The off-line planning mainly deals with the environment that does not contain moving obstacles and therefore, can provide the trajectory well in advance. The generated path can also be regarded as the globally optimised path. In contrast, the on-line planning is normally adopted in the dynamic environment where multiple moving obstacles are required to be avoided as well as the static ones. A trajectory will be generated in a more reactive way that environment information is continuously updated using the sensor suite and the trajectory is accordingly amended.

Figure 1.6 shows a typical flow chart of the path planning for a marine vehicle adopting both the on-line and off-line planning strategies. It can be seen that the main difference between these two strategies is that for the on-line strategy an additional collision risk assessment system is integrated with the path planning algorithm. The collision risk assessment is able to determine the potential collisions with other ships and if such a risk is high enough to trigger safety issue the path planning algorithm

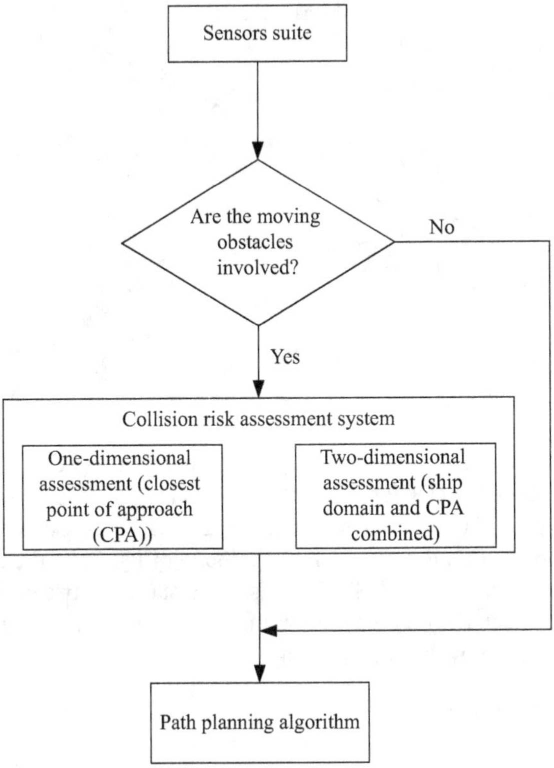

Figure 1.6 Flow chart for path planning of a marine vehicle

will be advised to re-plan the path and provide appropriate evasive actions. Typical collision risk assessment strategies include the one-dimensional (1D) approach (or *the closest point of approach (CPA)*) and the two-dimensional (2D) approach (or *the ship domain and CPA combined*).

In the following sections, basic theory behind the two different collision risk assessments will be reviewed, then a practical and feasible risk assessment strategy for marine vehicles will be described. In terms of the path planning, after reviewing and comparing a series of different but commonly adopted planning algorithms, the fast marching method (FMM) based path planning will be specifically introduced. Finally, algorithm validation on a practical USV platform will be given to demonstrate how the vehicle can autonomously navigate on a lake with the assistance of an autonomous navigation system.

1.3.1 Collision risk assessment strategies

1.3.1.1 Collision risk assessment based upon the CPA

An initial concept adopted for the collision risk assessment is based upon the CPA or the shortest instantaneous distance between vehicles involved. Parameters

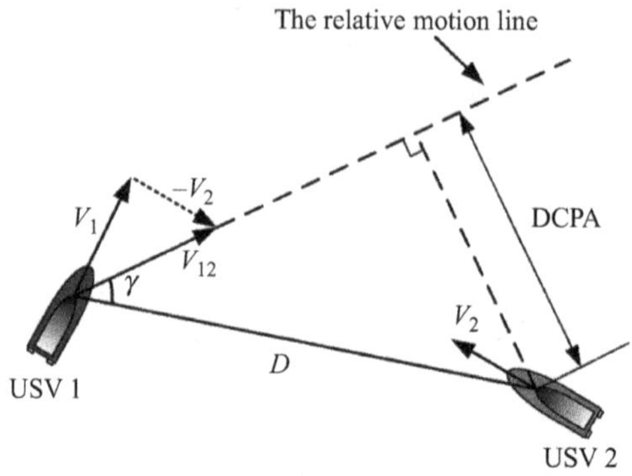

Figure 1.7 The collision risk assessment between two USVs based upon the DCPA

investigated include the time of closest point of approach (TCPA) and distance to closest point of approach (DCPA). Figure 1.7 illustrates the calculation of the TCPA and DCPA for a typical two vehicles encountering situation. Assume USV 1 is travelling with the velocity V_1 and USV 2 has the velocity of V_2. V_{12} is the relative velocity of the USV 1 with respect to USV 2, D is the distance between two vessels and γ is the angle between the relative motion line and the bearing angle of USV 1. The DCPA and TCPA can now be calculated as

$$d_{\mathrm{CPA}} = D * \sin(\gamma), \tag{1.21}$$

$$t_{\mathrm{CPA}} = \frac{d_{\mathrm{CPA}}}{V_{12}}. \tag{1.22}$$

In general, a potential collision risk between two vehicles can be viewed to be existed if following equations are satisfied as

$$d_{\mathrm{CPA}} < d_{\mathrm{CPA}}\_\mathrm{limit}, \tag{1.23}$$

$$0 \leq t_{\mathrm{CPA}} < t_{\mathrm{CPA}}\_\mathrm{limit}, \tag{1.24}$$

where both $d_{\mathrm{CPA}}\_\mathrm{limit}$ and $t_{\mathrm{CPA}}\_\mathrm{limit}$ are predefined parameters identifying the minimum safe distance and time value that should not be violated by the vehicles.

1.3.1.2 Collision risk assessment based upon ship domain

The CPA-based collision risk assessment is carried out using the 1D information (the distance information), and the vehicles involved in the assessment are normally viewed as point mass. However, due to the non-linear dynamic behaviour of vessels as well as the complexity of maritime traffic environment, it is more practical to represent the vessel with certain areas that can explicitly reveal the ship's behaviour and such an area is named as *ship domain*. The ship domain is a 2D safety area around vessel that 'prevents' all the other vehicles from entering. Its shape is largely constructed based on the navigational statistical data and normally consists of three different shapes, which are proposed by Fujii and Tanaka [15], Goodwin [16] and Davis *et al.* [17] respectively. As listed and compared in Table 1.2, it can been seen that Fujii and

Table 1.2 Comparison of three different ship domains

Ship domain	Developer	Shape and dimension	Environment
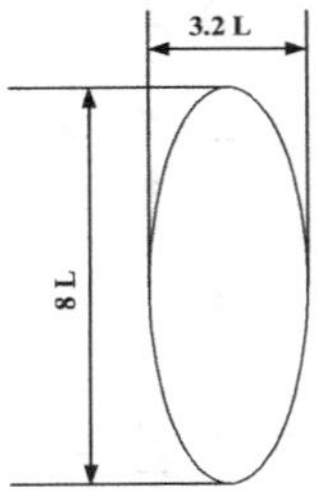	Fujii and Tanaka [15]	Ellipsoidal. 3.2 L for minor axis and 8 L for major axis (L is the ship length)	Confined areas such as channels.
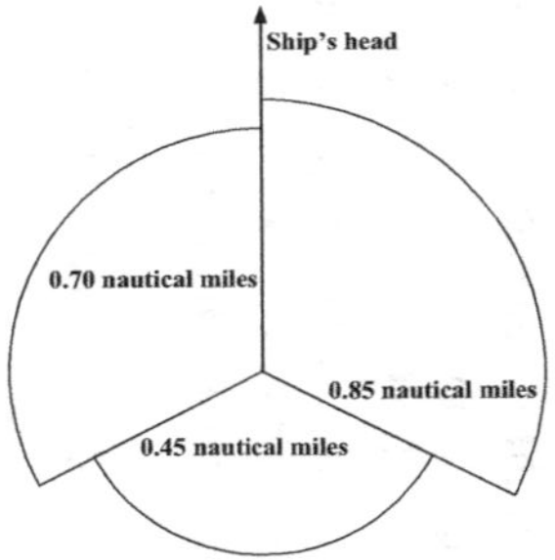	Goodwin [16]	The shape consists of three discrete zones. Each zone has the radius of 0.85 nautical miles, 0.70 nautical miles and 0.45 nautical miles, respectively.	Open sea area. The shape is compliant with the marine traffic regulation rules (COLREGS).
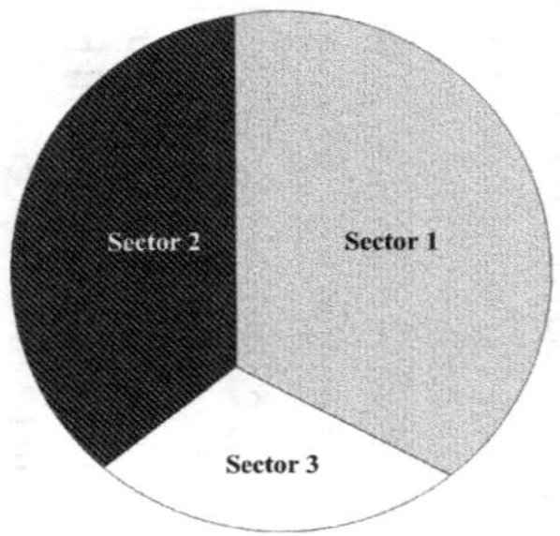	Davis *et al.* [17]	Circular shape. The shape is divided into three different sectors which are in accordance with the three sections defined in Goodwin [16].	Open sea area. Compliant with COLREGS. Easier to be mathematically modelled.

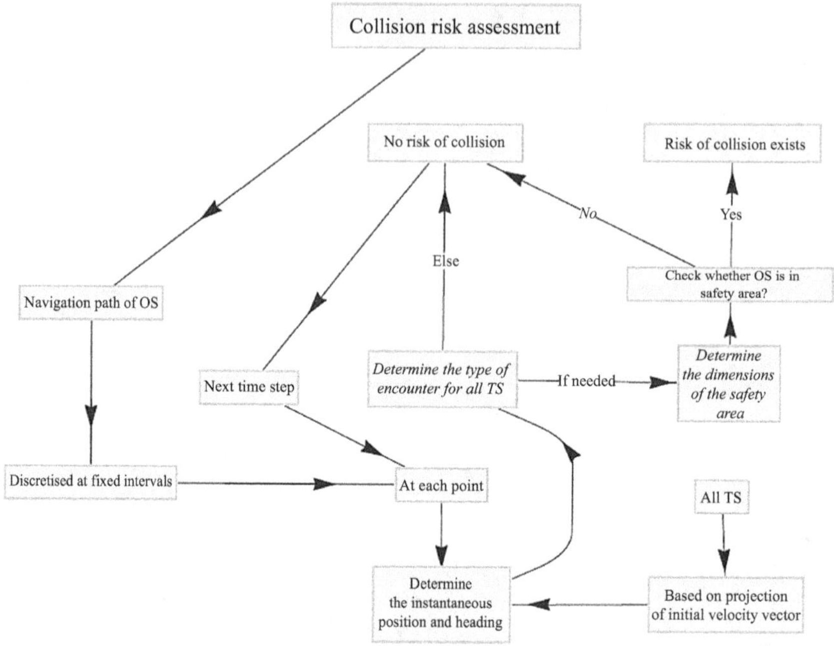

Figure 1.8 Flow chart of the collision risk assessment process

Tanaka's [15] shape is normally used in confined areas such as channels, while the other two shapes is more suitable for open sea area. These three shapes have become the fundamental shape when developing the ship domain, and have been modified and improved by subsequent studies using the methods, such as fuzzy logic [18] and artificial neural network [19] for particular environment or navigation requirements.

When performing the collision risk assessment using the ship domain, information regarding the CPA (mainly the DCPA value) is first calculated and if the DCPA has violated the ship domain, appropriate evasive manoeuvres are thus required. Based on this, a practical and feasible collision risk assessment strategy has been proposed by Tam and Bucknall [20] with the algorithm flow chart shown in Figure 1.8. It assesses the risk based upon the discretised trajectory of the OS and TS.[a] Along each fixed path intervals, the type of the instantaneous encounter and the collision risk are calculated and evaluated. The assessment mainly consists of two procedures, i.e., determine the type of the encounter based upon the OS and TS's positions and velocities, determine the dimension of the ship domain based upon the encounter types and TS's velocity. An assessment loop has been used in such a way that all the involved TS's will be considered and all of the discretised points along the navigational path will be evaluated for the risk. Based upon the collision risk assessment results, if evasive actions are needed the path will be accordingly amended using the path planning algorithms, which will be specifically discussed in the next section.

1.3.2 Motion planning for USV

The aim of the path planning for USV is to provide and generate a guidance trajectory such that by following this trajectory the vehicle is able to autonomously accomplish the tasks. When calculating the path, various costs need to be taken into account to ensure that the generated path is optimised. Commonly considered costs include the minimum distance, the least energy consumption, the maximum navigation safety, etc.

Favourite motion planning algorithms can be largely categorised into two general groups, i.e., the deterministic and the heuristic approach. The deterministic searching approach is done by following a set of defined steps with the advantages of completeness and consistency. A searching result can be guaranteed to be found as long as it exists; also, the output remains the same if there is no variation of searching environment. However, one of the major problems of deterministic searching is the long computation time in high dimension or complex space. Popular deterministic approaches include the artificial potential field, the roadmap based algorithm, the FMM and the optimisation method.

The heuristic searching approach is proposed to solve specifically the problem, which cannot be efficiently addressed by a deterministic approach. Also, it is able to provide an approximate solution when exact solutions are hard to find. However, since it only searches the subspace of the search space, the global optimality of the results cannot be guaranteed, i.e., only near optimal result can be obtained. In addition, because the algorithm stochastically searches within the space, the consistency of the results is not as good as the deterministic method [21]. Typical heuristic algorithms include the evolutionary algorithm (EA) such as the genetic algorithm, the particle swarm optimisation and the ant colony optimisation.

1.3.2.1 The FMM

The FMM was first proposed by J. Sethian in 1996 to iteratively solve the Eikonal equation to simulate the propagation of an interface [22]. The Eikonal equation has the form as

$$|\nabla T(\boldsymbol{x})|V(\boldsymbol{x}) = 1, \tag{1.25}$$

where $T(\boldsymbol{x})$ is the interface arrival time at point $\boldsymbol{x}$ and $V(\boldsymbol{x})$ is the interface propagating speed. The Eikonal equation belongs to the partial differential equation (PDE) and its numerical solution can be obtained via the upwind differential method when using the FMM. The solving process of the FMM is similar to Dijkstra's method but in a continuous way. Suppose (x, y) is the point that $T(x, y)$ needs to be solved. The neighbour of (x, y) is a point set containing four elements $(x + \Delta x, y)$, $(x - \Delta x, y)$, $(x, y + \Delta y)$, $(x, y - \Delta y)$, $T(x, y)$ can be obtained as

$$T_1 = \min(T_{(x-\Delta x,y)}, T_{(x+\Delta x,y)}), \tag{1.26}$$

$$T_2 = \min(T_{(x,y-\Delta y)}, T_{(x,y+\Delta y)}), \tag{1.27}$$

$$|\nabla T_{(x,y)}| = \sqrt{\left(\frac{T_{(x,y)} - T_1}{\Delta x}\right)^2 + \left(\frac{T_{(x,y)} - T_2}{\Delta y}\right)}, \tag{1.28}$$

$$\left(\frac{T_{(x,y)} - T_1}{\Delta x}\right)^2 + \left(\frac{T_{(x,y)} - T_2}{\Delta y}\right) = \frac{1}{(V_{(x,y)})^2} \tag{1.29}$$

where Δx and Δy are the grid spacing in x and y directions.

The solution of above equation is given by

$$T_{(x,y)} = \begin{cases} T_1 + \dfrac{1}{V_{(x,y)}}, & \text{if } T_2 \geq T \geq T_1, \\ T_2 + \dfrac{1}{V_{(x,y)}}, & \text{if } T_1 \geq T \geq T_2, \\ \text{Quadratic solution of Eikonal equation}, & \text{if } T > \max(T_1, T_2). \end{cases} \tag{1.30}$$

To further illustrate the FMM algorithm, a simple case representing how to update a 6*6 grid map is shown in Figure 1.9. Figure 1.9(a) shows the initial configuration of the algorithm with the middle point being the algorithm start point. The grid size is 1*1, and the interface propagating speed is set to be uniform as 1 at each grip point. When the FMM is executed, grid points are categorised into three different groups as follows:

- ***Far (marked as light blue):*** contains grid points with undecided arrival time value (T). In the first time step when running the FMM, all grid points except the start points belong to *Far*;
- ***Known (marked as red):*** contains grid points with decided arrival time values (T). Such values will not be changed when the algorithm is executed;
- ***Trial (marked as green):*** contains grid points with calculated arrival time values (T); however, values may be changed then the algorithm is running.

In Figure 1.9(b), the first step time of running the FMM is represented. The start point is currently the only *Known* point with T value as 0. Four neighbour points of the start point consist of the *Trial* set and therefore are marked as green with calculated arrival time T as 1.

For next time step, the point with minimum arrival time cost will be first selected from the *Trial* set to become the new *Known* point; however, four neighbours have the same cost, and the point below the start point is considered as the point with the minimum cost thereby being re-marked as *Known* with its neighbours identified as *Trial* as well (shown in Figure 1.9(c)). Figure 1.9(d) and (e) show how the grip map is updated by repeating such a process in time steps 3 and 4, and the final updating result is represented in Figure 1.9(f). In Figure 1.9(f), it is observed that the start point has the minimum arrival time 0; whereas, other points' arrival times are increasing proportionally to the distances to the start point, which forms a potential field where the potential field value is the interface arrival time with the minimum potential located at the start point.

1.3.2.2 The FMM-based path planning algorithms

The FMM-based path planning algorithm is described in Algorithm 1. Suppose that the planning space (***M***) shown in Figure 1.10(a), where the algorithm is performed

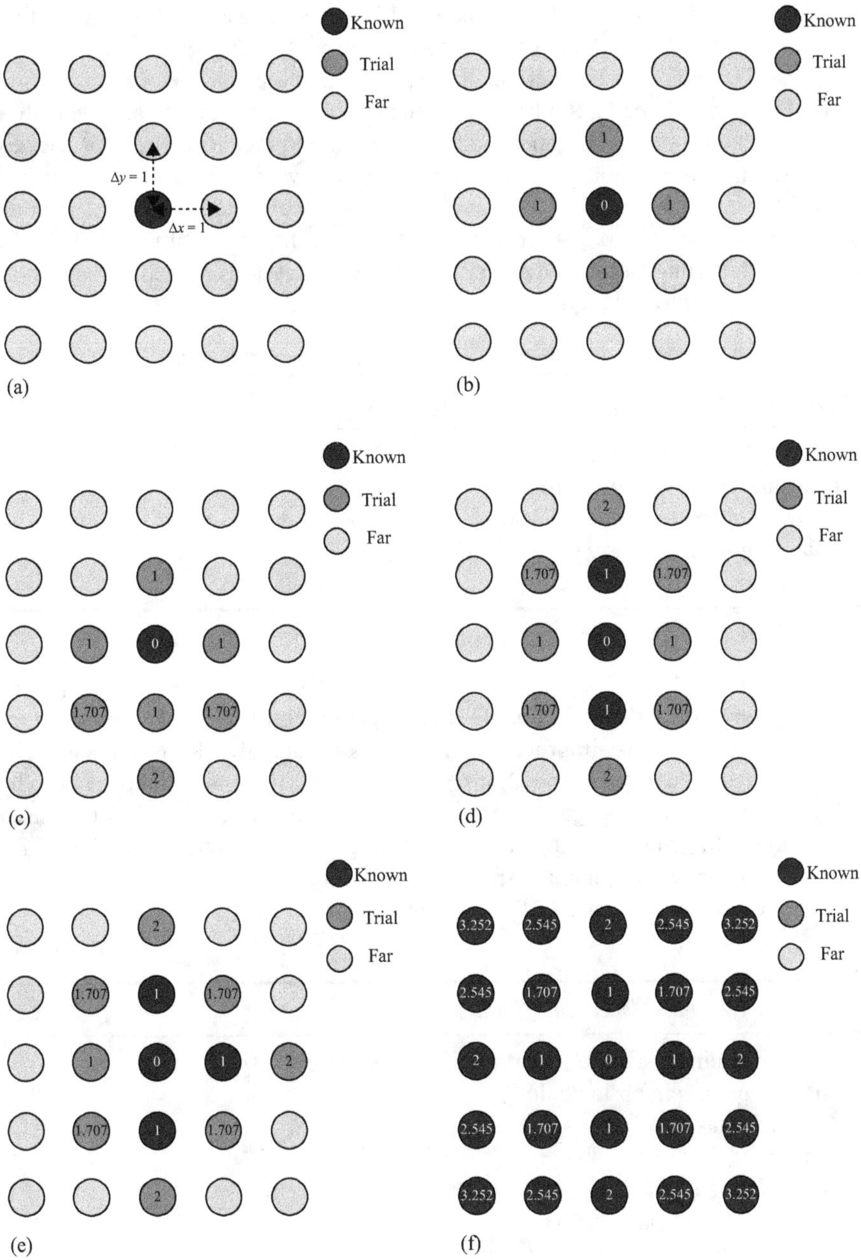

Figure 1.9 The updating process when using the FMM

on, has a representation of a binary map and is perfectly rasterised. The algorithm first reads in the M and calculates its speed matrix (V). The speed matrix (V) is the same size as the M and defines the interface propagation speed at each point in the $\boldsymbol{M}$. Based on the V, the FMM is executed to calculate an arrival time matrix $\boldsymbol{T}$ from the start point. The generated $\boldsymbol{T}$ (shown in Figure 1.10(b)) may be viewed as an arrival time potential field where the potential value represents local arrival time of the interface, which subsequently indicates local distance to the start point if a constant speed matrix is used. Then, based upon the arrival time matrix $\boldsymbol{T}$, the optimal path is finally searched by applying the gradient descent method (shown as the red line in Figure 1.10(c)).

Algorithm 1.1: *FMM Path_Planning* Algorithm

Require: planning space ($\boldsymbol{M}$), start point (p_{start}), end point (p_{end})
1: Calculate speed matrix $\boldsymbol{V}$ from $\boldsymbol{M}$
2: Arrival time matrix ($\boldsymbol{T}$) $\leftarrow$ *FMM* ($\boldsymbol{V}, p_{start}, p_{end}$)
3: *path* $\leftarrow$ *gradient Decent* ($\boldsymbol{T}, p_{start}, p_{end}$)
4: **return** *path*

Another FMM based path planning algorithm is the fast marching square (FMS) method proposed by Gomez *et al.* [23]. Compared with the FMM, the trajectory provided by the FMS has improved safety, as it is further away from obstacles. The FMS is represented in Algorithm 2. It first generates a safety potential map ($\boldsymbol{M}_s$) by applying the FMM to propagate interfaces from all the points in obstacle area. Based on $\boldsymbol{M}_s$, the FMM is executed again from the start point to generate the final path. By using the same previous planning space, the path generated by the FMS is shown as the green trajectory in Figure 1.10(c) with increased safety.

Algorithm 1.2: *Fast_Marching_Square* Algorithm

Require: planning space ($\boldsymbol{M}$), start point (p_{start}), end point (p_{end})
1: **for** each point a in obstacle **do**
2: obstaclePoints $\leftarrow a$
3: **end for**
4: $\boldsymbol{M}_s \leftarrow$ *FMM* (*obstaclePoints*)
5: $\boldsymbol{T} \leftarrow$ *FMM* ($\boldsymbol{M}_s, p_{start}, p_{end}$)
6: *path* $\leftarrow$ *gradientDecent* (T, p_{start}, p_{end})
7: **return** *path*

Finally, based on the kinematic motion of the USV and FMS algorithm, the angle-guidance fast marching method (AFMS) has been proposed by Liu and

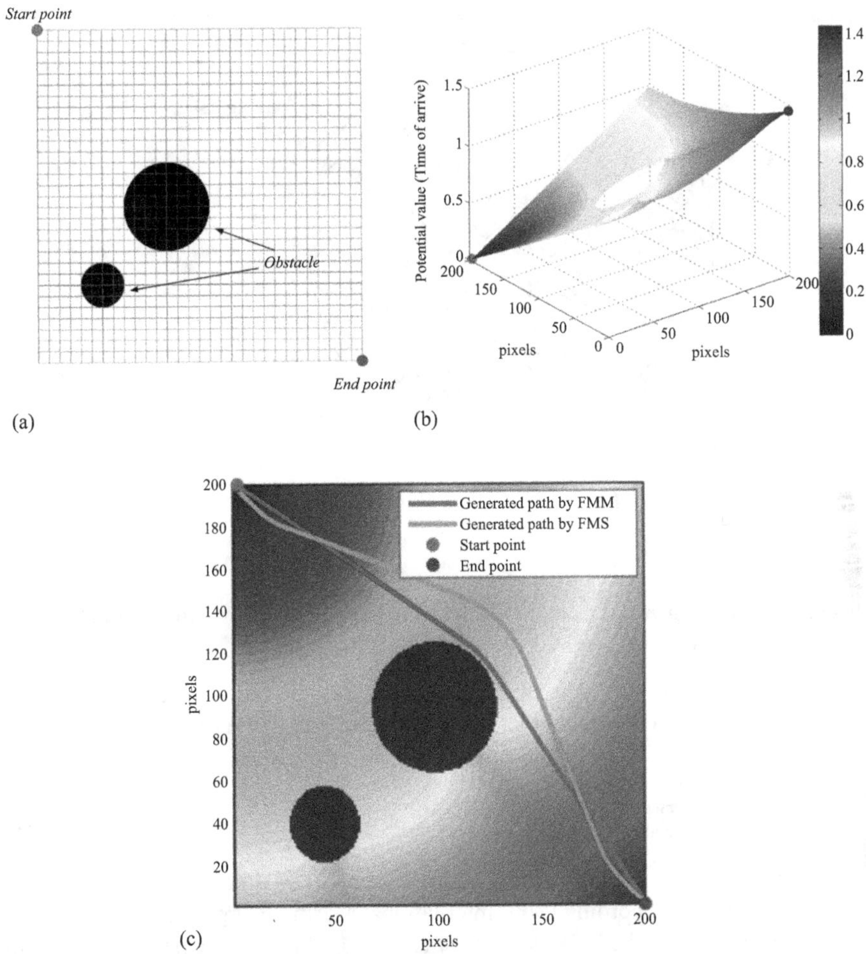

Figure 1.10 *(a) The grid map. (b) Potential field generated by using the FMM. Potential value at each point represents the local distance to the start point. Higher the potential is, longer the distance to the start point. (c) Path generated by using the FMM (red line) and the FMS (green line)*

Bucknall [24] with the aim of generating a trajectory compliant with vehicle's dynamics (pseudocode shown in Algorithm 3). It uses the FMS as the base algorithm, and in order to make the generated path compliant with USV's motion constraints, the core of the AFMS is to create a "guidance range" (GR) upon the planning space (***M***) (shown as Line 1 in Algorithm 3). The shape of the GR is constructed as graphically shown in Figure 1.11. It consists of two different sectors, i.e., the *Turning range* sector in white, and the *Obstacle region* sector in

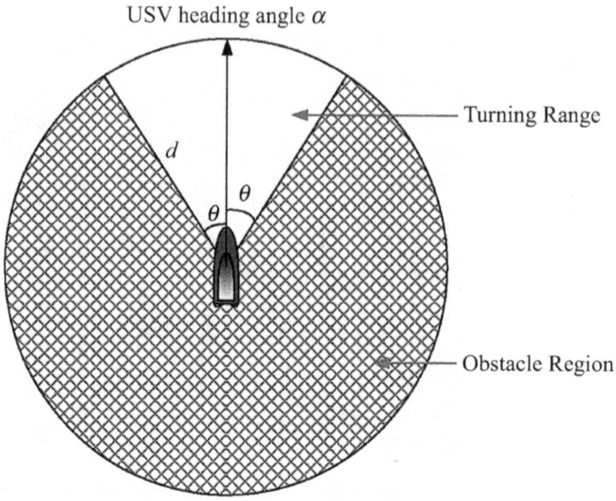

Figure 1.11 The illustration of GR

shade. The *Turning range* sector's dimension is controlled by three parameters as follows:

(1) Range distance (d): The range distance (d) is the radius of the cone shape, which is able to control the influence range affecting the path and is related to the surge speed of the USV as

$$d = \begin{cases} d_{\min} & \text{if } u < u_{\text{permit}}, \\ u * \text{rangeScalar} & \text{otherwise}, \end{cases} \tag{1.31}$$

where $d_{\min}$ is the predefined minimum distance (3 to 4 times the vessel's length is recommended according IMO (International Maritime Organisation) [25]) used in the case when the USV is at low speed. *rangeScalar* is the parameter used to regulate GR range size to prevent the algorithm from generating an oversized obstacle area such that the target point will be unnecessarily blocked.

(2) Heading angle (α): The heading angle (α) represents the USV's current heading angle and determines the direction of the GR.

(3) Turning angle (θ): The turning angle (θ) parameter is calculated according to vehicle yaw constraint as

$$\theta = r_{\max} * \Delta T, \tag{1.32}$$

where ΔT is the time step and $r_{\max}$ is the constraint of the yaw rate.

The weighting values inside the *Turning range* sector are assigned to remain the same as they are in the planning space $\boldsymbol{M}$. In addition, since it is desired that the path should be located within the *Turning range* sector making the *Obstacle sector* act like an obstacle, grid values of the *Obstacle sector* are assigned as 0.

Algorithm 1.3: *Angle_guidance FMS* Algorithm

Require: planning space ($\boldsymbol{M}$), start point (p_{start}), end point (p_{end}), heading angle (α), turning angle (θ), range radius (r)
1: range $\leftarrow$ *guidanceRange* ($\boldsymbol{M}\ r, \alpha, p_{\text{start}}, \theta$)
2: **for** each point p in *Obstacle sector* **do**
3: $M(p) = 0$
4: **end for**
5: $\boldsymbol{M}_s \leftarrow$ *FMS* ($\boldsymbol{M}$)
6: *path* $\leftarrow$ *pathgradientDecent* ($\boldsymbol{M}_{\text{s}}, p_{\text{start}}, p_{\text{end}}$)
7: **return** *path*

1.3.3 Autonomous and intelligent navigation of a USV

A simulation result is presented in this subsection to showcase the performance of a USV formation autonomously navigating in a practical maritime environment. More specifically, the primary aim of this test is to verify a path planning algorithm's ability to deal with collision avoidance requirement. Among different types of path planning algorithms, the FMS is adopted in this test as it is able to provide a path with guaranteed smoothness and safety. A number of static and dynamic obstacles are included in the simulation environment, especially the dynamic obstacles (or the moving ships) are configured to traverse in the environment with complex motions, i.e. the moving ships can change their velocities in both direction and speed, which place more demands on the algorithm's capability of high computational efficiency in order to avoid collisions.

In the simulations, it is assumed that identical USVs are used in formation. Speed of the leader USV is set constant such that it is easier for other USVs to follow. Followers, however, can vary their velocities with respect to their positions in the formation. For example, the follower USV needs to remain at the same velocity as leader's when it is moving at the desired formation position. If the current position of follower deviates from the desired position, the follower either speeds up to catch up or slows down to wait for the leader.

The simulation area used in this section is extracted from a practical environment near Portsmouth Harbour (Figure 1.12), which is a large natural water area and one of the busiest harbours in the UK. To facilitate the algorithm to search for a route, area with a 2.5 km × 2.5 km dimension has been transformed to a 500 pixels × 500 pixels binary map. Three target ships (TSs) are now included, and TS1 and TS3 are able to change their velocities during the simulation. The specific configurations of these TSs are listed in Table 1.3.

Simulation results are shown in Figure 1.12 with the movements of the formation in different steps being presented. Three TSs' trajectories have been plotted in different colours with TS1 in cyan, TS2 in green and TS in black. In addition, each TS's instantaneous heading has been displayed with a black arrow to give a better view of

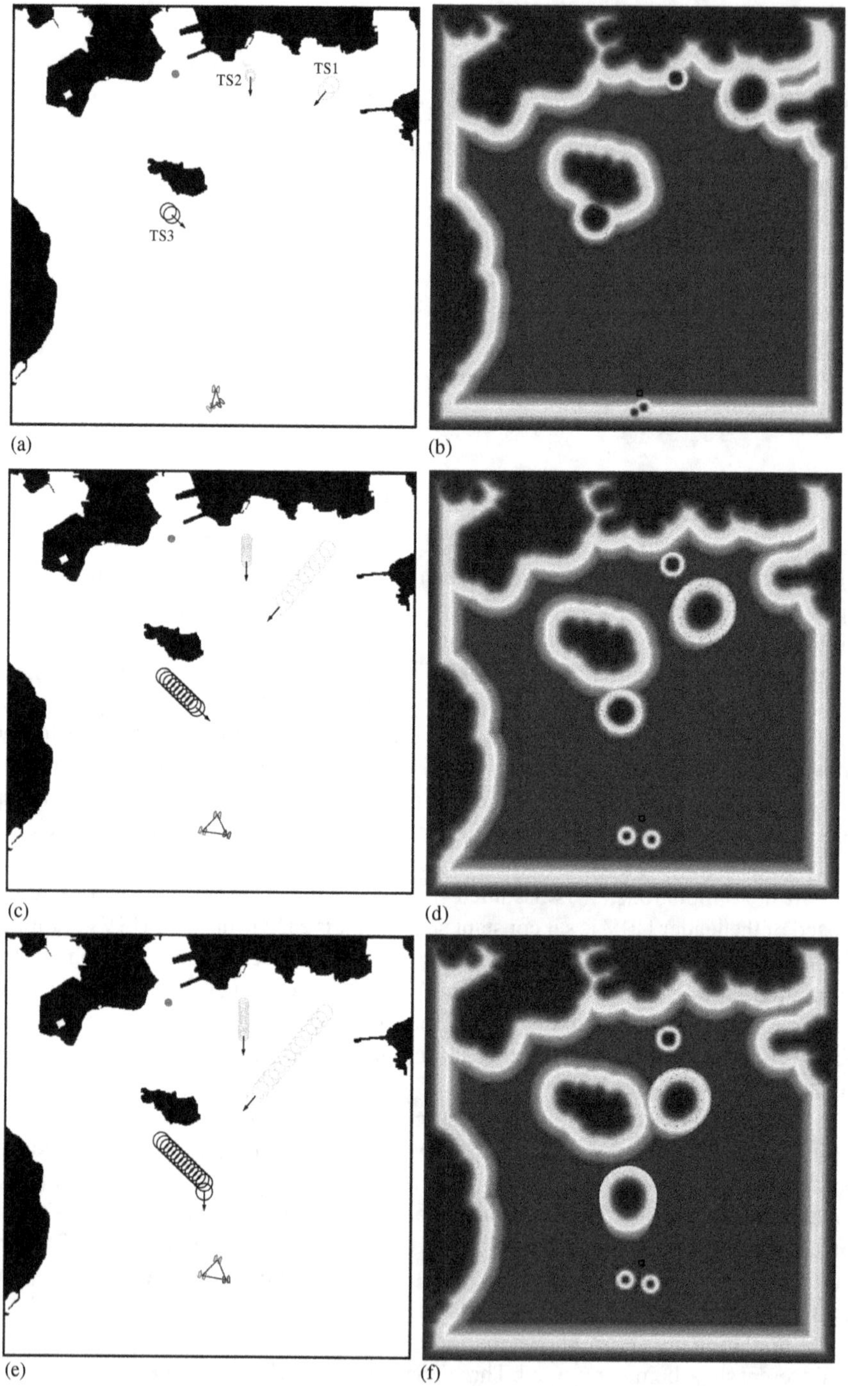

Figure 1.12 Formation movement sequences and corresponding potential maps for simulation near Plymouth Harbour with fully dynamic moving ships. (a) and (b) Time step = 2; (c) and (d) time step = 10; (e) and (f) time step = 14; (g) and (h) time step = 15; (i) and (j) time step = 19; (k) and (l) time step = 31; (m) and (n) time step = 46; and (o) and (p) time step = 64

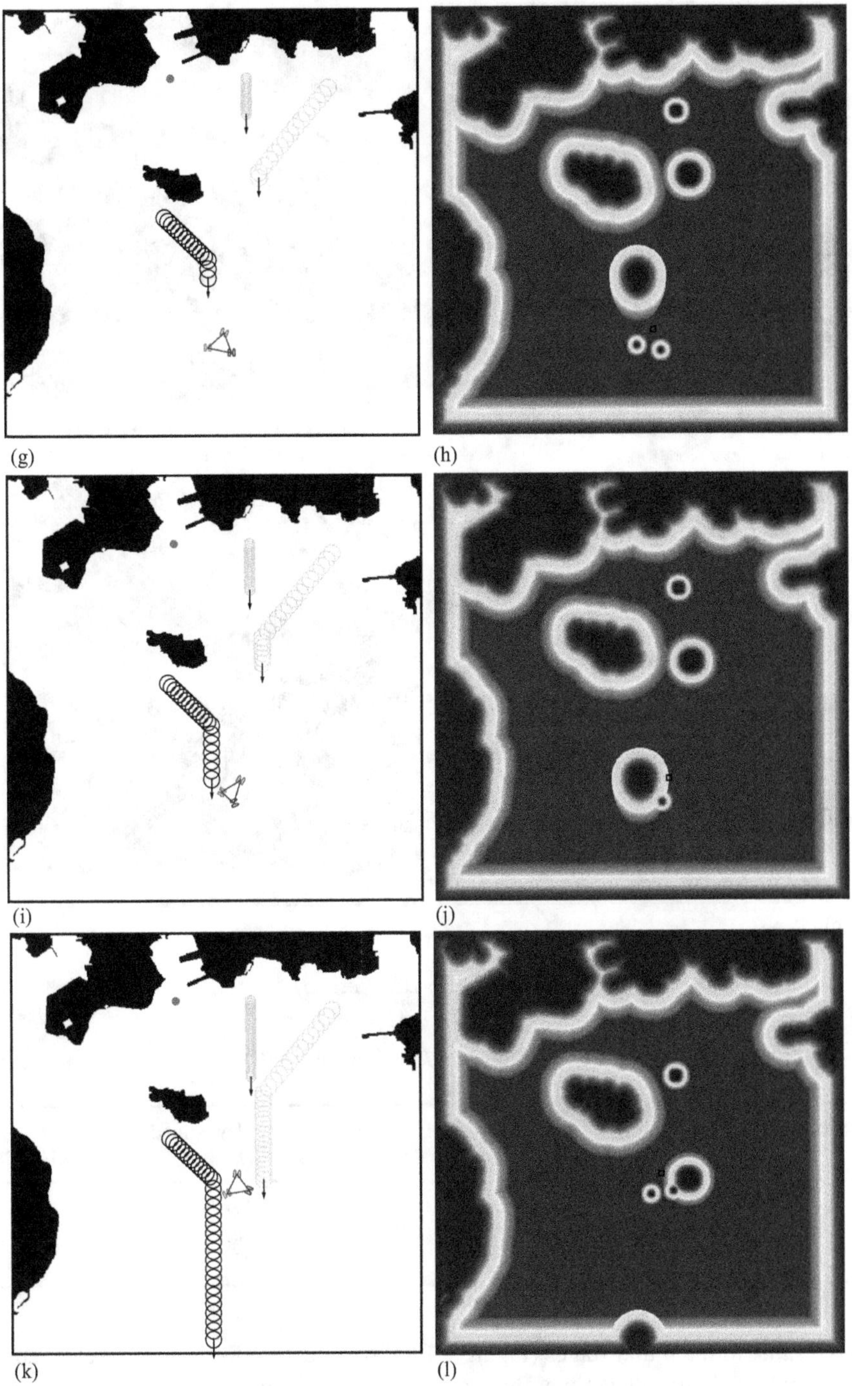

Figure 1.12 (Continued)

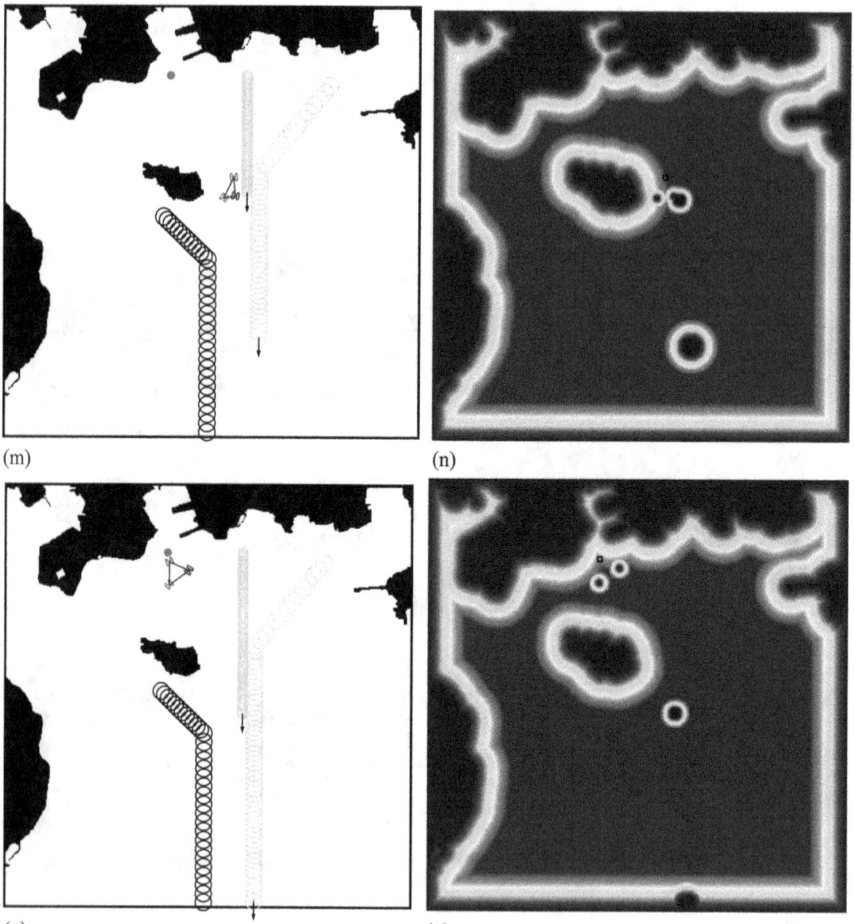

(m) (n) (o) (p)

Figure 1.12 (Continued)

Table 1.3 Parameters of the simulation of USV formation navigation

TS index	Start point	Initial speed (knots)	Initial heading (°)	New speed (knots)	New heading (°)
1	(399,415)	20	150	12	270
2	(295,422)	6	270	N/A	N/A
3	(191,260)	12	210	21	270

the results. In the initial stages (Figure 1.12(a)–(d)), three TSs' dynamic behaviours can first be well modelled. Since they so far have the same velocities as in the previous test, the same ship domain and collision avoidance areas have been generated for each TS to prevent the collisions.

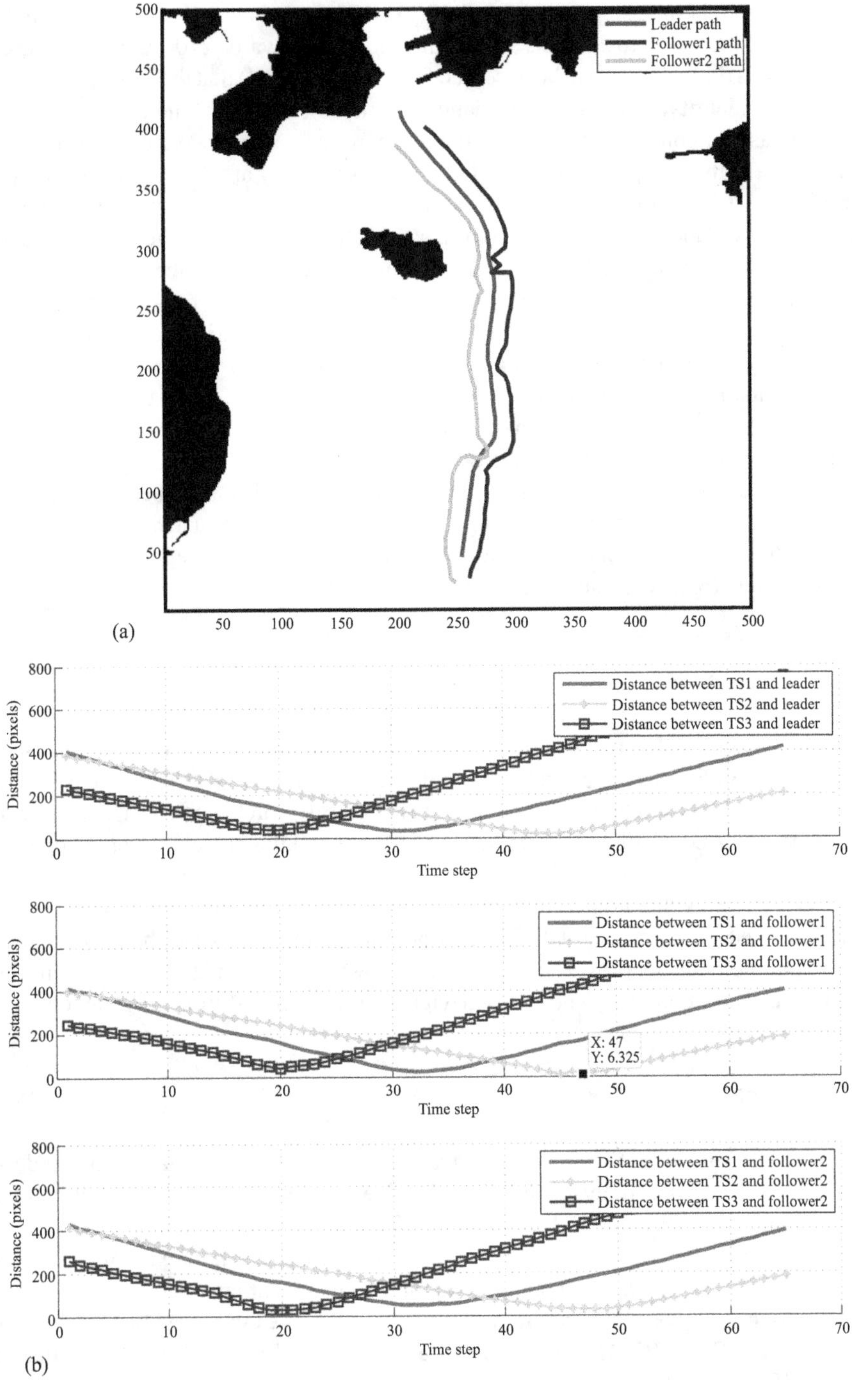

Figure 1.13 Evaluation results for simulation near Plymouth Harbour with fully dynamic moving ships. (a) Trajectories for formation and (b) distance between target ships and each USV in formation

However, at time step 14 (Figure 1.12(e)), TS3 first changes its velocity with the speed been increased to 21 knots and heading adjusted towards 270° (0° is configured in the x-axis direction). It can be observed from Figure 1.12(f) that the algorithm is able to quickly identify such a velocity change and new ship domain and collision avoidance areas can consequently be generated according to the new velocity. The same procedure is then followed as the TS1 changes its velocity at time step 15. As shown in Figure 1.12(h), since the speed of the TS1 has been reduced, its ship domain and collision avoidance areas are varied from a half-elliptical shape to the circular shape.

It should also be noted that because the algorithm can identify the velocity change of each TS, the collision avoidance with these TSs can still be retained in this simulation. As presented in Figure 1.12(i)–(p), the USV formation can avoid TSs by maintaining safe distances. Such a capability can be further proven from the algorithm evaluation results shown in Figure 1.13. In Figure 1.13(a), the overall trajectories of the USV formation have been plotted. It can be observed that each trajectory is well away from all the static obstacles meaning no collision occurs. In terms of the moving obstacles avoidance, as shown in Figure 1.13(b), distances between each TS and each USV in the formation are well above 0 pixels. The distance minimum value is approximately 6 pixels in the simulation area and about 30 m in the real environment, which is able to provide sufficient distance to avoid collisions.

1.4 Conclusion

In this chapter, modelling and control methods for autonomous marine vehicle, especially the USV, have been introduced and discussed. The 6 DOF as well as the 3 DOF mathematical modelling approaches are presented for vehicle control. The choice of the number of DOF is based upon the vehicle type and the vehicle's propulsion system. Then, path planning for USV is specifically introduced to show how to control a USV from the planning perspective. The FMS algorithm have been successfully implemented on a practical USV formation platform to showcase the performance of autonomous navigation. The results have shown that by using such a system, the vehicle can autonomously and safely navigate a practical maritime environment.

References

[1] Motwani, A. *A Survey of Uninhabited Surface Vehicles*. Marine and Industrial Dynamic Analysis, School of Marine Science and Engineering Plymouth University, Plymouth. Tech. Report, 2012.

[2] Bremer, R., Cleophas, P., Fitski, H., and Keus, D. *Unmanned Surface and Underwater Vehicles*. Technical Report, DTIC Document, 2007.

[3] Manley, J. Development of the autonomous surface craft aces. In *Proceedings of OCEANS'97. MTS/IEEE Conference*, Vol. 2. IEEE; Halifax, NS, Canada; 1997. pp. 827–832.

[4] Curcio, J., Leonard, J., and Patrikalakis, A. Scout—A low cost autonomous surface platform for research in cooperative autonomy. In *Proceedings of OCEANS 2005 MTS/IEEE Conference*, IEEE; Washington, DC, USA; 2005. pp. 725–729.

[5] Pascoal, A., Oliveira, P., Silvestre, C., *et al.* Robotic ocean vehicles for marine science applications: the European ASIMOV project. In *Oceans 2000 MTS/IEEE Conference and Exhibition*, Vol. 1. IEEE; Providence, RI, USA; 2000. pp. 409–415.

[6] Alves, J., Oliveira, P., Oliveira, R., *et al.* Vehicle and mission control of the delfim autonomous surface craft. In *2006 14th Mediterranean Conference on Control and Automation.* IEEE; Ancona, Italy; 2006. pp. 1–6.

[7] Sutton, R., Sharma, S., and Xao, T. Adaptive navigation systems for an unmanned surface vehicle. *Journal of Marine Engineering & Technology, 10*(3): 3–20, 2011.

[8] Liu, Y., Noguchi, N., and Yusa, T. Development of an unmanned surface vehicle platform for autonomous navigation in paddy field. *IFAC Proceedings Volumes*, 47(3):11553–11558, 2014.

[9] Majohr, J., and Buch, T. Modelling, simulation and control of an autonomous surface marine vehicle for surveying applications measuring dolphin Messin. *IEE Control Engineering Series, 69*:329, 2006.

[10] Caccia, M., Bono, R., Bruzzone, G., *et al.* Sampling sea surfaces with sesamo: An autonomous craft for the study of sea–air interactions. *IEEE Robotics & Automation Magazine, 12*(3):95–105, 2005.

[11] Twichell, D. C., Andrews, B. D., Lee Edmiston, H. and Stevenson, W. R. Geophysical mapping of oyster habitats in a shallow estuary, Apalachicola Bay, Florida. US Geological Survey Open-File Report 1381 (2007): 1–13.

[12] Ebken, J., Bruch, M., and Lum, J. Applying unmanned ground vehicle technologies to unmanned surface vehicles. *In Unmanned Ground Vehicle Technology VII*, Vol. 5804. International Society for Optics and Photonics, 2005. pp. 585–597.

[13] Caccia, M., Bibuli, M., Bono, R., and Bruzzone, G. Basic navigation, guidance and control of an unmanned surface vehicle. *Autonomous Robots, 25*(4): 349–365, 2008.

[14] Fossen, Thor I. *Guidance and Control of Ocean Vehicles*. New York: John Wiley & Sons Inc; 1994.

[15] Fujii, Y., and Tanaka, K. Traffic capacity. *The Journal of Navigation*, *24*(4): 543–552, 1971.

[16] Goodwin, E. M. A statistical study of ship domains. *The Journal of Navigation*, *28*(3):328–344, 1975.

[17] Davis, P. V., Dove, M. J., and Stockel, C. T. A computer simulation of marine traffic using domains and arenas. *The Journal of Navigation*, *33*(2): 215–222, 1980.

[18] Wang, N. An intelligent spatial collision risk based on the quaternion ship domain. *The Journal of Navigation*, *63*(4): 733–749, 2010.

[19] Zhu, X., Xu, H., and Lin, J. Domain and its model based on neural networks. *The Journal of Navigation*, *54*(1):97–103, 2001.

[20] Tam, C., and Bucknall, R. Collision risk assessment for ships. *Journal of marine science and technology*, 15(3):257–270, 2010.

[21] Tam, C., Bucknall, R., and Greig, A. Review of collision avoidance and path planning methods for ships in close range encounters. *The Journal of Navigation*, *62*(3):455, 2009.

[22] Sethian, J. A. A fast marching level set method for monotonically advancing fronts. *Proceedings of the National Academy of Sciences*, *93*(4): 1591–1595, 1996.

[23] Gómez, J. V., Lumbier, A., Garrido, S., and Moreno, L. Planning robot formations with fast marching square including uncertainty conditions. *Robotics and Autonomous Systems*, *61*(2):137–152, 2013.

[24] Liu, Y., and Bucknall, R. The angle guidance path planning algorithms for unmanned surface vehicle formations by using the fast marching method. *Applied Ocean Research*, *59*:327–344, 2016.

[25] Naeem, W., Xu, T., Sutton, R., and Tiano, A. The design of a navigation, guidance, and control system for an unmanned surface vehicle for environmental monitoring. *Proceedings of the Institution of Mechanical Engineers, Part M: Journal of Engineering for the Maritime Environment*, 222(2):67–79, 2008.

Chapter 2

Efficient optimal path planning of unmanned surface vehicles

Yogang Singh,[1] Sanjay Sharma,[1] Robert Sutton,[1] Daniel Hatton,[1] and Asiya Khan[1]

Abstract

The objective of the path planning of marine vehicles is to determine a collision-free path from the start to the goal point in a practical marine environment comprising static and moving obstacles. A good number of path-planning approaches have been adopted from the area of mobile robotics and extended in the area of marine robotics. The current chapter overviews application of grid-based path planners in the area of unmanned surface vehicles (USVs). The current chapter also proposes a novel and computationally efficient approach towards optimal path planning of USVs.

2.1 Introduction

The path-planning problem of marine vehicles, which is defined in terms of computation of collision-free paths in a marine environment consisting of static and moving obstacles, has been an important area of research in the last decade [1,2]. The most traditional approach is to have a marine robot in a known marine environment and find a collision-free path describing a motion that explains the robots transitions from its initial to the final state. The variants and modifications to this traditional approach are umpteen. The present chapter gives an account of the application of grid-based path planners in the area of optimal path planning of unmanned surface vehicles (USVs).

Marine vehicles can be broadly classified as shown in Figure 2.1. This classification is based on the displaced volume of the vehicles. Each class of these vehicles requires different levels of autonomy due to diverse nature of their missions and uncertainties involved in their operational environments. With advancements in the area of marine navigation, major research towards development of USVs technology started after successful implementation of the USVs in the 1990–91 Gulf war [3].

[1]Autonomous Marine Systems (AMS) Research Group, School of Engineering, University of Plymouth, United Kingdom

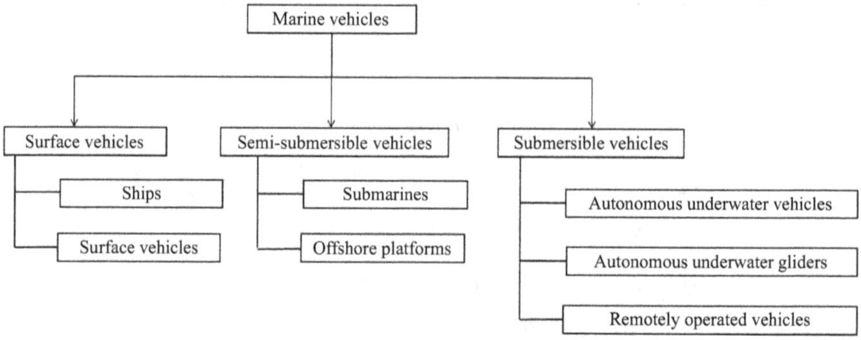

Figure 2.1 Classification of marine vehicles [4]

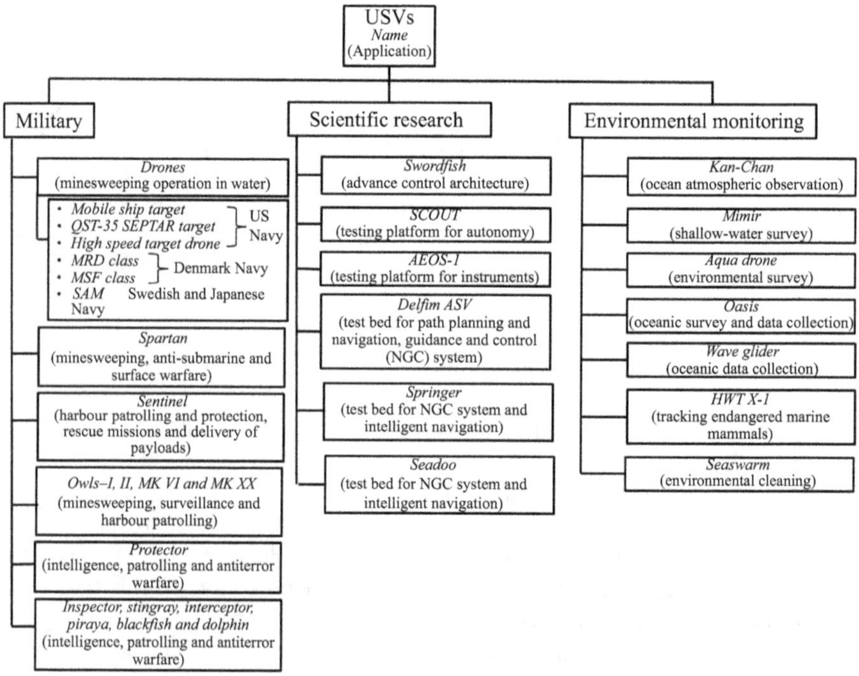

Figure 2.2 Classification of USVs based on application; Name (Application) [5]

USVs are watercraft of small (<1 t) or medium (100 t) size in terms of water displacement. Basic purposes of USVs are military, ocean and scientific research and exploration of hydrocarbons. Classification and developments of USVs based on their application has been explained by [5] and shown in Figure 2.2. A few USV prototypes are shown in Figure 2.3. The general architecture of USV operation in maritime environment has three basic systems, namely control and path planning,

Figure 2.3 A few USV prototypes: (a) Springer (Source: [6]); (b) Delfim (Source: [7]); (c) protector (Source: [8])

communication and monitoring and obstacle detection and avoidance (ODA) which are responsible for mission planning and execution as shown in Figure 2.4.

The present policies and legal framework constraint the use of a USV in a maritime environment due to the risk of injury and property damage. International Maritime Organisation (IMO) has suggested a list of regulations for the autonomous operation of USVs in a maritime environment [9–11] and with the current level of autonomy in the marine sector [12], such autonomous operations without human interference are not possible. This leads to the requirement of increasing autonomy in the operation of USVs. Path planning is an important subsystem of USV operation (from Figure 2.4), which has been an important area of research in the recent decade and fundamental to the objective of the autonomous functioning of USVs.

It has been proven, in general, that basic path-planning complexity increases exponentially with increase in degrees of freedom of the robot. For example, if we have *n* robots of *d* degrees of freedom each, the complexity becomes exponential in *nd* [13]. In terms of the marine environment, such complexity increases the manifold considering uncertainty of marine environment and convoluted nature of the obstacles. This problem of motion planning in robotics has led to the development of many complete planners [14–17]. For most of the practical problems, such planners are impossible to use and need simplification of the problem statement. This has led to several assumptions on the robot environment like fatness [15], bounded local complexity [18] and dispersion [19]. In order to move USVs autonomously in the

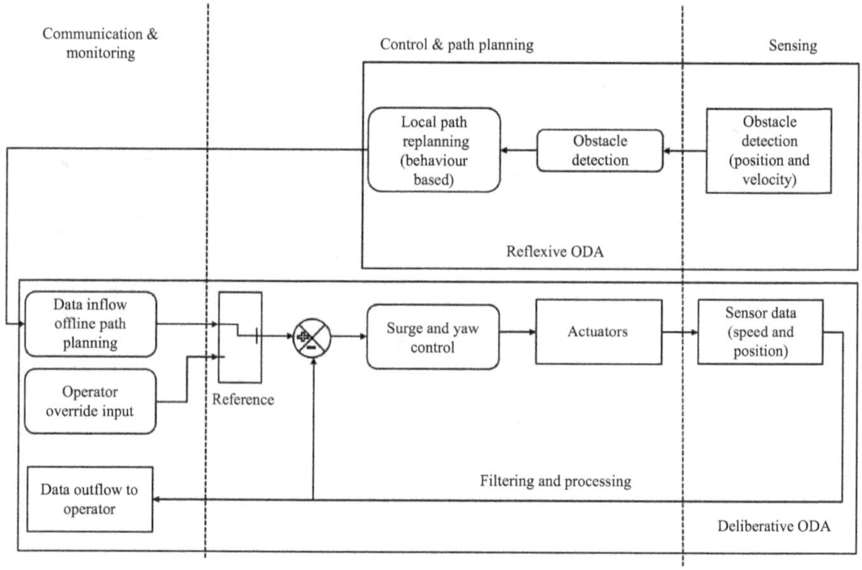

Figure 2.4 General architecture of a USV (Source: [3])

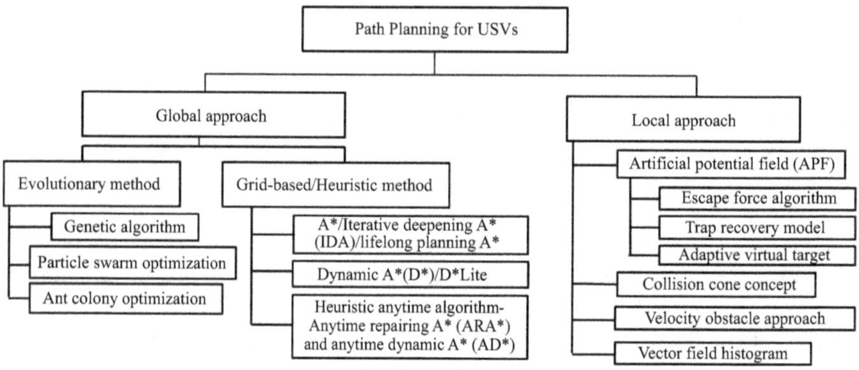

Figure 2.5 Path-planning approaches for USV (Source: [20])

maritime environment, it becomes imperative to develop practical path planners which can be implemented in the real world. Path-planning approaches for USVs can be divided into local (where complete information about environment is not known) and global (where complete information about environment is known) approaches, whose classification is shown in Figure 2.5. Grid-based global path planners have been developed that solves such complex path-planning problem with a substantial low computational time. Such heuristic approaches have been extensively used in robotic path planning for the last two decades and have been the first choice of path planners compared to local approaches (see Figure 2.6).

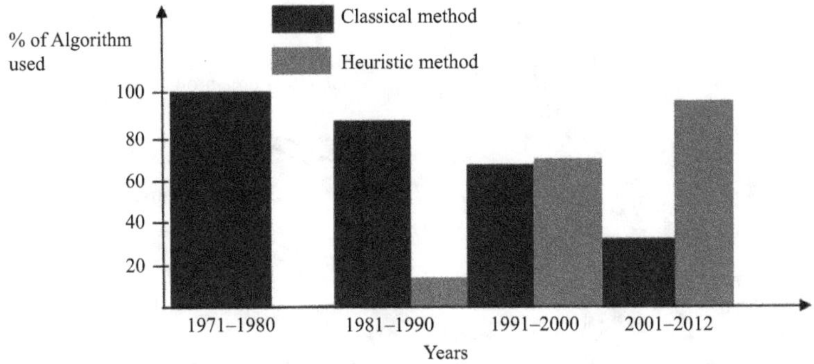

Figure 2.6 Application of classical (local approaches) and heuristic algorithms in the robotic path planning during last four decades (Source: [21])

Grid-based path planners are capable of finding solutions to the practical problems and satisfying objectives like optimality and resolution. Such heuristic methods work fine on configuration spaces like special Euclidean (*SE*) (2) and (3), which are the most popular mapping techniques being used in path planning of marine vehicles [20]. Two of the most successful such planners are Dijkstra [22] and A* algorithms [23], which have been extensively used in robotic path planning, and several of their modified versions have been the most popular approaches [24–26] in the last decade for path planning of USVs.

This chapter gives an overview of the feasibility of a constrained A* approach in the path planning of USVs. The method explores an A* approach with a USV enclosed by a circular boundary as a safety distance constraint on an optimal path curvature to resolve problem of motion planning for a USV as shown in Figure 2.7. The method is able to generate safer way points for a USV navigation with retention of optimality. This chapter also surveys the application of different heuristic methods developed in literature for USV navigation in a maritime environment.

2.1.1 Review of heuristic approaches in path planning of USVs

USVs are non-holonomic systems, characterized by differential equations constrained by time derivatives of state variables. Heuristic or grid-based methods guarantee solutions for finding a path for such systems, if there exist one. Dijkstra initiated the work in the area of heuristic methods being used for path planning by studying shortest path between two nodes specified on a grid map [22], which was later improved in the form of A* approach [23]. During the last decade, many improved versions of A* constrained in terms of computational effort, path length and heading angle have been proposed and applied in robotic path planning, i.e., A* with Post Smoothing [27], Field D* [28], Theta* [29] and D* Lite [26]. Owing to technical similarities between mobile robotics and USVs, some of these techniques have been extended in the path planning of USVs to understand the performance of USVs in the maritime

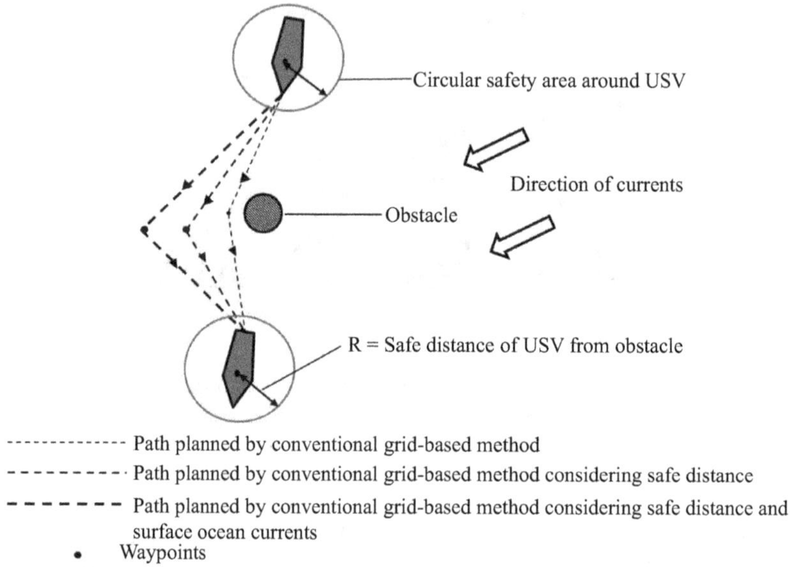

Figure 2.7 A schematic showing the path generated by a conventional grid based path planner compared against the path generated by a grid based path planner by considering safety distance and sea surface currents

environment. The first study towards this was proposed by combining A* approach with a local path planner for USV path planning in a constrained harbour [30]. The same approach was extended in the uncertain sea environment to understand the effectiveness of global path planner with uncertainty [31]. USVs need to comply with International Regulations for Preventing Collisions at Sea (COLREGs) in order to operate safely with IMO guidelines [9]. One of the initial efforts was made by combining A* approach with rule 14 of COLREGs in an environment cluttered with static and dynamic obstacles [32].

USVs are generally constrained by yaw rate and heading angle in real time manoeuvring. A modified A* algorithm, Theta*, for search in 3D Euclidean space at all orientation was implemented for USV path planning complying with heading angle of USV and compared with conventional grid-based 3D path planners [25]. In a further work, the Theta* algorithm was improved in terms of computational time and path length against conventional 3D path planners for USV path planning in the form of ARC-Theta* algorithm [24], which considers angular rate (yaw rate) of USV in path planning. Another improvement in the A* algorithm for USV path planning was proposed by a modifying heuristic for ocean environment with surface currents constrained to heading angle and diverse water depth [33]. Another novel work in area of optimal path planning of USVs has been conducted recently by using fast marching (FM) square (FM^2) algorithm, an optimal approach to FM method by

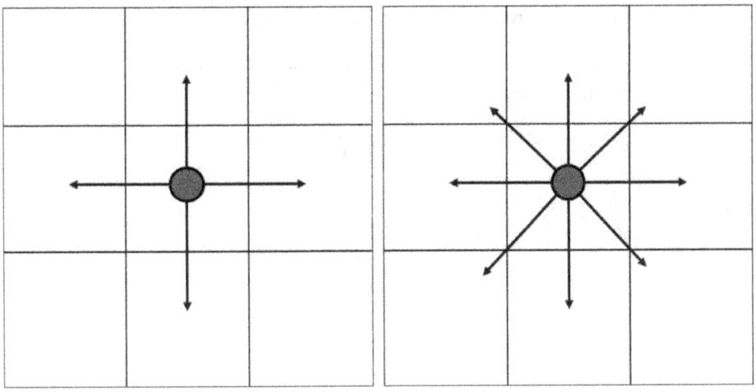

Figure 2.8 Schematic of 4-connectivity and 8-connectivity in C Space

considering environmental disturbances [34]. A modified A* approach, Finite Angle A* (FAA*), was proposed later to comply with the heading and safety requirements of the USV [35].

Until now in the literature, most studies related to path planning have been in the area of autonomous underwater vehicles [36,37], and very few studies have made an attempt to understand the effectiveness of global path planners on USV navigation. In addition to that, most global path-planning studies have not considered environmental effects and safety of USV into account while designing approaches. The proposed approach provides a computationally effective alternative for the optimal USV navigation compared to approaches adopted till now in the literature.

2.1.2 A approach*

The A* algorithm on a gridded map is restricted either to 4-connectivity or 8-connectivity, as shown in Figure 2.8, based on resolution required, where each cell in C Space is evaluated by the value

$$f(n) = h(n) + g(n) \tag{2.1}$$

where $h(n)$ is the heuristic distance of the cell to the goal state and $g(n)$ is the length of the path from initial state to goal state through selected sequence of cells. Each adjacent cell of actually reached cell is evaluated by value of $f(n)$ and the one with lowest value of $f(n)$ is chosen as the next one in sequence. This advantage of modifying distance in A* gives wide range of modifications which can be applied in the algorithm in the form of energy consumption and safety distance [38]. The present study considers the safety distance constraint to study the path planning of USV.

The following pseudocode describes the working of A* algorithm:

Algorithm 2.1: A* Algorithm

```
Data: start, goal(n), h(n), expand(n)
Result: path
Begin;
if goal(start) = true then
    return makePath(start)
end
open ← start
closed ← Ø
while open ≠ Ø do
    sort(open) ;
    n ← open.pop() ;
    kids ← expand(n) ;
    forall kid ∈ kids do
        kid.f ← (n.g + 1) + h(kid);
        if goal(kid) = true then
            return makePath(kid);
            if kid ∩ closed then
                open ← kid;
            end
        end
    end
    closed ← n
end
return Ø
```

2.2 Methodology overview

2.2.1 Environmental mapping

The abstraction of path planning for USVs is shown in Figure 2.9. In order to implement path-planning algorithms, mapping the environment becomes the initial step. Environmental mapping converts world space into C Space which helps in quick implementation of algorithms and manageable storage in computers. The C Space for USV are dynamic in nature with high spatial and temporal variability. This study adopts a popular mapping technique, namely regular occupancy grid due to its effective resolution in grid-based path planners [39]. Portsmouth harbour is among the busiest harbours in United Kingdom and is a perfect area for understanding path planning of USV. In this study, binary images of satellite images of Portsmouth harbour taken from *Google Maps* have been utilized as gridded maps for the proposed A* approach as shown in Figure 2.10. The C Space for the planar USV is considered

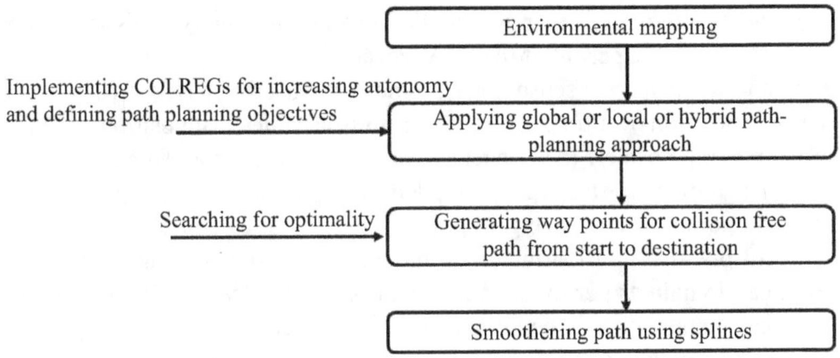

Figure 2.9 Path planning abstraction for USVs [20]

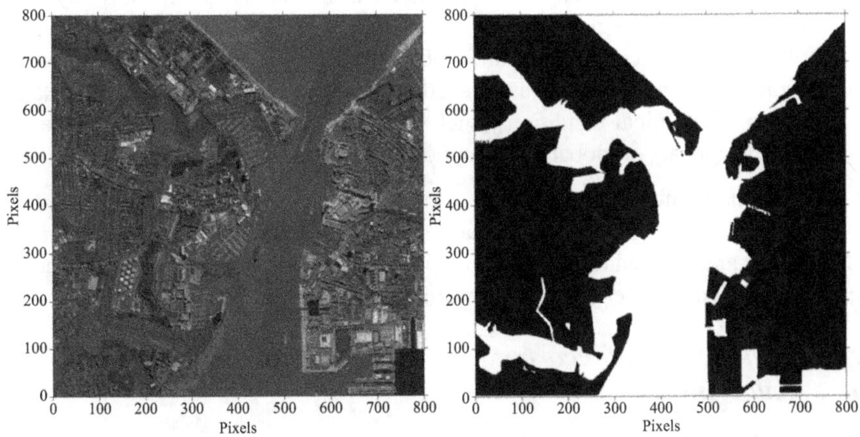

Figure 2.10 Satellite image of Portsmouth harbour and its corresponding binary image (Source: Google Maps)

as R^2, representing the planar positions of the USV where a USV is treated as a pixel point on the map. The map of the environment is the converted binary image where free space is considered as 1 (white) while obstacle is considered as 0 (black). The 800×800 binary image has a resolution of 3.6 m per pixel length.

2.2.2 Assumptions

The complexity of USV path planning is massive, and a number of simplifications have been recommended to reduce the intricacies of the problem [40]. Here, the following assumptions have been made:

1. The map (study area) is considered to be in a confined sea environment near to Portsmouth harbour. Henceforth, temporal and spatial variability in the chosen

study area in terms of environmental effects and moving vessels is considered quasi-static for the period of the USV voyage.
2. Kalman filter and other sensor measurements are used on a USV to determine the obstacle position over time. The current study assumes that position and velocity of the moving obstacle in C Space is known from a Kalman filter.
3. The given moving obstacles are modelled as ellipse on the grid map by combining two grid points, where each grid point comprises of a semi ellipse, since it is a standard practice in a marine environment to consider moving obstacles in an elliptical domain as per the recommendations of the IMO [41]. Overlapping of elliptical shape with grid cell boundary is neglected.
4. The USV is modelled as a particle under the assumption that an effective, robust controller quickly establishes the commanded velocity.
5. The USV is generally having a combination of deliberative and reactive systems on board for planning path in a marine environment. The deliberative systems help in determining global waypoints while reactive systems are responsible for collision avoidance when dynamic obstacles come in the USV safety domain described in Figure 2.7. It is assumed that such reactive collision avoidance takes over in off-nominal conditions, such as a case where a previously undetected obstacle appears or global path planner fails to generate a path.

A schematic of the path-planning system adopted in the current study is shown in Figure 2.11. Information of sea surface current, moving obstacles and topography of the study area is used to define the map in the form of a graph, and the proposed approach is used to generate safer waypoints for an optimal path.

2.2.3 Challenges of incorporating COLREGs in path-planning algorithms

The COLREGs serves as a handbook for selecting avoidance manoeuvres. It is a requirement suggested by IMO for all vessels moving in oceans to ensure operational safety. Recently, several efforts have been made to integrate COLREGs in path-planning algorithms for USVs [42,43]. However, these studies work safely in a scenario with very few complexities with an assumption that each vessel in the operational domain has the same amount of information about the current COLREGs situation and reacts in same way. This hypothesis does not hold true in real time where each sailor interprets COLREGs based on speed, size and heading of the other vessel [44]. In addition to that, various external factors such as limited field of view, ocean currents and seamanship in the case of breaching the COLREGs make it non-trivial to incorporate COLREGs rules into path-planning framework used in complex scenarios.

The present state of the approaches in COLREGs are local in nature and the present study assumes that reactive planner on the USV work satisfactorily in close encounter scenarios. Hence, the proposed approach in the present study do not consider incorporating COLREGs in path planning of USV and makes an effort to plan path in a computationally efficient manner so that local planners have enough time to respond to dynamically changing situations.

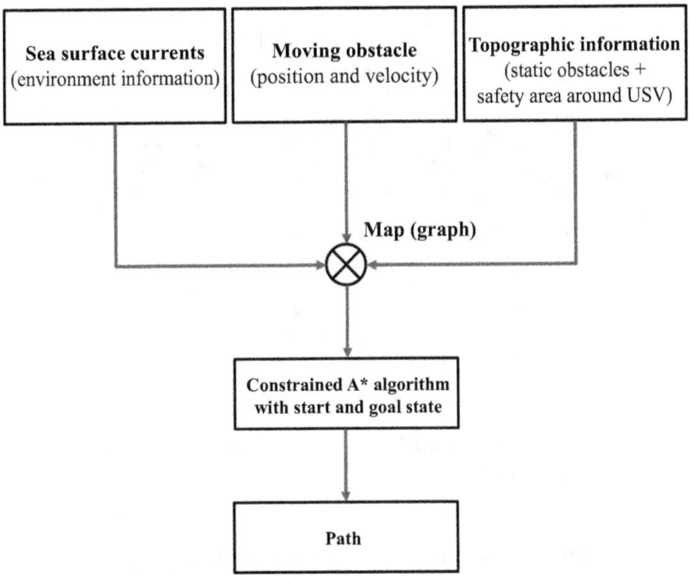

Figure 2.11 Schematic of the proposed path-planning system

2.2.4 Incorporating guidance and control system with path-planning algorithm

The general architecture of a USV operation in a maritime environment has basically three subsystems, namely control and path planning, ODA and communication and monitoring as shown in Figure 2.4.

Path planning is an important subsystem of this architecture responsible for generating waypoints within a desired environment. The current study proposes a computationally effective and safer approach for generation of optimal waypoints for USV navigation in the desired environment. In order to plan and execute a mission in real-time, it is hereby important to interface the guidance and control subsystems with navigation methods and provide quick feedback to the guidance and control subsystems for effective decision-making and higher autonomy.

Conventional waypoint guidance subsystems are designed by reducing surge, sway and yaw (3 DOF) to surge and heading (2 DOF) [45]. Guidance is responsible to achieve motion control objectives in the physical environment in which the vehicle moves [46]. The easiest way is to use a classical autopilot system, so that commanded yaw angle generated from a line-of-sight (LOS) guidance algorithm can be controlled (assuming sufficient bandwidth) and cross track error is minimized. The Figure 2.12 shows a waypoint tracking control system implemented with a standard proportional integral derivative (PID) autopilot in series with a LOS algorithm.

In terms of autopilot and control system development, a detailed review of studies conducted on USVs has been discussed by [48]. Many control techniques like H_∞ [49],

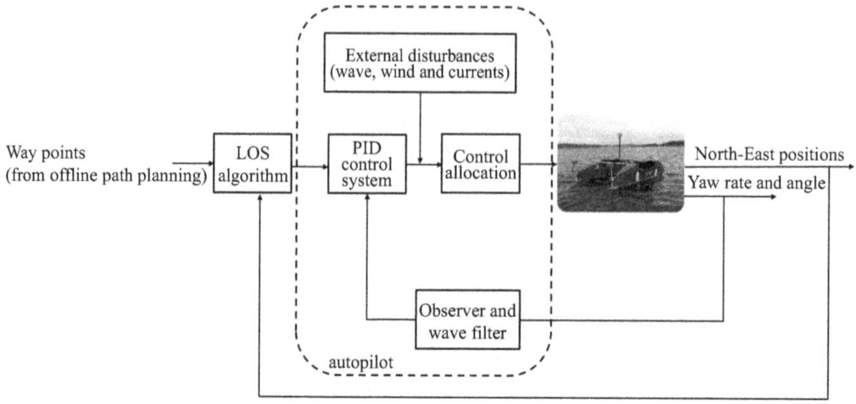

Figure 2.12 PID autopilot with a LOS projection algorithm for way-point tracking (Modified from [47])

linear quadratic Gaussian [50], model predictive control [51] have been proposed recently, together with development of an adaptive control system [52] towards making the controller effective for a range of USV speeds and operating sea conditions.

2.2.5 Collision avoidance in close encounter situation

The general architecture for a USV operation in a maritime environment described in Figure 2.4 shows that high-level planners send waypoints to low-level decision makers, i.e., local control systems and obstacle avoidance subsystems to execute the waypoint following task. When a time variant moving obstacle enters the working domain of the operating USV, it is expected that high-level planners quickly regenerate new set of way points based on the current information of the environment. Many other factors like relative velocity of the USV and the obstacle, the sensing horizon, etc. also play an important role in such regeneration process. In such transition, it is hereby required to have a quick response time from the high level planners which is one of the main objectives of the current study.

In real-time operations, collision avoidance is the most important objective. Since the current study considers inland UK water for operation of USVs, it is imperative to follow the local guidelines towards the development of a path planner and collision avoidance with moving obstacles. To enable the safe and secure operation of autonomous surface ships within the existing IMO requirement, a code of practice has been prepared by the UK Maritime Autonomous Systems Working Group (MASRWG) and published by Maritime UK through the Society of Maritime Industries [53]. Under this code of practice, all autonomous ships working within the UK waters have the following six levels of autonomy as developed by the European Defence Agency [53]:

1. Human on board
2. Operated

3. Directed
4. Delegated
5. Monitored
6. Autonomous

The current state of operation of USVs is either at level 3 or level 4 of the autonomy, where there is always a human-in-the loop towards monitoring the operation of USVs. In a case where an unknown obstacle of uncertain trajectory and nature enters into the domain of the USV and collision cannot be avoided, a few emergency actions like abort and stop are employed in response to fault conditions.

2.3 Simulation results

The proposed approach is simulated using C++ and OpenCV. All simulations are performed on a PC with *Microsoft* Windows 7 as OS with Intel i5 2.70 GHz quad core CPU and 16 GB RAM. The simulations were repeated for 500 times, especially in terms of computational time, to account for variable computational power in OS Windows. The average time from all repetitions was calculated for proper verification of the proposed approach.

2.3.1 Comparing A approach with and without safety distance*

The proposed study deals with inclusion of a safety distance criteria in the A* approach towards USV path planning. In order to benchmark the safety distance approach and to decide upon an optimum value of safety distance, four arbitrary values, 10, 20, 30 and 40 pixel are taken as safety distance on a grid map (as shown in Figure 2.10) and compared against an A* approach without safety distance in terms of computational time. The start and goal states used in the path-planning system are depicted on the binary map, as shown in Figure 2.13.

Figure 2.14 shows the comparison of A* approach with and without safety distance constraint in terms of computational time. The results show that on an R^2 grid map, a larger safety distance constraint produces computationally efficient path in a A* approach against paths produced without such constraint. This is due to the fact that search process explores lesser number of nodes with safety distance than without safety distance by pruning the search domain.

In terms of path length, simulations shows that the A* approach with and without safety distance constraint produces path of equal length, i.e., 1.043 km, although a difference in resultant path can be seen in Figure 2.15. This difference in resultant paths is less visible in smaller safety distance values, while a more noticeable difference is observed in paths produced with larger safety distance. This leads to the fact that optimality remains conserved in path planning with decrease in computational effort in the proposed approach unlike ones adopted in literature towards path planning of USV where an increase in computational cost has been observed with increase in path length for proposed approaches.

Since the current study considers a narrow channel of Portsmouth harbour for path planning of USV, it becomes necessary to choose a safety distance where a

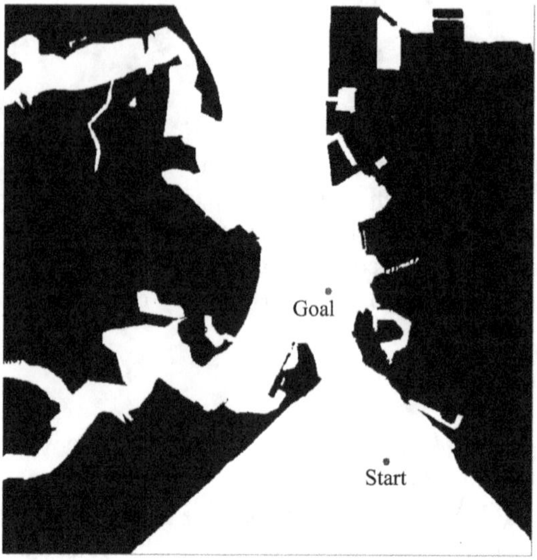

Figure 2.13 Binary map with start and goal states

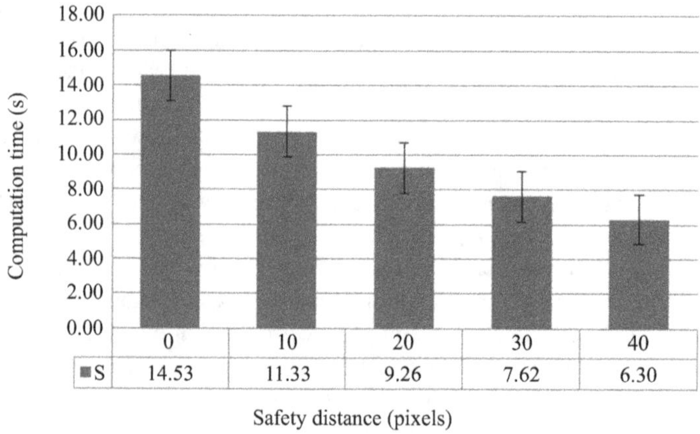

Figure 2.14 Compared computational time of A approach with and without safety distance constraint. The interval on each bar denotes the standard deviation of the computational time*

proper trade-off between computational time and safety distance from an obstacle can be maintained. Therefore, a 20 pixel safety distance (72 m on real map) has been chosen for the present study. This value also provides enough time for local reactive techniques for collision avoidance in the case where one or more moving obstacles are detected in the operational domain of the USV.

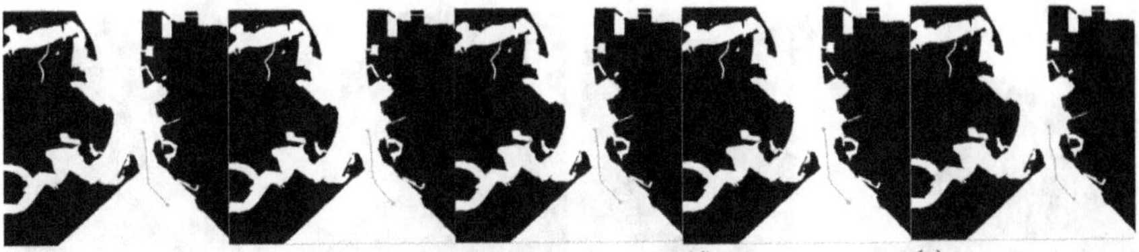

Figure 2.15 *Resultant path with safety distance of (a) 0, (b) 10, (c) 20, (d) 30 and (e) 40 pixel*

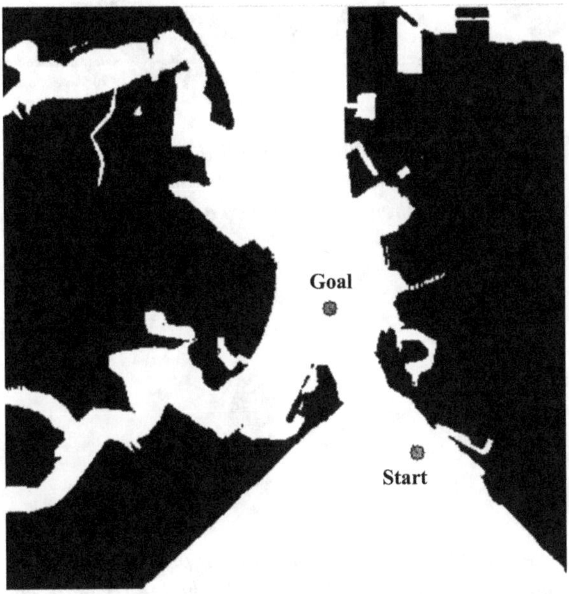

Figure 2.16 *Binary map with start and goal states for simulating A* approach under static and partially dynamic environment*

2.3.2 Constrained A* approach under static and partially dynamic environment

In order to understand the effectiveness of the proposed approach, simulations are conducted in binary maps of Portsmouth harbour comprising static obstacles as well as moving obstacles. Such an environment which consists of moving and stationary obstacles is termed as a partially dynamic environment. The effectiveness is defined in terms of path length and computational time obtained in simulations. In simulations, a stationary map with one and two moving obstacles for a constrained channel having start and goal points as defined in Figure 2.16 in Portsmouth harbour has been considered. A binary map of the simulation area with single and two moving obstacles is shown in Figure 2.17.

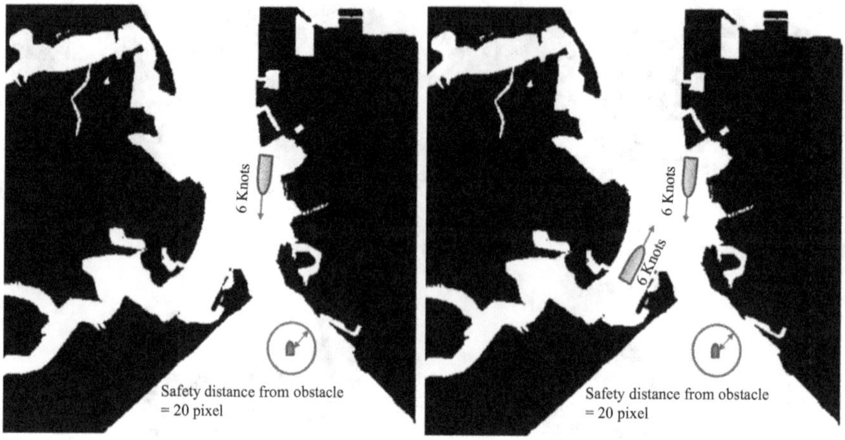

Figure 2.17 Binary map of the simulation area (Portsmouth harbour) showing velocity and direction of moving obstacles. In this study, 20 pixel has been chosen as the safety distance around a USV

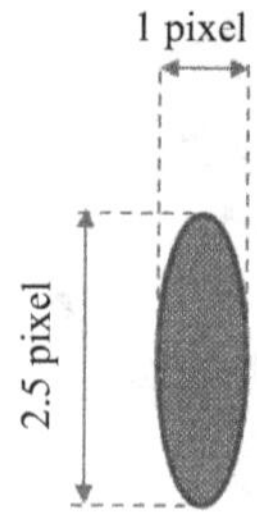

Figure 2.18 Dimension of the elliptical domain representing the encapsulation of a moving obstacle in a static one in the binary map. The dimensions of ellipse are chosen in accordance with the dimensions of high-speed craft having operational velocity range from 6 to 9 kn

Modelling of dynamic obstacles on a map for maritime path planning is defined in terms of the velocity of the moving obstacle in maritime environment. Reference [54] has suggested a circular shape for slow moving obstacles and elliptical shapes for fast-moving obstacles. Therefore, an elliptical shape as shown in Figure 2.18 has been adopted in the current study.

The results presented in Figure 2.19 show a path generated by the proposed approach in different scenarios of single moving obstacle. The scenarios presented in the figure shows a single moving obstacle moving in a straight line at a velocity of 6 kn, based upon its start point shown in Figure 2.16 and considers each instantaneous dynamic situation as static (based on the conventional method adopted in deliberative path planning by [55]). Path length and computational time are computed for each start time of the mission, and results are shown in Figures 2.20 and 2.21. It is found

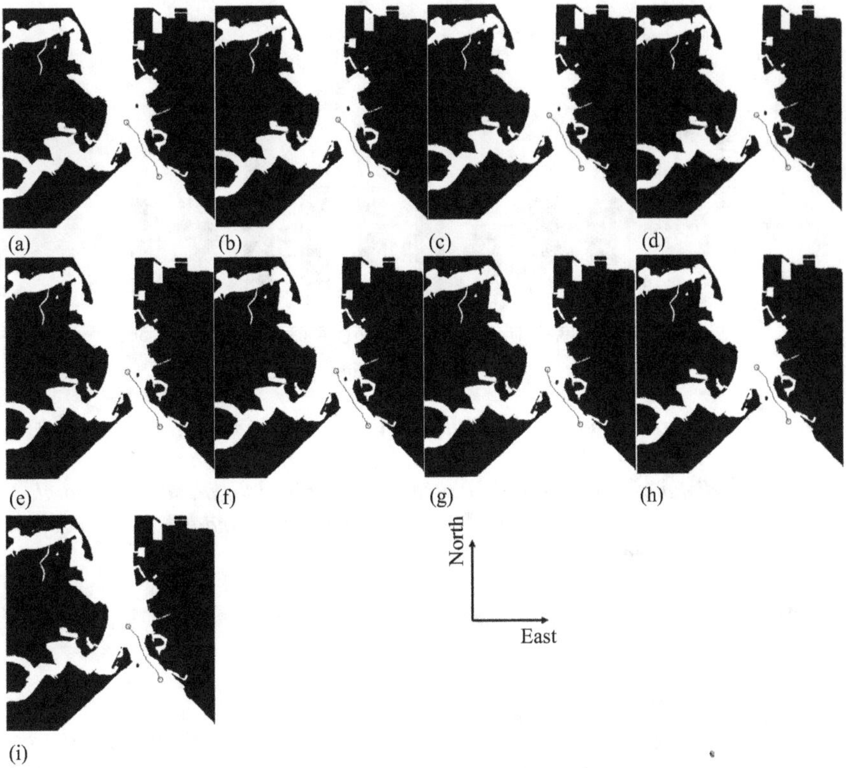

Figure 2.19 Comparison of paths obtained with different start time (a) 0, (b) 10, (c) 20, (d) 30, (e) 40, (f) 50, (g) 60, (h) 100 and (i) 120 s. Position of the single moving obstacle is plotted at each start time on the binary map, based on the velocity and direction mentioned in Figure 2.17

that as the moving obstacle approaches the safety domain of the USV, an increase in path distance is observed.

This is owing to the fact that vehicle moves further east, as shown in Figure 2.19(h), to maintain the constraint of keeping a safety distance of 20 pixel. In addition to that, a decrease in computational effort is observed with increase in path length once the moving obstacle is detected within the safety domain of USV. This is because the search space in the gridded map gets pruned in the proposed approach which leads to generation of longer path length with decrease in computational time. The computational time again increases once the moving obstacle escapes out of the safety domain of the USV.

In order to make the environment more complex and more cluttered, a scenario with two moving obstacles is considered for understanding the effectiveness of the proposed approach as shown in Figure 2.17 (right side). The results shown in Figure 2.22 show path generated by proposed approach in different scenarios of

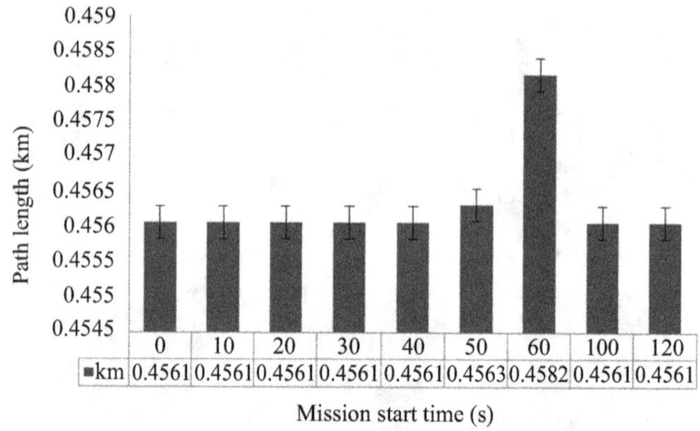

Figure 2.20 Comparison of path length obtained with different start time (a) 0, (b) 10, (c) 20, (d) 30, (e) 40, (f) 50, (g) 60, (h) 100 and (i) 120 s. A safety distance constraint of 20 pixel is maintained for all scenarios in the figure. The interval on each bar denotes the standard deviation of the path length

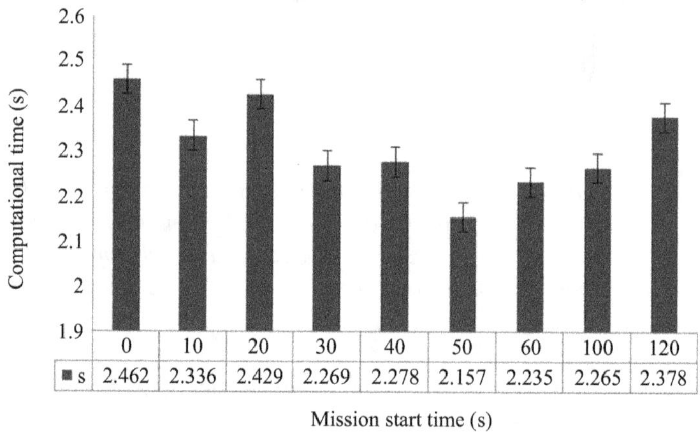

Figure 2.21 Comparison of computational time obtained with different start time (a) 0, (b) 10, (c) 20, (d) 30, (e) 40, (f) 50, (g) 60, (h) 100 and (i) 120 s. The interval on each bar denotes the standard deviation of the computational time

a maritime environment with two moving obstacles. The scenarios presented in the figure show both moving obstacles are moving in a straight line at a velocity of 6 kn, based upon their start points shown in Figure 2.16.

In this case also, the same pattern as found with the single moving obstacle scenario is observed. The comparison of path length and computational time is shown in

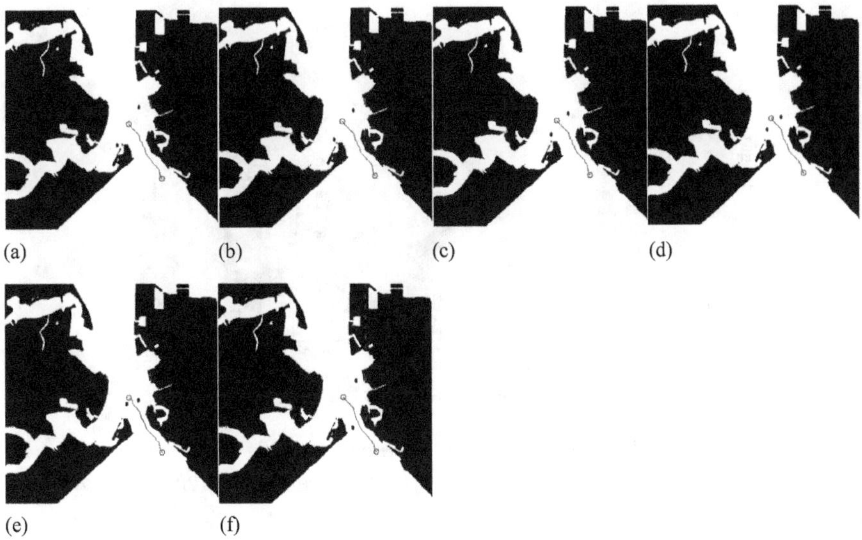

Figure 2.22 Comparison of paths obtained with different start time (a) 0, (b) 10, (c) 20, (d) 30, (e) 40 and (f) 100 s. A safety distance constraint of 20 pixel is maintained for all scenarios in the figure. Position of both moving obstacles is plotted at each start time on the binary map, based on the velocity and direction mentioned in Figure 2.17

Figures 2.23 and 2.24, respectively. The path length increases once the moving obstacles approaches the safety domain of the USV in order to maintain the safety distance constraint. This fact is reflected in the resultant paths obtained in different scenarios where a change in resultant path is obtained when moving obstacle approaches USV. With increase in path length, a decrease in computational time is observed. The computational time retains the increased value once the moving obstacles escape out of the safety domain of the USV.

2.3.3 Constrained A* approach with environmental disturbances

Ocean currents generated in the upper layer of the ocean environment by atmospheric wind system are referred as sea surface currents [56]. In the current study, the effect of steady non-uniform headwind and tailwind currents on USV navigation has been studied for the proposed approach. In general, ocean currents are provided in a NetCDF data format by various meteorological agencies around the world. Such data obtained from satellites have a resolution of 2 km [57], while the range of most navigation devices is less than 5 nmi which makes such data low in precision and not suitable for USV path planning. Hence, the synthetic vector field of moderate and strong intensity is created within the map to verify the effect of current on optimal path planning. Real ocean currents are multidirectional and irregular, spatially and temporally. In the

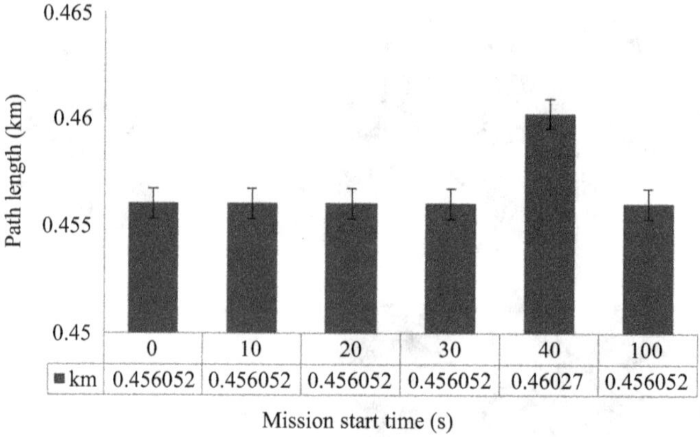

Figure 2.23 Comparison of path length obtained with different start time 0, 10, 20, 30, 40 and 100 s. The interval on each bar denotes the standard deviation of the path length

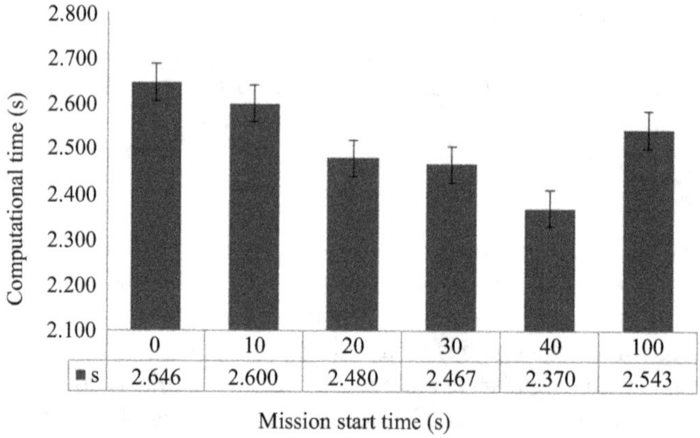

Figure 2.24 Comparison of computational time obtained with different start time of 0, 10, 20, 30, 40 and 100 s. The interval on each bar denotes the standard deviation of the computational time

present study, current effect on USV path planning is simplified as a constant disturbance by assuming the current to be unchanged over a period of time [58]. Two current scenarios, a moderate current intensity of 1.5 m/s and a strong current intensity of 2.5 m/s, are considered for the present study. These values are chosen on observation of high speed currents of 2–3 m/s in coastal regions [56].

In order to understand the steady non-uniform headwind and tailwind effects of current on path planning, clockwise and anticlockwise directions of chosen intensity

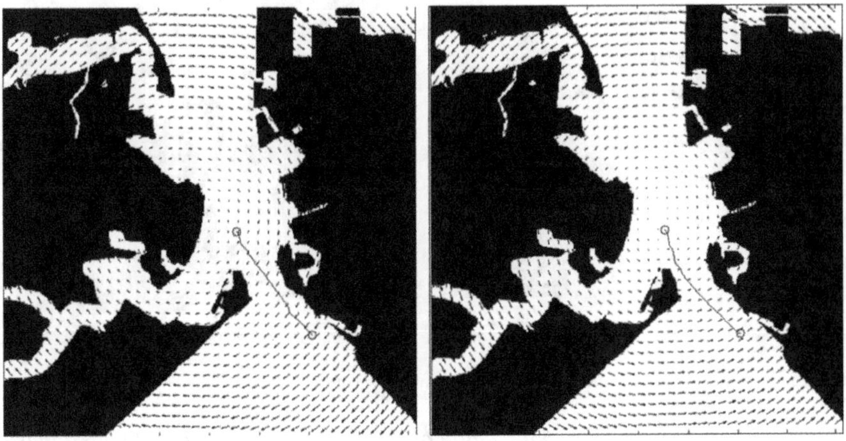

(a) (b)

Figure 2.25 Comparison of paths obtained for currents moving with intensity of 1.4 m/s in (a) anticlockwise and (b) clockwise direction. The start and goal states are same as shown in Figure 2.16. The safety distance constraint of 20 pixel is maintained for both scenarios

values are taken in the present study. Figure 2.25 shows the path obtained by the proposed approach with currents moving in anticlockwise and clockwise direction with intensity of 1.4 m/s. Path length and computational time are compared for both scenarios shown in Figure 2.25 and results are presented in Figures 2.26 and 2.27, respectively. The results show that when the USV operates in steady non-uniform tailwind currents, it has to cover a larger distance in current, while a smaller distance is observed in steady non-uniform headwind currents. This is due to the fact that presence of steady non-uniform tailwind currents in the USV voyage creates larger forces in the sway motion, directing the USV to move closer to the shore line (as seen in Figure 2.25(a)), which leads to generation of a path with a longer curvature. The current approach has been able to demonstrate a decrease in computational effort to find path when higher distance voyages are observed under influence of sea surface currents.

Along the same line, currents of 2.5 m/s are considered to understand the path-planning pattern of USV under influence of strong ocean currents. Figure 2.28 shows the path obtained by the proposed approach with currents moving in anticlockwise and clockwise direction with intensity of 2.5 m/s. Path length and computational time are compared for both scenarios shown in Figure 2.28 and results are shown in Figures 2.29 and 2.30, respectively. In this case also, a similar pattern as found with 1.4 m/s has been observed. In terms of current intensities of different magnitude moving in same direction (from a comparison of path length values for AC currents in Figures 2.26 and 2.28), it has been found that currents of higher magnitude are more favourable in minimizing energy usage for USV voyage with a no substantial

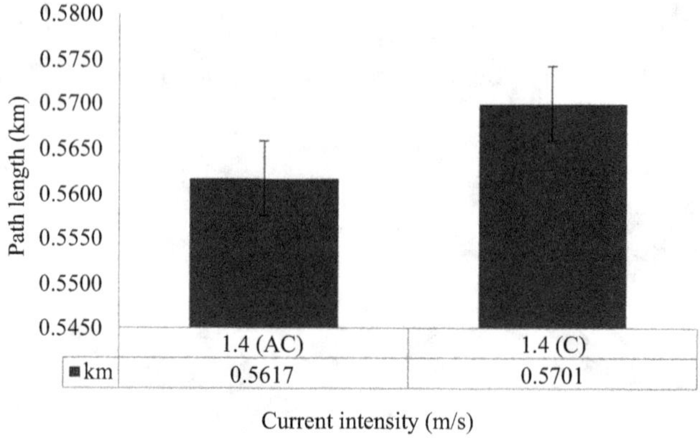

Figure 2.26 Comparison of path length obtained for currents moving with intensity of 1.4 m/s in anticlockwise (AC) and clockwise (C) directions. The interval on each bar denotes the standard deviation of the path length

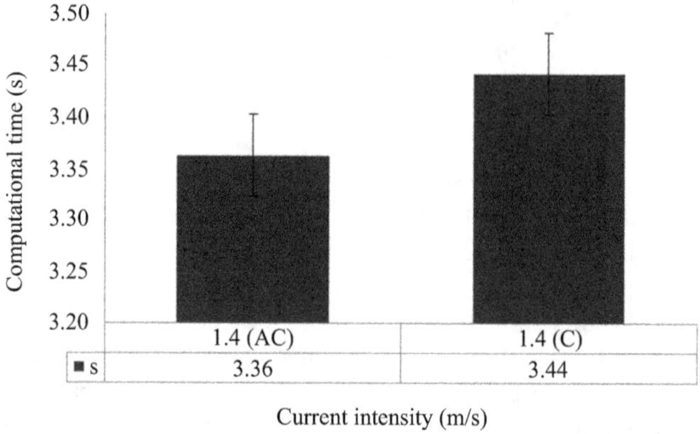

Figure 2.27 Comparison of computational time to determine path obtained for currents moving with intensity of 1.4 m/s in anticlockwise (AC) and clockwise (C) directions. The interval on each bar denotes the standard deviation of the computational time

increase in computational effort. This leads to the fact that proposed approach can assist USV in utilizing the ocean environment intelligently to minimize energy usage by integrating current information with path planner.

2.3.4 Constrained A* approach with single moving obstacle and environmental disturbance

In order to create a more complete picture of the operational environment near to Portsmouth harbour and to analyse the effectiveness of the proposed approach in

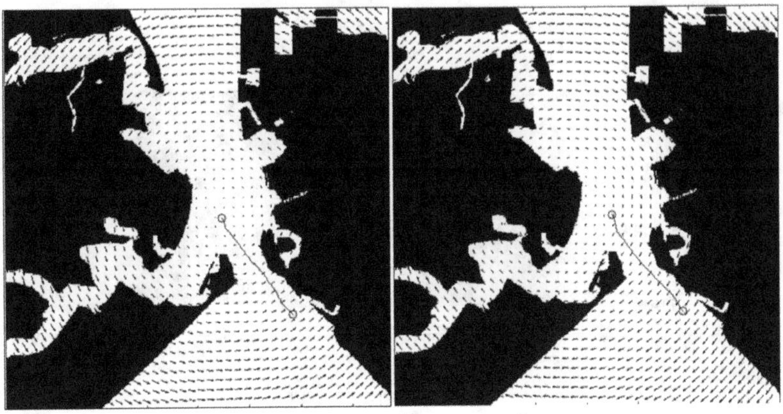

(a) (b)

Figure 2.28 Comparison of paths obtained for currents moving with intensity of 2.5 m/s in (a) anticlockwise and (b) clockwise direction. The start and goal states are same as shown in Figure 2.16. The safety distance constraint of 20 pixel is maintained for both scenarios

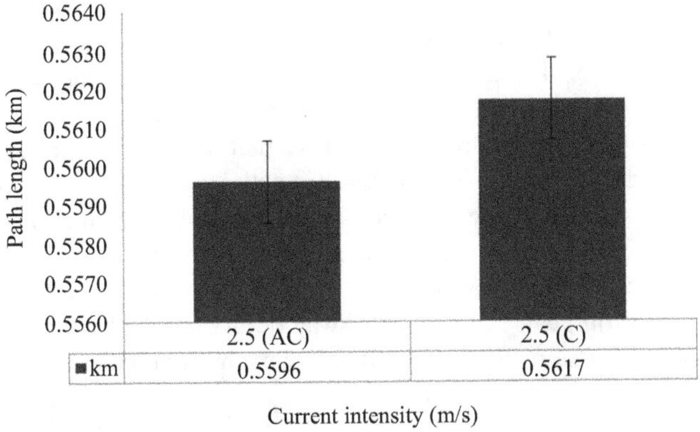

Figure 2.29 Comparison of path length obtained for currents moving with intensity of 2.5 m/s in anticlockwise (AC) and clockwise (C) directions. The interval on each bar denotes the standard deviation of the path length

cluttered environment, a single moving obstacle is introduced in the map in the presence of sea surface currents of moderate intensity as discussed in Section 2.3.3. Since the complexity of the environment has increased, a more flexible safety distance constraint of 15 pixel has been adopted for this study in order to keep a proper trade-off between optimal way points and environmental complexity. Figure 2.31 shows the generated paths for different start time in the environment comprising of static obstacle, sea surface currents of 1.4 m/s moving in anticlockwise direction and moving

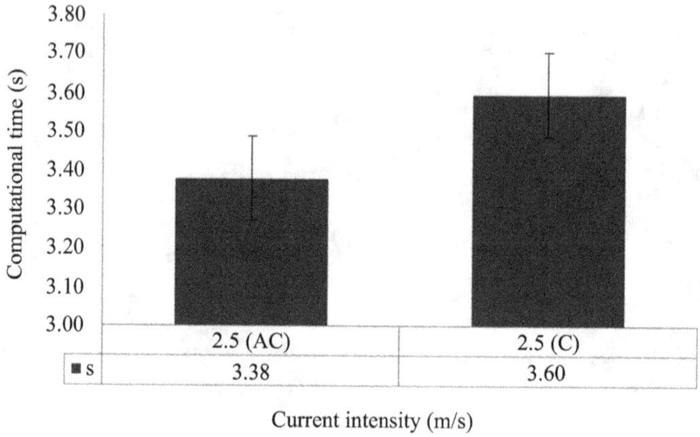

Figure 2.30 Comparison of computational time to determine path obtained for currents moving with intensity of 2.5 m/s in anticlockwise (AC) and clockwise (C) directions. The interval on each bar denotes the standard deviation of the computational time

obstacle (where each dynamic position is considered static). Comparison of path length and computational time for all scenarios presented in Figure 2.31 are shown in Figures 2.32 and 2.33, respectively.

From the results obtained, one can observe that as the obstacle approaches the safety domain of the USV, there is an increase in path length observed with a decrease in computational effort for cases (as found with mission start time of 30 and 60 s), where an increased path length is observed. In addition to that, most cases have been able to generate path within a reasonable computational time. These results show that the proposed algorithm can generate safer way points for the USV voyage for long and short duration missions in a cluttered complex environment.

2.4 Conclusions

In this chapter, a constrained A* approach for optimal path planning of USVs in a confined maritime environment is proposed. The objective of generating safer way points by keeping a safe distance from the obstacle was evaluated in simulations, conducted in various environments comprising static obstacle, moving obstacle and sea surface currents of different intensities. The upstream and effects of sea surface currents were evaluated, and effect of sea surface currents with moving obstacle was also analysed. The simulation results shows that the present approach generates safer way points for USV voyage in a computationally efficient manner against the conventional A* approach with no loss of optimality. The approach is found to be robust, computationally efficient and can be extended for real time path planning of USVs in

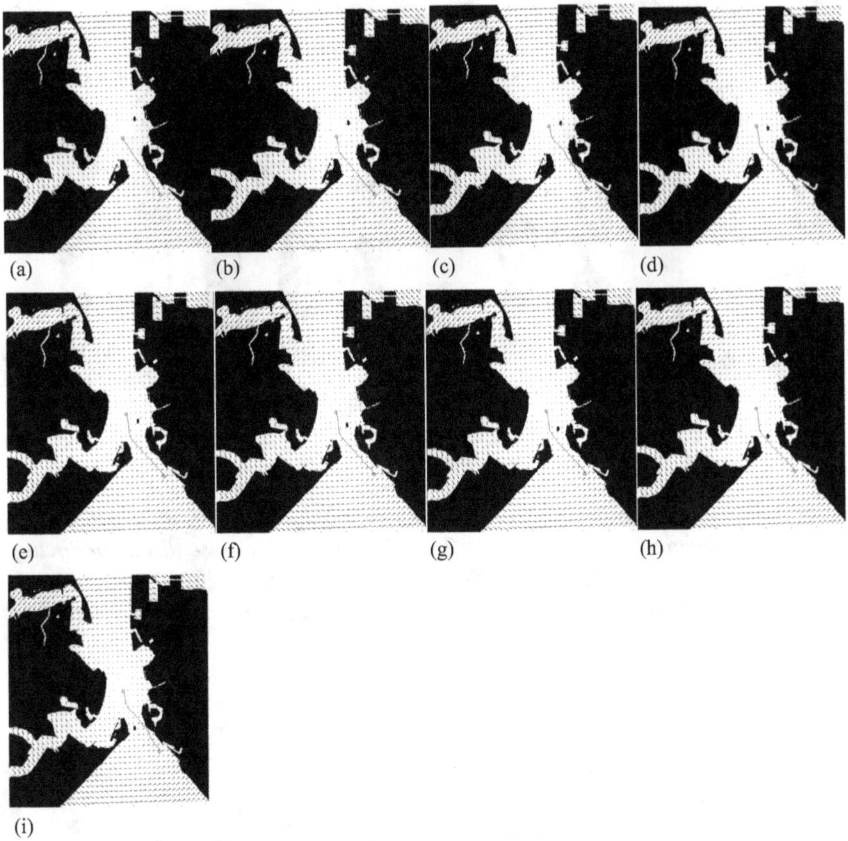

Figure 2.31 Comparison of paths obtained for scenario with sea surface currents of 1.4 m/s moving in anticlockwise direction having a moving obstacle for different start time of (a) 0, (b) 10, (c) 20, (d) 30, (e) 40, (f) 50, (g) 60, (h) 80 and (i) 100 s. The start and goal states are the same as shown in Figure 2.16. The safety distance constraint of 15 pixel is maintained for all scenarios

confined water. In conclusion, it is evident that such an optimal approach is suitable for global path planning of USV. In future work, it is planned to extend the work in development of a path follower approach working in conjugation with proposed approach for a reactive path planning in scenarios involving close encounters. Another extension of the present work lies in considering heading angle constraint for USV, in such cases, where path length is more important than computational time. This converts the problem from an R^2 to an *SE(2)* path-planning approach.

Most leading companies in USV operations are looking for the integration of COLREGs with optimal path planners to abide the working guidelines of the IMO. A challenging extension of the current work lies in the fact of finding a heuristic cost

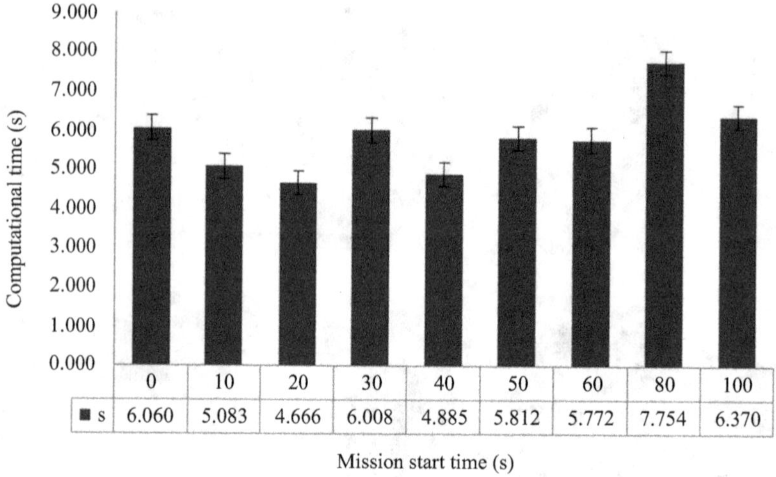

Figure 2.32 Comparison of computational time obtained for scenario with sea surface currents of 1.4 m/s moving in anticlockwise direction having a moving obstacle for different start time of 0, 10, 20, 30, 40, 50, 60, 80 and 100 s. The interval on each bar denotes the standard deviation of the computational time

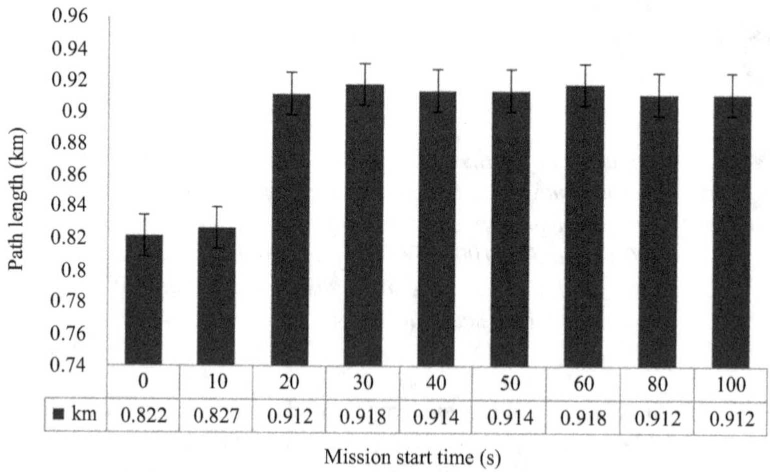

Figure 2.33 Comparison of path length obtained for scenario with sea surface currents of 1.4 m/s moving in anticlockwise direction having a moving obstacle for different start time of 0, 10, 20, 30, 40, 50, 60, 80 and 100 s. The interval on each bar denotes the standard deviation of the path length

function which can take into account rules of the COLREGs without compromising the optimality and computational effort required to find a feasible trajectory.

References

[1] Latombe JC. Robot Motion Planning. vol. 124. Stanford, CA: Springer Science & Business Media; 2012.

[2] Laumond JP, Jacobs PE, Taix M, *et al.* A motion planner for nonholonomic mobile robots. IEEE Transactions on Robotics and Automation. 1994;10(5):577–593.

[3] Campbell S, Naeem W, and Irwin GW. A review on improving the autonomy of unmanned surface vehicles through intelligent collision avoidance manoeuvres. Annual Reviews in Control. 2012;36(2):267–283.

[4] El-Hawary F. The Ocean Engineering Handbook. Boca Raton, FL: CRC Press; 2008.

[5] Motwani A. A Survey of Uninhabited Surface Vehicles. Technical Report. Marine and Industrial Dynamic Analysis, School of Marine Science and Engineering, Plymouth University; 2012. p. 1–54.

[6] Naeem W, Sutton R, Sharma SK, *et al.* Experimental results on the development of a pollutant tracking unmanned surface vessel. Control Applications in Marine Systems. 2007;7:361–366.

[7] Alves J, Oliveira R, Oliveira P, *et al.* Vehicle and mission control of the Delfim autonomous surface craft. In: 14th Mediterranean Conference on Control and Automation (MED'06); 2006.

[8] Caccia M, Bruzzone G, and Bono R. A practical approach to modelling and identification of small autonomous surface craft. IEEE Journal of Oceanic Engineering. 2008;33:133–145.

[9] IMO. Code on Intact Stability for All Types of Ships. A749 (18), Amended by Resolution MSC75l; 1988. p. 69.

[10] IMO. Guidance to the Master for Avoiding Dangerous Situations in Following and Quartering Seas. MSC/Circ707; 1995.

[11] IMO. Revised Guidance to the Master for Avoiding Dangerous Situations in Adverse Weather and Sea Conditions. MSC/Circ1228; 2007.

[12] UK Maritime and Coastguard Agency. The Maritime Autonomous Surface Ships Code of Practice. MCA; 2017.

[13] Laumond JP. Robot Motion Planning and Control. Stanford, CA: Springer; 1998.

[14] Canny J. The Complexity of Robot Motion Planning. Boston, MA: MIT Press; 1988.

[15] van der Stappen AF, Halperin D, and Overmars MH. The complexity of the free space for a robot moving amidst fat obstacles. Computational Geometry. 1993;3(6):353–373.

[16] van der Stappen AF, Overmars MH, de Berg M, *et al.* Motion planning in environments with low obstacle density. Discrete & Computational Geometry. 1998;20(4):561–587.

[17] van der Stappen AF, and Overmars MH. Motion planning amidst fat obstacles. In: Proceedings of the Tenth Annual Symposium on Computational Geometry. ACM; 1994. p. 31–40.

[18] Schwartz JT, and Sharir M. A survey of motion planning and related geometric algorithms. Artificial Intelligence. 1988;37(1–3):157–169.

[19] Lazard S. Planification de trajectoires de robots mobiles non-holonomes et de robots à pattes. Université Pierre et Marie Curie-Paris VI; 1996.

[20] Singh Y, Sharma S, Hatton D, *et al.* Optimal path planning of marine vehicles. Indian Journal of Geo-Marine Sciences. 2018;47(7):1325–1334.

[21] Mohanty PK, and Parhi DR. Controlling the motion of an autonomous mobile robot using various techniques: a review. Journal of Advanced Mechanical Engineering. 2013;1(1):24–39.

[22] Dijkstra E. A note on two problems in connexion with graphs. Numerische Mathematik. 1959;1:269–271.

[23] Hart PE, Nils JN, and Bertram R. A formal basis for the heuristic determination of minimum cost paths. IEEE Transactions on Systems Science and Cybernetics. 1968;4:100–107.

[24] Kim H, Kim D, Shin J, *et al.* Angular rate-constrained path planning algorithm for unmanned surface vehicles. Ocean Engineering. 2014;84(Supplement C):37–44. Available from: http://www.sciencedirect.com/science/article/pii/S0029801814001292.

[25] Kim H, Lee T, Chung H, *et al.* Any-angle path planning with limit-cycle circle set for marine surface vehicle. In: Robotics and Automation (ICRA), 2012 IEEE International Conference on. IEEE; 2012. p. 2275–2280.

[26] Koenig S, and Likhachev M. D* Lite. AAAI/IAAI. 2002;476–483.

[27] Rabin S. A* speed optimizations. Game Programming Gems. 2000;1:272–287.

[28] David F, and Anthony S. The Field D* Algorithm for Improved Path Planning and Replanning in Uniform and Non-Uniform Cost Environments. CMU-RI-TR-05-19. Pittsburgh, PA: Carnegie Mellon University; 2005.

[29] Nash A, Daniel K, Koenig S, *et al.* Theta*: Any-angle path planning on grids. In: AAAI; 2007. p. 1177–1183.

[30] Casalino G, Turetta A, and Simetti E. A three-layered architecture for real time path planning and obstacle avoidance for surveillance USVs operating in harbour fields. In: Oceans 2009-Europe. IEEE; 2009. p. 1–8.

[31] Svec P, Schwartz M, Thakur A, *et al.* Trajectory planning with look-ahead for Unmanned Sea Surface Vehicles to handle environmental disturbances. In: 2011 IEEE/RSJ International Conference on Intelligent Robots and Systems; 2011. p. 1154–1159.

[32] Naeem W, Irwin G, and Yang A. COLREGs-based collision avoidance strategies for unmanned surface vehicles. Mechatronics. 2012;22(6):669–678. Available from: http://www.sciencedirect.com/science/article/pii/S0957415811001553.

[33] Lee T, Kim H, Chung H, *et al.* Energy efficient path planning for a marine surface vehicle considering heading angle. Ocean Engineering. 2015;107: 118–131.

[34] Garrido S, and Moreno L. Cálculo de trayectorias para vehıculos autónomos marinos (USV) utilizando el método Fast Marching Square sometido a campo vectorial. In: XXXVII Jornadas De Automatica, Madrid; 2016.

[35] Yang J, Tseng C, and Fan C. Collision-free path planning for unmanned surface vehicle by using advanced A* algorithm. In: 26th Asian-Pacific Technical Exchange and Advisory Meeting on Marine Structure; 2012. p. 251–256.

[36] Alvarez A, Caiti A, and Onken R. Evolutionary path planning for autonomous underwater vehicles in a variable ocean. IEEE Journal of Oceanic Engineering. 2004;29(2):418–429.

[37] Kruger D, Stolkin R, Blum A, *et al.* Optimal AUV path planning for extended missions in complex, fast-flowing estuarine environments. In: Robotics and Automation, 2007 IEEE International Conference on. IEEE; 2007. p. 4265–4270.

[38] Duchoň F, Babinec A, Kajan M, *et al.* Path planning with modified a star algorithm for a mobile robot. Procedia Engineering. 2014;96:59–69.

[39] Mooney P, Corcoran P, and Winstanley A. Towards quality metrics for Open-StreetMap. In: Proceedings of the 18th SIGSPATIAL International Conference on Advances in Geographic Information Systems. ACM; 2010. p. 514–517.

[40] Azariadis P, and Aspragathos N. Obstacle representation by Bump-surfaces for optimal motion-planning. Robotics and Autonomous Systems. 2005; 51(2–3):129–150.

[41] Tam C, Bucknall R, and Greig A. Review of collision avoidance and path planning methods for ships in close range encounters. The Journal of Navigation. 2009;62(3):455–476.

[42] Svec P, Shah B, Bertaska I, *et al.* Dynamics-aware target following for an autonomous surface vehicle operating under COLREGs in civilian traffic. In: Intelligent Robots and Systems (IROS), 2013 IEEE/RSJ International Conference on. IEEE; 2013. p. 3871–3878.

[43] Kuwata Y, Wolf M, Zarzhitsky D, *et al.* Safe maritime autonomous navigation with COLREGS, using velocity obstacles. IEEE Journal of Oceanic Engineering. 2014;39(1):110–119.

[44] Shah B. Planning for Autonomous Operation of Unmanned Surface Vehicles. University of Maryland, College Park, MD; 2016.

[45] Healey AJ, and Marco D. Slow speed flight control of autonomous underwater vehicles: experimental results with NPS AUV II. In: The Second International Offshore and Polar Engineering Conference. International Society of Offshore and Polar Engineers; 1992.

[46] Bibuli M, Bruzzone G, Caccia M, *et al.* Path-following algorithms and experiments for an unmanned surface vehicle. Journal of Field Robotics. 2009;26(8):669–688.

[47] Fossen TI, Breivik M, and Skjetne R. Line-of-sight path following of underactuated marine craft. IFAC Proceedings Volumes. 2003;36(21):211–216.

[48] Roberts G. Trends in marine control systems. Annual Reviews in Control. 2008;32(2):263–269.

[49] Lefeber E, Pettersen KY, and Nijmeijer H. Tracking control of an underactuated ship. IEEE Transactions on Control Systems Technology. 2003;11(1):52–61.

[50] Sharma S, Naeem W, and Sutton R. An autopilot based on a local control network design for an unmanned surface vehicle. The Journal of Navigation. 2012;65(2):281–301.

[51] Annamalai K, Sutton R, Yang C, *et al.* Robust adaptive control of an uninhabited surface vehicle. Journal of Intelligent & Robotic Systems. 2015;78(2):319.

[52] Sharma SK, Sutton R, Motwani A, *et al.* Non-linear control algorithms for an unmanned surface vehicle. Proceedings of the Institution of mechanical engineers, Part M: Journal of Engineering for the Maritime Environment. 2014;228(2):146–155.

[53] Maritime UK. The Maritime Autonomous Surface Ships Code of Practice; 2017. Available from: https://www.maritimeuk.org/media-centre/publications/being-responsible-industry-industry-code-practice/.

[54] Liu Y, and Bucknall R. Path planning algorithm for unmanned surface vehicle formations in a practical maritime environment. Ocean Engineering. 2015;97:126–144.

[55] Borenstein J, and Koren Y. The vector field histogram-fast obstacle avoidance for mobile robots. IEEE Transactions on Robotics and Automation. 1991;7(3):278–288.

[56] Fossen T. Guidance and Control of Ocean Vehicles. New York, NY: Wiley; 1995.

[57] Bonnett R, and Campbell J. Introduction to Remote Sensing. 3rd Edition. New York: CRC Press; 2002.

[58] Antonelli G, Fossen T, and Yoerger D. Underwater Robotics. In: Springer Handbook of Robotics. Berlin, Heidelberg: Springer; 2008. p. 987–1008.

Chapter 3
Collision avoidance of maritime vessels

Wasif Naeem[1], Sable Campbell de Oliveira Henrique[1], and Mamun Abu-Tair[1]

Abstract

This chapter presents an artificial potential field (APF)-based online collision avoidance system for manned and unmanned maritime vehicles, which is capable of reacting to static and dynamic obstacles in the vicinity. The standard marine 'rules of the road' are integrated into the collision avoidance framework. A risk assessment module is also introduced which is based on the standard closest point of approach (CPA) method. A decision maker then selects appropriate rules based on relative heading and positions of the vessels. For the detection part, an integrated vision and laser-based system has been developed to provide sensing functionality for multiple obstacles.

Autonomous craft are preferred over manned vessels in scientific operations as they are more suited for long enduring and tedious missions in dangerous or hazardous environments. The technique presented is fairly generic and is applicable to a general class of marine vehicles ranging from a small/medium-sized craft to a large freighter or an oil tanker. The small/medium-sized vehicles have been widely employed in surveillance and scientific missions. For surveillance missions, deploying autonomous vehicles will maximise the coverage and reduce the number of personnel involved in the operation area. Simulation results are provided to include the three fundamental collision encounter scenarios, that is, overtaking, head-on and crossing. It is also shown that Dubin circles could be successfully employed to take the dynamics of the craft into account, which provides a general method independent of the craft size.

3.1 Introduction

A major challenge and ongoing efforts for current generation of unmanned surface vehicles (USVs) lie in the design of their intelligent decision making, whereby USVs can operate (semi or fully) autonomously with minimal or no human intervention.

[1]Energy, Power and Intelligent Control Research Cluster, School of Electronics, Electrical Engineering and Computer Science, Queen's University Belfast, UK

This would require, in addition to a typical onboard navigation and control system and standard sensor suite, an intelligent guidance strategy, which is capable of detecting risk of collisions as early and accurately as possible and then replans a safe path accordingly in good time. An intelligent onboard risk assessment and collision avoidance module is thus, a vital and mandatory aspect of any fully autonomous craft operating in an area littered with other manned or unmanned vessels and objects. Another important aspect of the collision avoidance system is that it must obey the standard marine 'rules of the road' defined by the International Maritime Organisation (IMO) so that a USV is not perceived as a threat for other marine traffic in the vicinity.

A number of USVs are now in operation, with a variety of civilian applications, such as surveying, mapping, collection of oceanographic data for analysis to name a few. Some of the well-known scientific USVs include *C-Eduro*[a] and *Springer* craft [1]. In military and defence strategies, they play a more significant role, such as in surveillance, coastal patrolling and mine countermeasures. Examples include the DARPA's *Sea Hunter*[b] trimaran USV and Atlas Electronik UK's *ARCIMS*[c] vessel. However, most of these advanced vessels are not commercially usable on a large scale due to the absence of a reliable collision detection and avoidance system. This means that most vessels still rely heavily on manned supervision for lookout and collision avoidance. Moreover, the legislation for the deployment of USVs remains a grey area in terms of traffic regulations, liability and avoidance protocols. It is believed that without such a legal framework, the demand for the mass manufacture of commercial (and even some defence) USVs will remain slow or stagnant.

Studies have revealed that collisions involving manned vessels account for more than 60% of casualties at sea [2]. According to European Maritime Safety Agency, 3,669 vessels have been involved in maritime incidents in 2016 involving 976 casualties in Europe alone [3]. Of these, human error is attributed to 63% of those incidents. Some of the primary factors include increasingly smaller crew size, fatigue and/or incorrect interpretation of rules. Thus, automating the rules will help alleviate some of the challenges associated with human factors by providing a consistent and deterministic decision making and path replanning. The use of both manned and unmanned vessels at sea in the same environment will pose another challenge; however, it would be even more important in those scenarios to follow the rules consistently and accurately so that no differentiation could be made between the two craft by the *officer on the watch* on the manned ship or indeed by the risk assessment module on the USV.

3.1.1 Motivation and background

Unmanned underwater vehicles (UUVs) including both remotely operated and autonomous underwater vehicles are often used to carry out subsurface operations; therefore, they seldom pose a direct threat to ambient surface traffic. Despite this, there is a great deal of research in sonar-based collision detection and avoidance strategies being developed for UUVs in contrast to USVs (which are lesser investigated), primarily due to a surge in interest in underwater exploration. For USV guidance, the well-known Line-of-Sight (LOS) guidance law is generally adopted [4]. In this method, the vessel's heading follows the LOS angle between the vehicle's current position and the goal location. Several other guidance laws are also based on this methodology, such as pure pursuit guidance and waypoint guidance by LOS. In the absence of a collision detection system, the unmanned vehicle will follow the reference path regardless of the presence of any objects along the way. This could lead to a collision as the vehicle may run into an obstacle, thus damaging the on-board components and in the worst case, cause it to sink.

For fully autonomous operation, a standardised, reliable and robust collision detection and avoidance system is needed which translates standard IMO marine traffic rules into automatic collision avoidance protocols. These guidelines or maritime rules of the road are known as Collision Regulations (COLREGs), the International Regulations for Preventing Collisions at Sea 1972 produced by the IMO. The COLREGs define typical collision scenarios including their variants such as crossing, head-on and overtaking, to mention a few and suggest actions to avoid a collision. Although the rules do provide a set of guidelines for safe manoeuvring at sea, they were written for human navigators, who, based on their experience, alter the course of the ship accordingly. This subjective nature of COLREGs is one of the major causes of ship collisions to this day. For USVs, it is even more vital to abide by the regulations as any deviation from the standard rules could result in a catastrophe. It is important to note that it is ultimately down to a mariner's experience and good seamanship practice that will avoid incidents at sea as humans continue to make error judgements, whilst misinterpreting the rules. Thus, having an automatic COLREGs-compliant collision avoidance system will help in reducing those incidents by providing some consistency in rule interpretation and following. In light of the increasing presence of autonomous vessels, the Navigation Safety Advisory Council (NAVSAC) have called for an update of COLREGs to include additions for unmanned vessels, for example, an amendment to the lookout rule [5]. Full COLREGs compliance is clearly challenging to implement for an autonomous vehicle, as it must exhibit intelligent behaviour in order to interpret and execute these rules designed for humans. The USV's decision-making abilities are only as accurate as the data provided by the sensors regarding the state of the ambient traffic, and the estimations of their respective trajectories. Recent developments in sensor technology have allowed fully autonomous USVs to take one step closer towards realisation, without the need for such laborious and costly operator supervision and intervention.

A description of some of the most common and successful algorithms developed such as D*, D* Lite, ARA*, etc. can be found in a survey by [6]. However, as with all existing path planning techniques, further modifications are required to incorporate navigation regulations based on situational awareness. Other types of path planning techniques with consideration to COLREGs in recent years include evolutionary algorithms [7], fuzzy logic [8–10], interval programming [11], velocity obstacle method [12], model predictive control [13] and heuristic A* method [14], to name a few. A review paper by Campbell [15] outlines a number of existing techniques that take COLREGs into account. However, most, if not all of the existing strategies do not scale well to multiple target ships and multiple/conflicting COLREGs rules, and usually a single objective is considered when using these techniques. Alternatively, a number of multiobjective evolutionary algorithms have also been developed [16–20] to tackle multiple objectives simultaneously in path planning. These results are developed either as general robotic motion planning algorithms [16,18], or applied in other kinds of autonomous vehicles including aerial and ground vehicles [17,19,20].

In this chapter, an artificial potential field (APF)-based path planner is presented based on the work published in [21] which is integrated with an image/LIDAR-based detection system mounted on a pan and tilt platform. The image/LIDAR system provides USV a visual reference of the surrounding area and range to any visible objects. Further details of the detection system can be found in [14]. In other work, researchers have successfully employed vision-based detection systems in their vehicles. SSC San Diego, for instance, have used a monocular vision-based system for obstacle detection and range finding. The algorithm utilises the Earth's radius and the altitude of the camera above the water to estimate the distance to the object. The same group of researchers have also implemented a stereo vision technique in their USV that was originally developed for NASA Mars Rovers and some other unmanned ground vehicles [22]. A coastline following controller has been devised and implemented in [23] in the *DelfimX* vehicle developed by the Instituto Superior Tecnico, Lisbon, Portugal which uses an onboard laser range finder to detect and track the coastline. The *ARCIMS* USV sensing equipment consists of a radar and electro-optic cameras, whereas the *C-Enduro* USV has broadband satellite communications and an optional suite of cameras.

Marine radar is commonly employed together with an automatic radar plotting aid (ARPA) for detection and tracking for a general class of marine vessels. This system in practice provides an excellent range and other benefits; however, a radar system may fail to detect those objects with a smaller signature above the water. Also, the radar system requires a significant amount of time (in order of minutes) for an obstacle presence to be confirmed. For surveillance missions, for example, this cannot be overlooked as a boat with low signature, which may be tracked with a delay or may not be tracked at all by the patrolling vessel. Also, for pop-up obstacles at a close range, the radar may completely miss the obstacle. Moreover, the ARPA system only tracks moving objects, which limits the capability of the unmanned craft.

The modified potential field algorithm presented in this chapter, works on the principle of attractive and repulsive fields to goal and obstacles, respectively. The method can be suitably adjusted for multiple approaching vehicles, requires minimum or no dependency on communication between vessels, does not need complex layered reactive and deliberative architecture, able to handle complex scenarios effectively and allow for real-time execution. The standard APF approach is commonly employed in robotics-related research for obstacle avoidance purposes; however, the basic technique needs to be modified to cater for COLREGs.

APFs have been applied for USV collision avoidance in a limited context and primarily without including COLREGs. In [24], for instance, the authors implemented APF-based COLREGs-compliant method on an autonomous vessel. The proposed technique was developed for ship route-finding and collision avoidance in real-time, as opposed to solving an optimisation problem for a pre-defined set of waypoints. When a vessel breaches the pre-defined safety zone, the current path is automatically replanned using a repulsive potential field around the target. For path smoothing, generated waypoints were interpolated using cubic splines. However, the well-known local minima issue of APFs is not addressed.

Reference [25] defines a unique Modified Virtual Force Field approach with fuzzy logic for track-keeping and obstacle avoidance for both stationary and dynamic obstacles. The authors presented arguments for adopting artificial intelligence (AI) approaches and soft computing techniques over methods based on modern control theory, which require precise mathematical models of the specific vessel and environment. Simulations are provided for single static and mobile obstacles. More recently, [26] proposed a modified APF approach to USV collision avoidance which takes COLREGs into account. The proposal strategy is shown to work for stationary and moving hazards and is also effective for non-compliant targets, however, only simulation results are shown.

The chapter is organised as follows. Section 3.2 presents the key IMO collision avoidance regulations which is one of the major subjects of this chapter. This is followed by a brief overview of the standard APF technique in Section 3.3 followed by the proposed risk assessment methodology in Section 3.4, which is based on the well-known CPA method. The decision-making system is presented in Section 3.5, whilst the modification to the APF methodology to take COLREGS into account is presented in Section 3.6. Simulation results are provided in Section 3.7 followed by concluding remarks in Section 3.8.

3.2 COLREGs

The IMO regulations, that is, COLREGs consists of three parts (A, B and C) where Part A contains general sailing guidelines, Part B provides steering and sailing rules, whilst Part C includes lights and shapes. Although all sections are applicable, the Steering and Sailing Rules in Section B, in particular, of vessels in sight of one another present perhaps the greatest challenge in achieving full autonomy. Section I of Part B specifies rules for ships in any visibility conditions, i.e., general practice. The key points from [27] are as follows:

- Rule 5 – *Look-out*: Every vessel must at all times maintain a proper look-out by sight and hearing, as well as by all available means in order to make a full assessment of the situation and risk of collision. USVs fulfil this criterion by an operator monitoring the radar or camera, in addition to an automatic identification system (AIS). An autonomous vessel is required to do this independently by being 'aware' of its proximity to other obstacles.
- Rule 6 – *Safe speed*: Any vessel is required to proceed at a safe speed at which she can take action to avoid collision and be able to stop within a distance suitable to the prevailing conditions. These conditions include the visibility; traffic density; her manoeuvrability (e.g., stopping distance and turning ability); background lights on shore, dazzle and backscatter from her own lights; the state of the wind, sea, current and nearby hazards; and draft in relation to the available water. A safety

buffer should also be included when navigating to allow for these phenomena such as drift, tidal currents, waves, limitations in the accuracy of equipment and the vessel's and surrounding vessels' own dynamic constraints.

- Rule 7 – *Risk of collision*: All sea going vessels must use all available means to determine the risk of a collision, including the use of radar (if available) to obtain early warning of the risk of collision by radar plotting or equivalent systematic observation of detected objects (e.g., ARPA, AIS). As an autonomous USV cannot rely on human judgement; therefore, there must be a means of assessing the risk. According to the rules, if in any doubt, avoidance action should be taken.
- Rule 8 – *Action to avoid collision*: Actions taken to avoid collision should be: positive, obvious and made in good time. This rule is very important, as any ambiguity in manoeuvring decisions could be hazardous. Very often, navigators use VHF communications to convey their intentions, particularly within ports and harbours which is not possible for unmanned vehicles.

Rules 9 and 10 describe protocols for navigation in narrow channels and describe traffic separation schemes (TSS), for example, crossing at right angles to the TSS and keeping towards starboard and crossing lanes as quickly as possible so as not to impede other vessels. The practicalities of deploying USVs in highly dense marine traffic in busy ports or harbours where such lanes exist remains questionable, due to the subjective nature, signalling and verbal negotiation required between mariners. However, that does not preclude one in designing an automatic collision detection and avoidance system for those environments.

On the other hand, Section B, Part II details regulations for ships in sight of one another. The rules defined in Part II are the essence of COLREGs and applicable in almost any encounter scenario at sea.

- Rule 13 – *Overtaking*: Any vessel overtaking any other shall keep out of the way of the vessel being overtaken. See Figure 3.1. The overtaking manoeuvre can either be carried out from port or starboard side with proper signalling. Starboard-side preference is maintained whenever and wherever possible.
- Rule 14 – *Head-on*: When two power-driven vessels are meeting on reciprocal or nearly reciprocal courses so as to involve risk of collision, each shall alter her course to starboard so that each shall pass on the port side of the other. See Figure 3.2.
- Rule 15 – *Crossing*: When two power-driven vessels are crossing so as to involve risk of collision, the vessel which has the other on her own starboard side shall keep out of the way and shall, if the circumstances of the case admit, avoid crossing ahead of the other vessel. See Figure 3.3.
- Rule 16 – *Action by give-way vessel*: Every vessel which is directed to keep clear of another vessel shall, so far as possible, take early and substantial action to keep well clear.
- Rule 17 – *Action by stand-on vessel*: Where one of two vessels is to keep out of the way, the other shall keep her course and speed. The latter vessel may however take action to avoid collision by her manoeuvre alone, as soon as it becomes apparent to her that the vessel required to keep out of the way is not taking appropriate action in compliance with these rules.

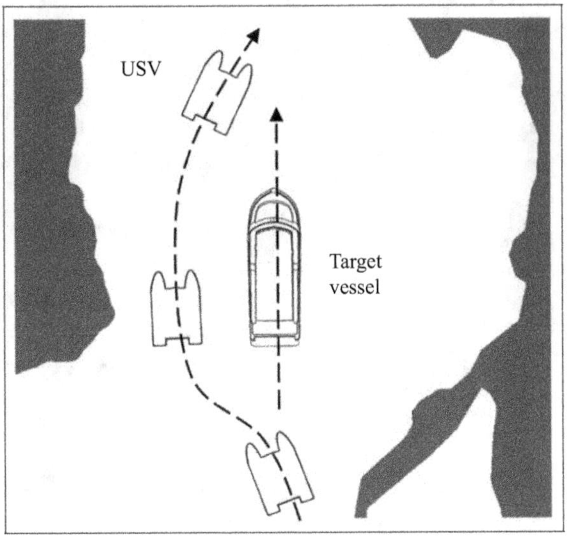

Figure 3.1 COLREGs Rule 13: Overtaking scenario

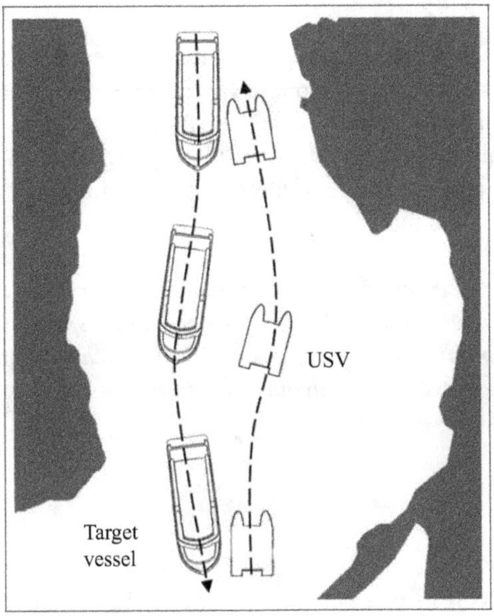

Figure 3.2 COLREGs Rule 14: A typical head-on scenario

Rule 18 specifies which type of vessels must be prioritised when giving way, depending on the type of craft in question, for example, a power-driven ship must always keep out of the way of a sailing boat. The guidelines in Part A must also be taken into consideration, for example maintaining a safe speed and taking action

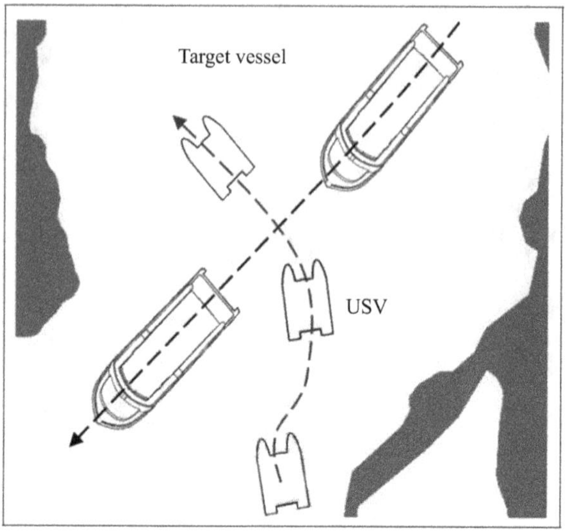

Figure 3.3 COLREGs Rule 15: A typical crossing scenario

when entering narrow channels. The sound and lighting protocols detailed in Part C can be integrated within the obstacle avoidance routines for indicating and issuing warnings, as well as emergency action, or fail-safe methods when another vessel or dynamic objects behave in an unpredictable manner.

Finally, it should be pointed out that Rule 2 *in extremis* takes precedence over other rules in case of non-compliant targets or when imminent risk of collision is detected. Rule 2(b) states that

> *In construing and complying with these Rules due regard shall be had to all dangers of navigation and collision and to any special circumstances, including the limitations of the vessels involved, which may make a departure from these Rules necessary to avoid immediate danger.*

3.3 APFs

Unlike grid-based heuristics such as A*, APFs are designed to operate in a continuous environment. APFs consist of a potential energy gradient, determined by the sum of individual potential functions. In this case, the USV is considered to be a positively charged particle moving towards a negatively charged goal location. The USV will follow the gradient of steepest descent in the virtual field, being repelled by positively charged obstacles along the way. An illustration of this concept is presented in Figure 3.4. The aspects which make this methodology particularly attractive as an online, real-time path planner are the rapid computation times, reactive avoidance

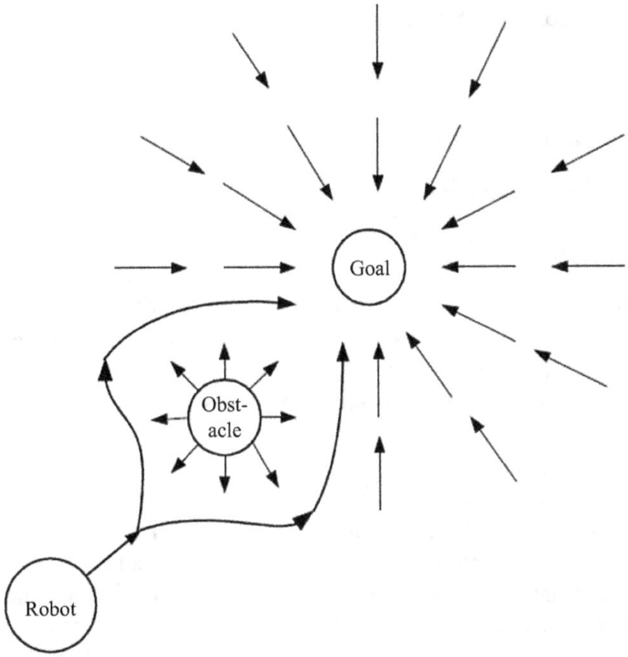

Figure 3.4 APF illustration

capability, smooth path generation, ability to handle multiple objects with ease and also the ability to incorporate COLREGs constraints by adapting the potential function. Limitations however, include cluttered and evolving environment handling, with tendency to get trapped in local minima, the standard of modelling objects as points and the USV considered to be a point source, and failure to execute where no solution can be found. The success of APFs largely depends on the initial set of assumptions made and the level of control desired.

Potential functions represent any obstacle, hazard, constraint (repulsive, positively charged), as well as a global or local waypoint (attractive, negatively charged). The sum of forces at any point is equal to the vectorial addition of the sum of attractive forces, F_{att} and the sum of repulsive forces, F_{rep}. F_{att} and F_{rep} being the negative gradients of the attraction and repulsive functions respectively as given by (3.1)–(3.3).

$$F(q)_{\text{att}} = -\nabla U_{\text{att}}, \tag{3.1}$$

$$F(q)_{\text{rep}} = -\nabla U_{\text{rep}}, \tag{3.2}$$

$$F(q) = -\nabla U_{\text{att}} - \Sigma \nabla U_{\text{rep}}. \tag{3.3}$$

The attractive potential U_{att} monotonically increases towards the goal and may be modelled as a quadratic or linear function depending on the desired field shape. Both will result in a convex environment, analogous to a marble inside a bowl with a sink where the goal is situated. The corresponding functions are provided in (3.4) and (3.5).

Conical potential function:

$$U(q)_{\text{att}} = K_G d_G(q, q_{\text{goal}}). \tag{3.4}$$

Quadratic potential function:

$$U(q)_{\text{att}} = \tfrac{1}{2} K_G d_G^2(q, q_{\text{goal}}), \tag{3.5}$$

where d_G is the Euclidean distance between the each set of coordinates in the field, q and the global goal coordinates, $U(q)$ is the magnitude of the potential field at coordinate, q and K_G is the attractive gain constant for the goal. The advantage of using a conical function in this systemic context is that it is fast to compute due to its linearity. However, there is a discontinuity at the goal location. The quadratic function yields a parabolic shaped field, which avoids the discontinuity problem at the singularity and shall thus be selected for implementation. The non-linearities do not affect the system's performance as the computation time is only microseconds slower than the linear function, which is not critical in this application.

On the other hand, for the repulsive function, it is noted that as the USV shall essentially be a point under influence, whilst the attractive pull towards the goal should be constant and influential at any given location, repulsion from obstacles should only be in effect within a given radius of the object. There is no need to repel objects which are far away, therefore there should be a discrete and defined region of influence, with the strength of this field diminishing radially outwards to encourage flow around the obstacle. This is achieved by setting the repulsive potential function as follows [28]:

$$U(q)_{\text{rep}} = \begin{cases} \tfrac{1}{2} K_{\text{ob}} \left(\dfrac{1}{d(q, q_{\text{obs}})} - \dfrac{1}{D_o} \right)^2 & \text{if } d \le D_o, \\ 0 & \text{if } d > D_o \end{cases} \tag{3.6}$$

where K_{ob} is the repulsive gain constant and D_o is the limit distance of influence of the obstacles. Whilst all obstacles in general robotic path planning may be assigned common adjustable parameters, here individual gain parameters shall be assigned to each type of obstacle, differentiating between static obstacles, moving vessels and coastlines for a more flexible and optimised design. The APF programme and set-up are designed based on this foundation theory.

As outlined earlier, a collision avoidance system, in general, consists of risk assessment, decision making and path replanning subsystems. The risk assessment is a key element of the collision avoidance strategy as it dictates how early and how accurate the path is to be replanned. The decision-making subsystem provides COLREGs guidelines in terms of capability of the overall system in comparison to human decision making. Finally, the path replanner modifies the current path of the USV for give-way manoeuvres when required by generating one or more additional waypoints in order to avoid the approaching vessel in compliance with the COLREGs. These subsystems are covered in the following sections.

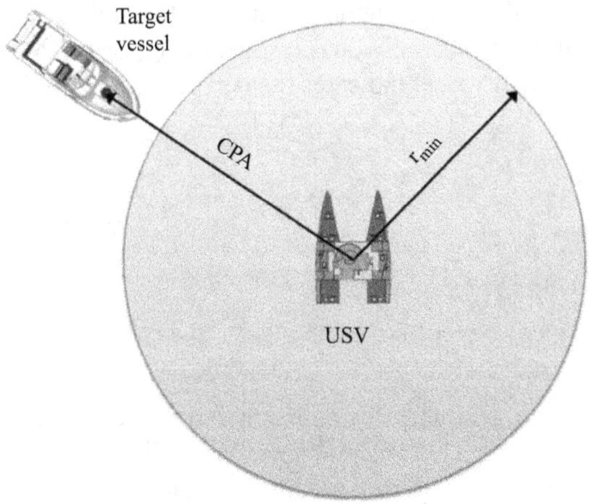

Figure 3.5 Circle of acceptance and CPA

3.4 Collision risk assessment

The path planner's responsibility is to calculate a waypoint or a series of waypoints towards each path subgoal in order to avoid any static or dynamic obstacles. However, it is possible that although a target vessel is detected on the map, it may not pose a direct threat at its current distance, speed and heading and thus, any computation in replanning the path is a waste of processing effort and memory resource. In order to avoid this, a threshold can be set for the minimum acceptable proximity between vessels, commonly known as a 'circle of acceptance,' $r_{\min}$ (see Figure 3.5). The actual minimum distance between the USV and target vessel, based on the assumption that both vessels continue at their current speeds and headings, is referred to as the CPA.[d] The time to CPA or TCPA is another important parameter used by mariners which is closely related to $r_{\min}$. A combination of CPA and TCPA is typically used to determine a risk of collision; however, here the focus will be on CPA and $r_{\min}$ mainly.

For any vessel inside the detection zone, the USV must automatically determine:

(a) own surge velocity, V_{USV};
(b) target vessel's surge velocity, V_{Tar};
(c) USV heading in earth-fixed reference frame (absolute in radians), θ_{USV};
(d) target vessel heading in earth-fixed reference frame (absolute in radians), θ_{Tar}.

The following parameters are then calculated:
Relative velocity components in earth-fixed x- and y-axes

$$V_{x\text{vess}} = V_{\text{USV}} \cos(\theta_{\text{USV}}), \tag{3.7}$$

$$V_{y\text{vess}} = V_{\text{USV}} \sin(\theta_{\text{USV}}), \tag{3.8}$$

$$V_{x\text{Tar}} = V_{\text{Tar}} \cos(\theta_{\text{Tar}}), \tag{3.9}$$

$$V_{y\text{Tar}} = V_{\text{Tar}} \sin(\theta_{\text{Tar}}). \tag{3.10}$$

This then enables the determination of the relative velocity by calculating the vector difference, in terms of individual resolved components. The velocity of the USV relative to the target vessel indicates the perceived velocity in the rest frame of the observer.

$$V_{x\mathrm{rel}} = V_{x\mathrm{vess}} - V_{x\mathrm{Tar}}, \tag{3.11}$$

$$V_{y\mathrm{rel}} = V_{y\mathrm{vess}} - V_{y\mathrm{Tar}}, \tag{3.12}$$

$$\angle V_{\mathrm{rel}} = \mathrm{arctan2}\left(\frac{V_{y\mathrm{rel}}}{V_{x\mathrm{rel}}}\right), \tag{3.13}$$

$$V_{\mathrm{rel}} = \sqrt{(V_{x\mathrm{rel}}^2 + V_{y\mathrm{rel}}^2)}. \tag{3.14}$$

In [29], a collision geometry for collision avoidance of a UAV and aircraft in shared air space is defined. This definition can be applied to the scenario for a USV and target vessel in proximity. Using a collision triangle as in Figure 3.6, the CPA can be identified. This is done by pre-calculating the relative velocity vector, V_{rel}, relative to the position of the USV itself. A direct LOS between the two vessels at their current locations is adjoined to the origin of the relative velocity vector, and V_{rel} is then extended, intersecting at a point to form a right angle with the position of the target vessel. This perpendicular line is the CPA.

CPA equation from collision geometry from Figure 3.6 can be found as follows:

$$\mathrm{CPA} = \delta_{\mathrm{LOS}} \sin(\alpha), \tag{3.15}$$

where α is the angle between the relative velocity, V_{rel}, and the LOS, and δ_{LOS} is the shortest LOS distance between vessels.

The calculated CPA distance is then compared to r_{min}, the radius of the pre-defined circle of acceptance. If the CPA distance is greater than r_{min}, then no pre-emptive action is necessary. This is of course reviewed upon the next sampling interval. However, if the CPA distance breaches the circle of acceptance by falling short of the minimum radius, r_{min}, then the decision maker is invoked to determine the COLREGs rule applicable (if any). This is further explained in the next section.

3.5 COLREGs decision maker

As mentioned in the above, a decision maker is required to assess whether or not any COLREGs rules apply and to select an appropriate one, if applicable. This judgement is based on logic and on the current data available, and is updated online at every sampling interval to ensure continuous situational awareness. When suitable COLREGs rule(s) have been identified, the path can be modified accordingly.

The intelligent, decision-making capabilities of a human navigator are coded here by designing an automated COLREGs decision maker. Upon the confirmation of an identified risk, it is necessary to pre-determine if any marine traffic rules are likely to apply to the encounter before an avoidance path is calculated. There are several considerations for the design of this subsystem. It is a time-critical process, as any lag in information processing and output could cause a hazardous close-encounter

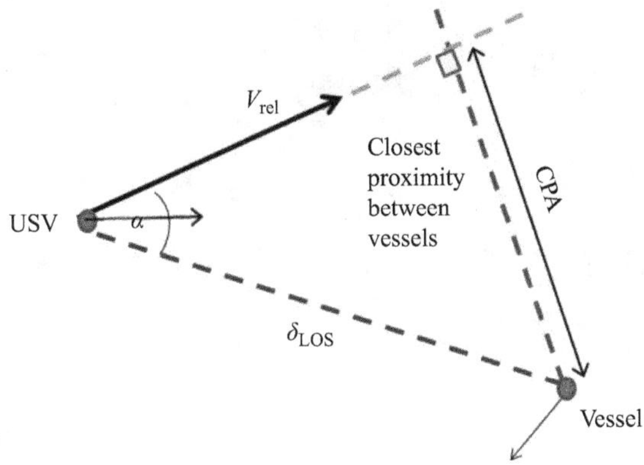

Figure 3.6 CPA as derived from collision geometry

Table 3.1 Ship domain behaviours for defined regions

Range	COLREGs rule	Action
$0° \geq \gamma \geq 15°$ **or** $345° \geq \gamma \geq 359°$	Head-on	Starboard manoeuvre
$15° < \gamma \leq 160°$	Crossing; give-way	Starboard manoeuvre
$200° < \gamma > 345°$	Crossing; stand-on	Starboard manoeuvre
$160° < \gamma \leq 200°$	Overtaking	Starboard or portside manoeuvre

scenario. It could also cause the USV to stand-on when it should in fact be taking evasive actions. However, an incorrect assumption could also be problematic, resulting in avoidance action(s), which could be confusing and misleading to other mariners. If COLREGs rule selection is ambiguous, there is a potential for causing additional risks for the target vessels resulting in unwanted vehicle motion. This probability is reduced by the fact that there is always a starboard-bound bias as defined in the IMO regulations.

The principle on which the decision shall be made depends on the relative angle of approach between the USV and target vessel or vessels. The relative angle of approach is obtained from the relative velocity vector, V_{rel}, as previously calculated in Section 3.4 (see Figure 3.6). The incident angle which V_{rel} makes with the USV in its current orientation lies within a COLREGs defined domain. Those domain regions are described in Table 3.1, which are not standardised, but are defined here as indicative figures. A clear illustration is provided in Figures 3.7 and 3.8, whereby the incident angle clearly falls within the crossing region, denoted by C.

Once a decision is taken on what COLREGs rule to apply, the next step is to generate a new path in the form of a waypoint or a set of waypoints in order to avoid the encountering vessel. For scenarios where the ownship is required to stand-on to

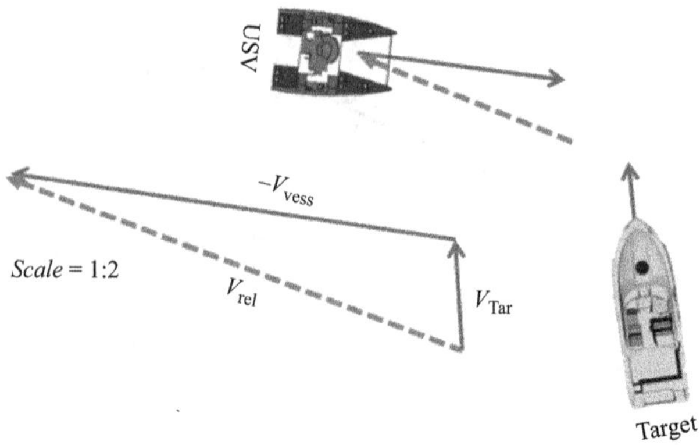

Figure 3.7 Relative velocity example for determining COLREGs encounter type

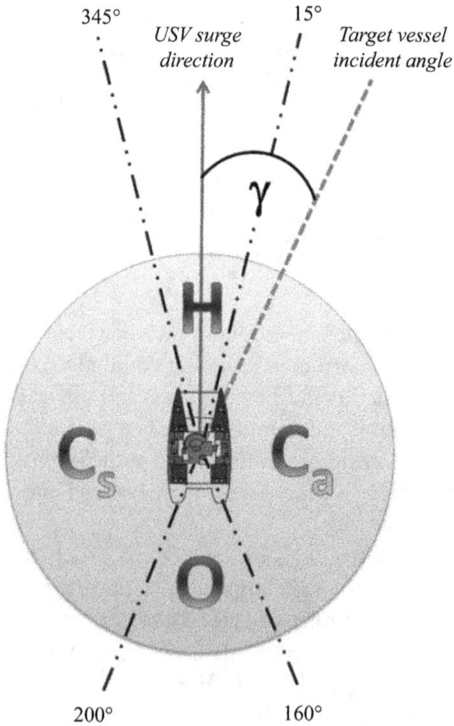

Figure 3.8 The USV domain, H activates head-on avoidance, O represents the overtaking region, C_a indicates a crossing scenario with give-way avoidance and C_s indicates a crossing scenario with stand-on behaviour

a target vessel, that is, being overtaken or crossing from the port side, the ownship will not generate any new path but will maintain a constant lookout in case the target vessel does not take the required action to avoid her. The path planner presented here is the APF method which was already described in Section 3.3. The next section presents the modification that must be made to the standard APF technique in order to generate COLREGs-compliant paths.

3.6 COLREGs zones for APF adaptation

An introduction to the fundamentals of the COLREGs rules appeared in Section 3.2, along with illustrations of the applicable avoidance guidelines. The detection system provided in [14] is integrated with a risk assessment module, which identifies any potential threats and generates suitable actions in order to alleviate it. The detection system is paired with an online APF-based path planning algorithm, which yields fast COLREGs-compliant paths for safe navigation in open seas. The attractive and repulsive functions having been defined in Section 3.3.

When a collision risk is detected, the decision maker determines applicable COLREGs rules and attempts to enforce them by ascertaining which region proximal to the approaching vessel is forbidden and assign virtual obstacles, which generate their own repulsive field individually and collectively. Rather than merely adding a string of discrete objects which could cause discontinuities in the path or trapping in local minima, these virtual objects are connected (see Figure 3.9) to exhibit continuity in the field generated. In the figure, cli is the minimum clearance of the ownship to the target vessel. The placement of these obstacles on the map with suitably tuned gains achieves a tapered avoidance region. This would encourage natural flow during avoidance and prevent the USV approaching too close to the target vessel ship.

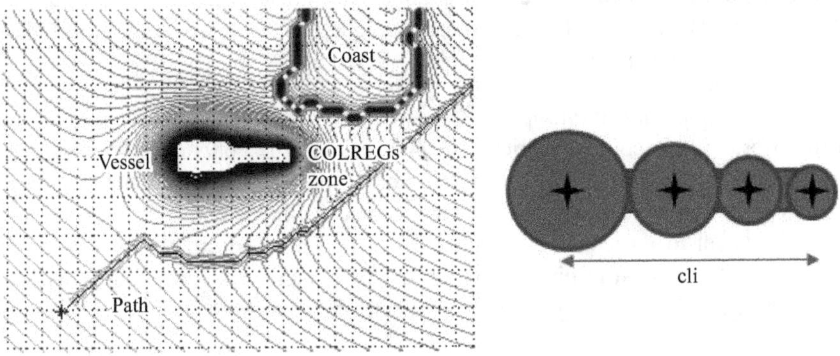

Figure 3.9 Illustration of the COLREGs zone of virtual obstacles [30]

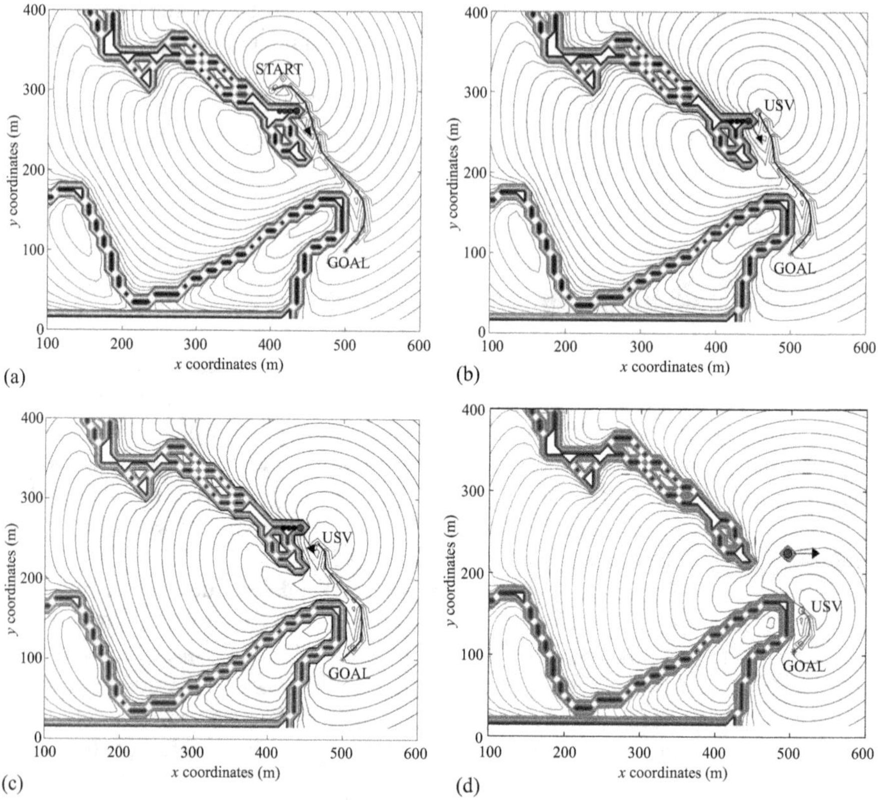

Figure 3.10 APF implementation to overtake vessel in real-time

3.7 Simulation results

In this section, results of the application of the modified APFs to take into account COLREGs are presented. The three fundamental collision scenarios, that is, overtaking (Rule 13), head-on (Rule 14) and crossing (Rule 15) as well as Rule 16 which covers action by the give-way vessel, are all presented.

(1) Rule 13: *overtaking encounter*

In the overtaking encounter depicted in Figure 3.10, the collision risk is detected when the CPA is less than the permissible range. As potential fields are continuous in nature, the 'no go' region is merged with the coastline because of the proximity of target vessel to the coastline (Figure 3.10(a)). This is desirable to prevent the emergence of any local minima in which the USV could become trapped. Having overtaken the slower-moving vessel on the port side (to avoid grounding), it is no longer flagged as a risk.

From the visual plots in Figure 3.10(a)–(d), APFs give satisfactory performance for this overtaking scenario. The path is COLREGs-compliant, the closest LOS

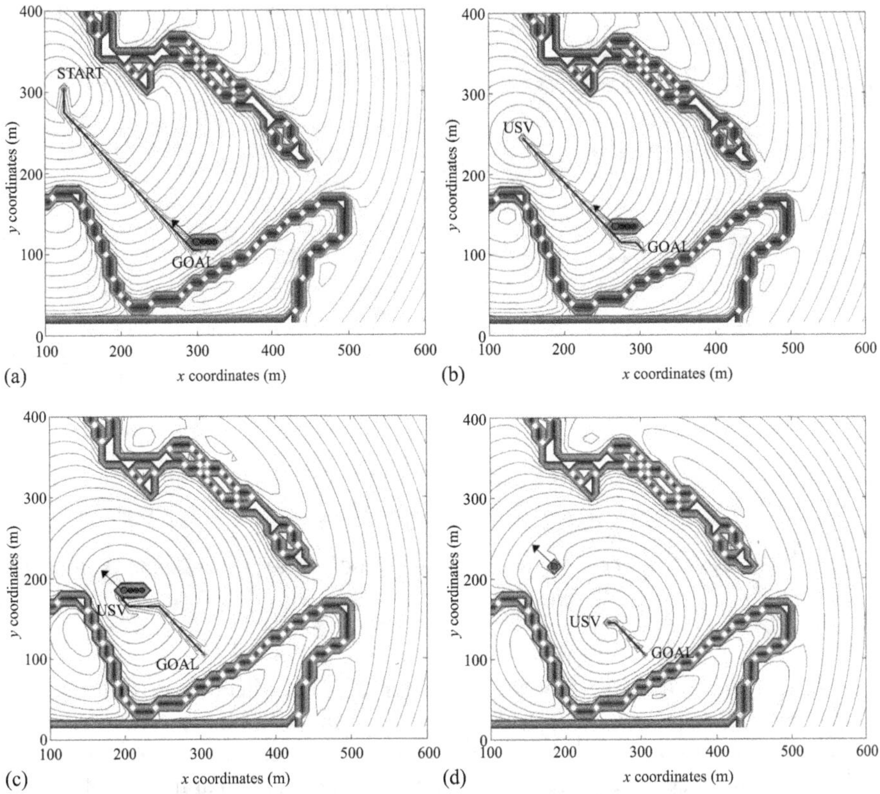

Figure 3.11 APF implementation for avoiding an oncoming vessel in real-time

distance is observed as 12 m (more than the minimum allowable of 10 m), and the computation time was 0.1363 s on average which is more than sufficient in practical terms with a 10 s sample time in between iterations. A head-on encounter is considered next.

(2) Rule 14: *head-on encounter:*

The head-on encounter simulations are depicted in Figure 3.11. The APF method takes give-way avoidance action almost immediately as in Figure 3.11(a), so that the two vessels are no longer approaching head-on and continue on their respective courses in opposite directions. Please note that in this scenario, the target vessel fails to yield and make a mutual starboard turn. This early avoidance behaviour could be changed by altering the minimum CPA limit, thus the vessel would not be deemed a threat until a closer approach is achieved. The consequences of avoiding the vessel too soon include potentially ambiguous perceived behaviour (not obvious enough) or a less efficient path by spending a significant proportion of time off the original course.

This APF head-on encounter simulation exhibits satisfactory performance, safely avoiding the oncoming vessel port-to-port as per COLREGs guidelines, which are automatically implemented by the decision maker. Once the vessels have passed

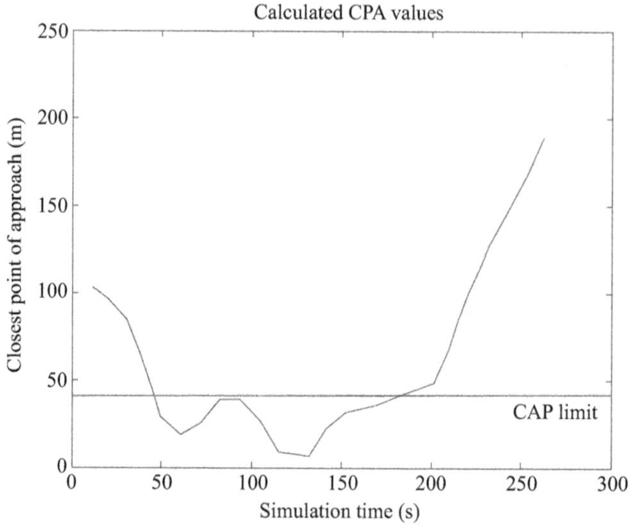

Figure 3.12 CPA distances throughout single vessel crossing avoidance manoeuvre

and are moving away from each other, the target is no longer deemed a threat (see Figure 3.11(d)). The average simulation time per replanning iteration is 0.1296 s. The minimum LOS distance was 10 m exactly, equal to the minimum allowable for safe avoidance. This distance would have been larger had the target vessel also manoeuvred to her starboard-side as required by the IMO guidelines. The final individual rule; a crossing scenario follows.

(3) Rule 15: *crossing encounter:*
As can be seen in Figure 3.12, the target vessel is not initially considered a threat, as the CPA distance is not lower than the CPA limit. This is evident in Figure 3.13, as there are no virtual vessels proximal to the target, representing a COLREGs zone. The USV safely gives-way to the crossing vessel by avoiding the COLREGs zone in front of the target, and instead passing from aft. After the encounter, there is no longer any risks identified and the USV proceeds towards the goal.

The resulting path exhibits some jaggedness (although not excessive, taking into account the map scale). This would imply that there may be a need to introduce smoothing in some instances when using APFs. This is further explored in the next subsection. Safe COLREGs execution is observed for crossing in this scenario, with an average computation time of 0.1628 s and a large clearance of 51 m.

3.7.1 Path dynamics

A USV is an underactuated system, which means that it has fewer actuators than output degrees of freedom. Hence it is vital to consider maximum speed, turning radius, drift when designing a path so that it can be traversed and the USV is able to

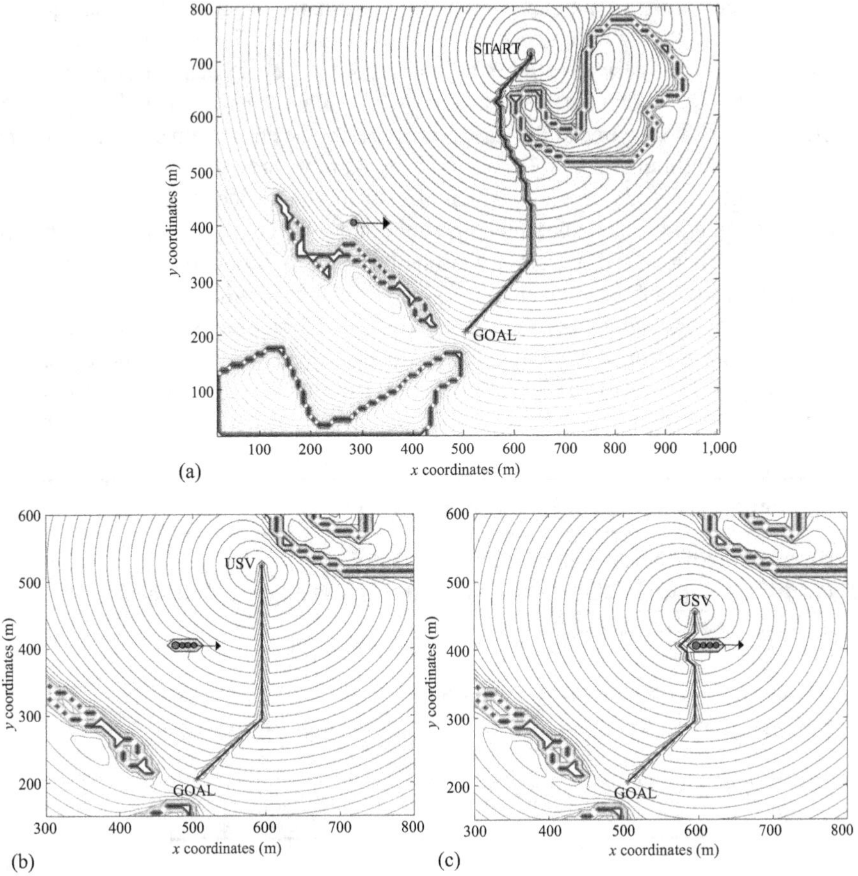

Figure 3.13 APF implementation for giving way to a crossing vessel in real-time [30]

stop or turn at will. Additionally, each vehicle has a unique dynamics which are specific to its individual design. Instead of incorporating advanced non-linear dynamics of an individual ship into the navigation and control system design, these features are generalised into algorithms and buffers are introduced. This is much more representative of navigation in practice, as a human navigator does not know the exact dynamics of the vessel, but can predict the behaviour as an approximation and make allowances for this. Also, a navigation, guidance and control system which is tailored only to a specific vessel without universal applicability is unlikely to be particularly commercialised, as it cannot be easily transferred to different platforms.

Smoothing the path is the first step of post-processing, eliminating any jagged edges and discontinuities. By attempting to follow this path in its raw form, the USV would exhibit jerky, non-continuous motion with high turning rates, which have the potential to damage actuators. Rather than attempting to fit an entire path to a curve, it is dealt with in segments or subpaths to make the problem more manageable. These paths can then be merged to form the whole path. To guarantee that a path is smooth, the first derivative must be continuous. The use of splines is common throughout related studies for the purpose of polynomial fitting. Also, with polynomial interpolation, changing one of the waypoints requires the entire path coefficients to be recomputed. Continuity is established at the breakpoints between individual subpaths using the Cubic Hermite method, which effectively determines the coefficients of a polynomial function, $F(x)$, guaranteeing a continuous first derivative, $F'(x)$. Although the second derivative, $F''(x)$ is not continuous, the monotonicity of the subpath is preserved (i.e., the subpath is neither non-increasing nor non-decreasing indicated by the first derivative being consistently negative or non-negative). As a consequence of smoothing the path, collision avoidance is no longer guaranteed and requires a further collision-check to verify that no ship is within CPA distance in the following procedure in Algorithm 1.

Algorithm 1: Final path collision-check for $F'(x)$

1: For all pairs of coordinates of smoothed path
2: **for all** $[x_j, y_j]$ **do** If LOS distance to vessel is less than CPA distance
3: **if** $\sqrt{(x_{jn} - x_v)^2 + (y_{jn} - Y_v)^2} < CPA$ **then**
4: Else if path nodes are members of occupied static obstacle nodes
5: **else if** floor$(x_{jn}, y_{jn}) \in [x_{\text{static}}, y_{\text{static}}]$ **then**
6: Display 'Vessel Collision Detection' Notify operator for manual override
7: **else**
8: Collision check OK
9: **end if**

It is assumed that the vessel's speed remains constant (a fair assumption for open sea navigation) or is externally controlled. The issue of incorporating turning radii, for example, in excess of 500 m for a large draft vessel, is solved by Dubins circles [31] applied after spline smoothing to non-conforming curve sections. Where the angle between consecutive smoothed waypoints is less than permissible, the path is altered. Assuming constant speed, the discontinuous region of the path is replaced by an arc of radius $r_{\min}$, calculated by simple geometry for the circle which intersects both LOS tangents at a single point. It would also imply that the end pose is pointed towards the next waypoint. The process logic is as follows in Algorithm 2.

The application of Dubins circles is vital when the USV is less able to manoeuvre caused by slow rudder response, handling quality, control system performance and vessel geometry. See Figure 3.14 for an illustration of Dubin circles application.

It is important to note that Dubins circles are not necessarily needed for highly manoeuvrable vessels such as the *DELFIM* catamaran, which is essentially able to rotate about its own axis at low speeds. Thus, Dubins act as a secondary post-processing technique to offer a universal solution, which would be important when dealing with cargo vessels, oil tankers, etc.

Algorithm 2: Dubins implementation

For all calculated headings between sequential waypoints;
If the angle between three successive waypoints is less than the accepted value;
if $\angle abc \leq \angle abc_{\text{accepted}}$ **then**
Apply Dubins circles to expand the applicable waypoints;
Display 'call Dubins'
$|a - c|^2 = |b - a|^2 + |c - b|^2 - 2(|b - c|^2 |c - b|^2 \cos(2\beta))$
➔$r = l \tan(\beta)$
else
➔ Current path OK
end if=0

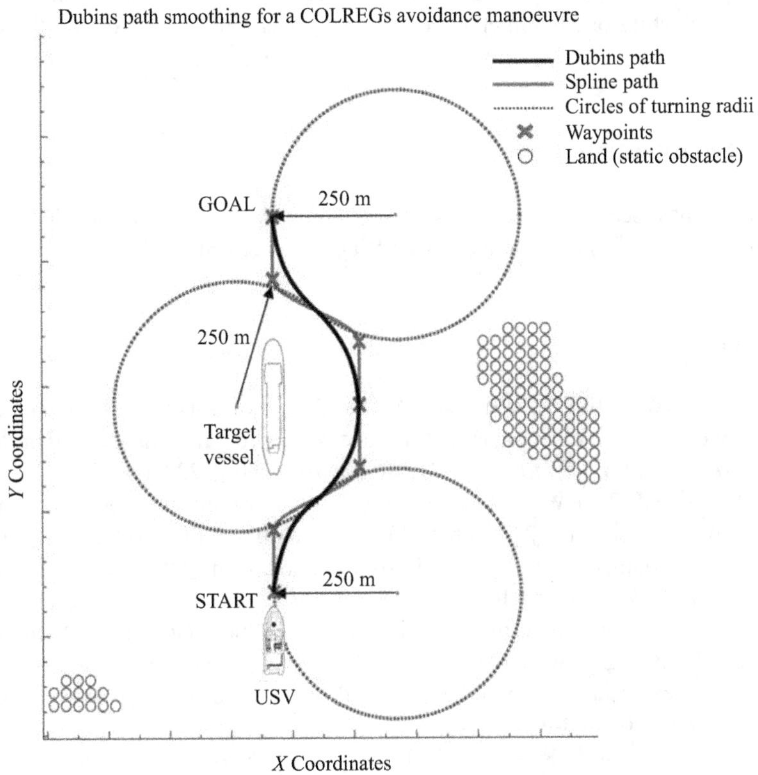

Figure 3.14 Dubins path smoothing incorporates the minimum turning radius of the vessel to yield traversable paths where splines are insufficient

3.8 Discussion and concluding remarks

Simulation results presented here demonstrated the applicability of the APF-based collision avoidance technique to generate suitably qualified COLREGs-compliant paths. The three fundamental COLREGs rules, that is, overtaking (Rule 13), head-on (Rule 14) and crossing (Rule 15), are all successfully simulated. However, it should be pointed out that on several occasions during the validation and verification process of APFs, the path failed to generate a resulting solution. This is due to the 'moving particle' getting trapped in local minima. A fast solution to this problem was proposed, which involves the current solution backstepping iteratively along the path in reverse until a valid solution is found. A virtual obstacle is then placed at the point of local minima to deter trapping and the algorithm is re-executed. This had generally resulted in a desirable path to goal; however, it does reduce the accuracy of the field and may conflict with other repulsive obstacles which may present themselves.

In general, the APF technique presented in this chapter is able to handle static, dynamic and multiple obstacles simultaneously with low computational cost, but there is still a potential to get trapped in local minima. Also, the resulting raw path may require smoothing to avoid jaggedness and parameter tuning may not be trivial as adjusting one parameter independently has a relative effect on the entire field.

Acknowledgement

This work has been supported mainly by Grant EP/I003347/1 awarded by the Engineering and Physical Sciences Research Council (EPSRC), UK.

References

[1] Naeem W, Sutton R, and Tiano A. The design of a navigation guidance and control system for an unmanned surface vehicle for environmental monitoring. *Proc. IMechE, Part M J Eng. Marit. Environ.* 2008;222(2):67–79.

[2] Jingsong Z, Price WG, and Wilson PA. Automatic collision avoidance systems: Towards 21st century [Internet]. Vol. 1, Department of Ship Science, University of Southampton, 2008. p. 1–10. Available from: https://eprints.soton.ac.uk/51087/1/acas21st.pdf.

[3] EMSA. Annual overview of marine casualties and incidents 2014 [Internet]. 2014. p. 76. Available from: http://www.emsa.europa.eu/news-a-press-centre/external-news/item/2303-annual-overview-of-marine-casualties-and-incidents-2014.html.

[4] Naeem W, Sutton R, Ahmad SM, and Burns RS. A review of guidance laws applicable to unmanned underwater vehicles. *J. Navig*. 2003;56(1):15–29.

[5] Norris CA. Legal issues relating to unmanned maritime systems monograph [Internet]. 2013. Available from: https://www.hsdl.org/?view&did=731705.

[6] Ferguson D, Likhachev M, and Stentz A. A guide to heuristic-based path planning. In: *Proceedings of theWorkshop on Planning under Uncertainty for Autonomous Systems at the International Conference on Automated Planning and Scheduling*. 2005.

[7] Szlapczynski R. Evolutionary sets of safe ship trajectories: A new approach to collision avoidance. *J. Navig*. 2011;64(1):169–81.

[8] Kao S-L, Lee K-T, Chang K-Y, and Ko M-D. A fuzzy logic method for collision avoidance in vessel traffic service. *J. Navig*. 2007;60(1):17–31.

[9] Perera LP, Oliveira P, and Soares CG. Maritime traffic monitoring based on vessel detection, tracking, state estimation, and trajectory prediction. *IEEE Trans. Intell. Transp. Syst*. 2012;13(3):1188–200.

[10] Perera LP, Carvalho JP, and Guedes Soares C. Solutions to the failures and limitations of mamdani fuzzy inference in ship navigation. *IEEE Trans. Veh. Technol*. 2014;63(4):1539–54.

[11] Benjamin MR, Leonard JJ, Curcio JA, and Newman PM. A method for protocol-based collision avoidance between autonomous marine surface craft. *J. Field Robot*. 2006;23(5):333–46.

[12] Kuwata Y, Wolf MT, Zarzhitsky D, and Huntsberger TL. Safe maritime autonomous navigation with COLREGs, using velocity obstacles. *IEEE J. Ocean. Eng*. 2014;39(1):110–19.

[13] Johansen TA, Perez T, and Cristofaro A. Ship collision avoidance and COLREGS compliance using simulation-based control behavior selection with predictive hazard assessment. *IEEE Trans. Intell. Transp. Syst*. 2016;17(12): 3407–22.

[14] Campbell S, Abu-Tair M, and Naeem W. An automatic COLREGs-compliant obstacle avoidance system for an unmanned surface vehicle. *Proc. Inst. Mech. Eng. Part M J. Eng. Marit. Environ*. 2014;228(2):108–21.

[15] Campbell S, Naeem W, and Irwin GW. A review on improving the autonomy of unmanned surface vehicles through intelligent collision avoidance manoeuvres. *Annu. Rev. Control*. 2012;36(2):267–83.

[16] Ahmed F, and Deb K. Multi-objective optimal path planning using elitist non-dominated sorting genetic algorithms. *Soft Comput*. 2013;17(7): 1283–99.

[17] McEnteggart Q, and Whidborne J. A multiobjective trajectory optimisation method for planning environmentally efficient trajectories. In: *The UKACC International Conference on Control (CONTROL)*. 2012. p. 128–35.

[18] Oral T, and Polat F. MOD* Lite: An incremental path planning algorithm taking care of multiple objectives. *IEEE Trans. Cybern*. 2016;46(1):245–57.

[19] Lee K-B, Myung H, and Kim J-H. Online multiobjective evolutionary approach for navigation of humanoid robots. *IEEE Trans. Ind. Electron*. 2015;62(9):5586–97.

[20] Hu L, Naeem W, Rajabally E, *et al*. COLREGs-compliant path planning for autonomous surface vehicles: A multi-objective optimization approach. In: *Proceedings of the 20th International Federation of Automatic Control (IFAC) World Congress*. Toulouse, France; 2017;50(1):13662–13667.

[21] Campbell S, and Naeem W. A rule-based heuristic method for COLREGS-compliant collision avoidance for an unmanned surface vehicle. In: *9th IFAC Conference on Manoeuvring and Control of Marine Craft,* Vol 45, Issue 27. Arenzano, Italy; 2012. p. 386–91.

[22] Larson J, Bruch M, Halterman R, Rogers J, and Webster R. Advances in autonomous obstacle avoidance for unmanned surface vehicles. In: AUVSI unmanned systems North America, Curran Associates. Washington; 2007. p. 1–15.

[23] Gomes P, Silvestre C, Pascoal A, and Cunha R. A coastline following preview controller for the DELFIMx vehicle. In: *Proceedings of the 17th international offshore and polar engineering conference (ISOPE)*, July. Cupertino, CA; 2007. p. 1080–7.

[24] Xue Y, Lee BS, and Han D. Automatic collision avoidance of ships. *Proc. Inst. Mech. Eng. Part M J. Eng. Marit. Environ*. 2009;223(1):33–46.

[25] Lee S-M, Kwon K-Y, and Joh J. A fuzzy logic for autonomous navigation of marine vehicles satisfying COLREGS guidelines. *Int. J. Control Autom. Syst*. 2004;2(2):171–81.

[26] Lyu H, and Yin Y. Ship's trajectory planning for collision avoidance at sea based on modified artificial potential field. In: *2017 2nd International Conference on Robotics and Automation Engineering (ICRAE)*. Shanghai, China; 2017. p. 351–357.

[27] Cockcroft AN, and Lameijer JNF. *Guide to the Collision Avoidance Rules*. Oxford: Elsevier Butterworth-Heinemann; 2003.

[28] Khatib O. Real-time obstacle avoidance for manipulators and mobile robots. *Int. J. Robot. Res*. 1986;5(1):90–8.

[29] Tsourdos A, White B, and Shanmugavel M. *Cooperative Path Planning of Unmanned Aerial Vehicles*, Advances in Unmanned Marine Vehicles. Chichester, UK: John WIley & Sons, Ltd.; 2011. 133 p.

[30] Naeem W, de Oliveira Henrique S, and Hu L. A Reactive COLREGs-compliant navigation strategy for autonomous maritime navigation. In: *10th IFAC Conference on Control Applications in Marine Systems CAMS 2016*, Vol 49, Issue 23. Trondheim; 2016. p. 207–13.

[31] Dubins LE. On curves of minimal length with a constraint on average curvature, and with prescribed initial and terminal positions and tangents. *Am. J. Math*. 1957;79(3):497–516.

Chapter 4

Sliding mode control for path planning guidance of marine vehicles

Shashi Ranjan Kumar[1], Ashwini Ratnoo[2], and Debasish Ghose[2]

Abstract

This chapter presents a path planning algorithm for marine vehicles using the sliding mode paradigm to generate paths to a moving target point where it is required that the vehicle approaches the destination with a certain angle and proposes implementation of the guidance commands. In many underwater scenarios, information about the target point is unavailable after some time. These vehicles use sonar to extract the information, which fails as it comes close to the target. In such cases, the open-loop control method is the only way to reach the targets. A natural way of its implementation is sample and hold technique which do not yield satisfactory performance with its usual sampling method. This chapter proposes different sampling techniques to implement the guidance law in open loop. These methods are based on sampling the guidance command generated using closed-loop feedback with last known estimates about the target. To ensure robustness, a sliding mode control based impact angle guidance law is used. Performance of the guidance scheme using proposed sampling methods are evaluated through extensive numerical simulations for different engagement scenarios and shown to work well.

4.1 Introduction

Marine vehicles like autonomous underwater vehicles (AUVs), automated surface vehicles, among others, are quite often called upon to operate for short distances after being launched from a bigger marine vehicle. The applications of such vehicles include surveillance of a particular spot on the ocean bed or a particular point on the hull of a ship for maintenance or monitoring purposes. This kind of applications requires the surveillance vehicles to arrive at the desired location at a certain angle to

[1]Department of Aerospace Engineering, Indian Institute of Technology Bombay, India
[2]Department of Aerospace Engineering, Indian Institute of Science, India

make observation more accurate. The applications can also include automated ocean vehicles for transportation or AUVs for intercepting hostile vehicles. The path of the vehicle is quite often pre-computed and programmed into the vehicle, as sensor-based information availability and subsequent data processing is slow for such short trajectories or may not even be available.

A classical approach to such path planning problems is that of Dubins' paths [1] which has been used extensively in the literature for this class of vehicles, both for marine vehicles and unmanned ground or aerial vehicles. Several other variants of these path planning algorithms for marine vehicles exist in the literature. However, most of these techniques use the classical Dubins formulation where the final destination point is stationary. But, in the scenario we consider, the target point itself may be moving. Only very recently, this class of problems has been addressed in the literature [2–4].

In recent years, sliding mode techniques have been used to design guidance commands for automated aerial vehicles that result in desirable trajectories [5–8]. This chapter proposes a path planning algorithm for marine vehicles using the sliding mode paradigm to generate paths to a moving target point where it is required that the vehicle approaches the destination with a certain angle and proposes implementation of the guidance commands. The inaccuracies resulting from the implementation are also presented along with a comparison with an ideal closed-loop implementation.

In all air-based engagement scenarios, the information about the targets are available with a reasonable accuracy with the use of modern radar systems. This facilitates the designer to use various control techniques to develop guidance laws which can be easily implemented using closed-loop feedback. However, in the underwater scenarios, the main problem lies in the availability of the information of the ships. During the terminal phase of engagement, the AUV is very close to the target. Therefore, the AUV, which uses sonar for the determination of the positions and heading angles of the target, fails to determine the information. In such scenarios, the guidance laws based on closed-loop feedback cannot be implemented directly. Hence, the implementation of guidance laws in open loop is the only approach which can be used to intercept target in such situations. The usual adopted approach is to use sample and hold technique and apply the obtained sampled guidance command. But, as will be shown later in this chapter, this method suffers as the sampling interval becomes larger. This generates a need to develop an efficient sampling technique to obtain the guidance command for implementation in the open-loop sense. In order to get satisfactory performance in such cases, the impact angle guidance strategy, used for generating guidance commands, should also have robustness against uncertainties.

In general, the AUV may also have limited available memory to store the guidance commands. This limitation motivates the design of technique which minimises the total number of guidance commands needed to be stored. In this chapter, the open-loop methodology is used to generate guidance commands, and the memory requirement to store these commands is equal to the ratio of total travel time and sampling interval. Note that the number of storage point reduces with increase in the sampling interval which motivates the use of sampling method which allows to use large sampling intervals.

In the guidance literature, there exist several papers addressing impact angle constraints using various control techniques. Some of these guidance laws are the variants of well-known proportional navigation guidance (PNG) law [9–16]. The guidance laws in [9,10] are designed as multistage PNG to satisfy impact angle constraints. In [11–16], the guidance laws are derived as PNG with an additional bias term to satisfy the impact angle requirement. Some of them are also designed using optimal control theory [17–25], polynomials [26,27], sliding mode control theory [5–8,28] and other control techniques [29–31]. The optimal-control-based guidance laws are usually designed after linearising the engagement dynamics and are also dependent on time-to-go information. These guidance may not provide satisfactory performance when these linearisations remain no longer valid in large heading error cases. The guidance laws in [26,27] also depends on time-to-go information. The guidance laws in [5–8,28] provide robustness against uncertainties due to the inherent nature of sliding mode control [32]. The guidance laws in [29–31] are based on geometrical approach.

All the above guidance laws are designed for a closed-loop feedback implementation. However, in the problem considered here, this kind of closed-loop implementation is not feasible due to lack of information. In this chapter, the different methods of generating the guidance command for implementation in open-loop sense is proposed. The guidance law proposed in [5], which has robustness against disturbances, is used to generate the closed-loop guidance command via computer simulations.

Organisation of the chapter is as follows. Section 4.2 discusses the problem formulation. Section 4.3 discusses the design of sliding mode control based guidance law briefly. Section 4.4 presents the different methods to sample the guidance command, generated using closed-loop feedback through simulation, for open-loop implementations. The efficacy of the proposed techniques is evaluated with extensive simulations in Section 4.5. Finally, Section 4.6 concludes the chapter with possible future directions of research.

4.2 Problem statement

Consider a planar engagement between an AUV and a ship as shown in Figure 4.1, where the AUV has a speed and flight path angle denoted by V_T and γ_T, respectively, and the ship has a speed and flight path angle denoted by V_S and γ_S, respectively. The AUV has a lateral acceleration a_T, and the ship is assumed to have a constant velocity. Without loss of generality, it can be assumed that $\gamma_T = 0^\circ$. We assume that the origin of the frame of reference is at the AUV's launch point. The relative distance of the ship to the AUV is denoted by r and line-of-sight (LOS) angle of the ship to the AUV is θ. The engagement dynamics based on the kinematic model are given by

$$\dot{r} = V_r = V_S \cos\theta_S - V_T \cos\theta_T \tag{4.1a}$$

$$r\dot{\theta} = V_\theta = V_S \sin\theta_S - V_T \sin\theta_T \tag{4.1b}$$

$$\dot{\gamma}_T = \frac{a_T}{V_T};\ \theta_T = (\gamma_T - \theta),\ \theta_S = (\gamma_S - \theta). \tag{4.1c}$$

Here, V_T and V_S are assumed to be constant throughout the engagement.

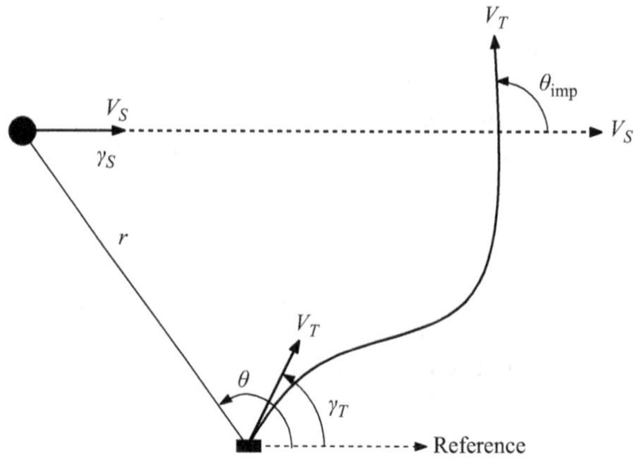

Figure 4.1 Underwater engagement scenario

The impact angle θ_{imp}, as shown in Figure 4.1, is defined as the angle between the velocity vectors of the AUV and ship at the time of interception. Note that the word interception is used here in a broad sense to imply that the AUV reaches the ship (which is the intended destination or the target). In order to achieve interception with the desired impact angle, it is first defined in terms of the final LOS angle by the relation given by [5–7]:

$$\theta_F = \gamma_{Sf} - \tan^{-1}\left[\frac{\sin\theta_{imp}}{\cos\theta_{imp} - \nu}\right], \tag{4.2}$$

where $\nu = (V_S/V_T)$ is the ratio of speed of ship to AUV. For the constant velocity ship, γ_S is fixed, and without losing generality, it can be assumed to be equal to 0°. For a desired impact angle θ_{imp}, the final desired LOS angle θ_F is fixed and is given by

$$\theta_F = -\tan^{-1}\left[\frac{\sin\theta_{imp}}{\cos\theta_{imp} - \nu}\right]. \tag{4.3}$$

To ensure interception, it is required for the AUV to change its course in such a way that the AUV and ship should be on collision course, that is, $\dot{\theta} = 0$. Now, the objectives of intercepting the ship with desired impact angle turns into a problem of satisfying $\dot{\theta} = 0$ and $\theta = \theta_F$.

4.3 Design of impact angle guidance

In this section, the guidance law which enables the AUV in achieving desired impact angle will be discussed briefly [5]. From the engagement dynamics, given by (4.1a)–(4.1c), it can be easily seen that the dynamics of LOS angle θ has relative degree 2 with respect to the AUV's lateral acceleration a_T and is given by

$$\ddot{\theta} = -\frac{2\dot{r}\dot{\theta}}{r} - \frac{\cos\theta_T}{r}a_T. \tag{4.4}$$

Note that the lateral acceleration a_T is multiplied with $\cos\theta_T$ and hence, if $\cos\theta_T$ is zero then the it is not possible to control LOS angle θ. In [5–7], it has been shown that $\theta_T = \pi/2$ is not an attractor. In this work, it is assumed that $\cos\theta_T \neq 0$ during the engagement.

In order to design sliding mode control (SMC)-based guidance law [5], the switching surface, which ensure finite time convergence of error and its derivative, is defined as

$$S = \dot{\theta} + C|\theta - \theta_F|^{\alpha}\text{sign}(\theta - \theta_F); \quad C > 0,\ 0 < \alpha < 1. \tag{4.5}$$

where θ_F is the desired final LOS angle. Using equivalent control approach, the equivalent control component of a_T can be obtained, by making $\dot{S} = 0$, as

$$a_T^{\text{eq}} = \frac{\dot{\theta}}{\cos\theta_T}\left[-2\dot{r} + Cr\alpha|\theta - \theta_F|^{(\alpha-1)}\right] \tag{4.6}$$

The discontinuous component is selected as

$$a_T^{\text{disc}} = \frac{\tilde{M}\text{sign}(S)}{\text{sign}(\cos\theta_T)}, \qquad \tilde{M} > 0 \tag{4.7}$$

where $\tilde{M}$ is the gain of discontinuous controller. The guidance law is given by

$$\begin{aligned} a_T &= a_T^{\text{eq}} + a_T^{\text{disc}} \\ &= \frac{\dot{\theta}\left[-2\dot{r} + Cr\alpha|\theta - \theta_F|^{(\alpha-1)}\right]}{\cos\theta_T} + \left[\frac{\tilde{M}\text{sign}(S)}{\text{sign}(\cos\theta_T)}\right]. \end{aligned} \tag{4.8}$$

Now, consider a Lyapunov function candidate V given by

$$V = \frac{1}{2}S^2 \tag{4.9}$$

On differentiating V and substituting for $\dot{S}$, we get

$$\dot{V} = S\left[-\frac{2\dot{r}\dot{\theta}}{r} + C\alpha|\left(\theta - \theta_F\right)|^{(\alpha-1)}\dot{\theta} - \frac{\cos\theta_T}{r}a_T\right] \tag{4.10}$$

On substituting for a_T, we get

$$\dot{V} = -\frac{\tilde{M}|\cos\theta_T|}{r}|S|. \tag{4.11}$$

From (4.11), it can be easily seen that $\tilde{M} > 0$ is the sufficient condition to make $\dot{V}$ negative definite. Hence, the sliding mode will occur and as a consequence $\theta = \theta_F$ and $\dot{\theta} = 0$ within a finite time.

Note that the sliding mode dynamics, which is given by

$$\dot{S} = -\frac{\tilde{M}|\cos\theta_T|}{r}\text{sign}(S) \tag{4.12}$$

with the above selected discontinuous control, is sluggish when the r is very large. In order to improve the transient performance of the sliding mode dynamics, the discontinuous control component can be selected as

$$a_T^{\text{disc}} = \frac{\tilde{M} r \,\text{sign}(S)}{\text{sign}(\cos\theta_T)}, \qquad \tilde{M} > 0. \tag{4.13}$$

This choice of a_T^{disc} will improve the transient performance and the condition on the gain $\tilde{M}$ to ensure Lyapunov stability remains the same.

Note that the term $|\theta - \theta_F|^{(\alpha-1)}$ in (4.8) gives rise to singularity when $\theta \approx \theta_F$. To avoid this singularity, β is introduced in a_T^{eq} as

$$a_T^{\text{eq}} = \frac{r\dot{\theta}}{\cos\theta_T}\left(C\alpha\beta|\theta - \theta_F|^{(\alpha-1)} - \frac{2\dot{r}}{r}\right), \tag{4.14}$$

where

$$\beta = \begin{cases} 0 & \text{if } |S| < \varepsilon_1, \ \& \ |\theta - \theta_F| \le \varepsilon_0 \\ \dfrac{|S| - \varepsilon_1}{\varepsilon_2 - \varepsilon_1} & \text{if } \varepsilon_1 \le |S| \le \varepsilon_2, \ \& \ |\theta - \theta_F| \le \varepsilon_0\,, \\ 1 & \text{otherwise} \end{cases} \tag{4.15}$$

and ε_1 and ε_2 are positive constants. The correction term β can be represented graphically as shown in Figure 4.2. In order to ensure satisfactory performance even in case where closing speed becomes negative, the modifications as suggested in [5] can be made to the guidance law. The equivalent control component can also be rewritten as

$$a_T^{\text{eq}} = -\frac{AV_r\dot{\theta}}{\cos\theta_T}, \qquad A = 2 + \frac{Cr\alpha\beta|\theta - \theta_F|^{(\alpha-1)}}{-V_r}. \tag{4.16}$$

In presence of very large heading error, $V_r > 0$ then $A < 2$ which is not desirable for any guidance law. To ensure $A > 2$ during engagement, the equivalent control component of the guidance law, (4.16), can be further modified as

$$a_T^{\text{eq}} = \frac{|V_r|\dot{\theta}}{|\cos\theta_T|}\left[2 + \frac{Cr\alpha\beta|\theta - \theta_F|^{(\alpha-1)}}{|V_r|}\right]. \tag{4.17}$$

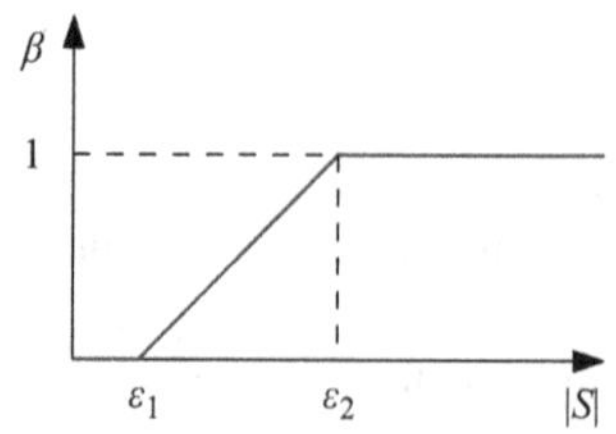

Figure 4.2 Graphical representation of term β

Finally, the guidance law can be written as

$$a_T = \frac{|V_r|\dot{\theta}}{|\cos\theta_T|}\left[2 + \frac{Cr\alpha\beta|\theta - \theta_F|^{(\alpha-1)}}{|V_r|}\right] + \frac{\tilde{M}r\text{sign}(S)}{\text{sign}(\cos\theta_T)} \tag{4.18}$$

4.4 Application of guidance scheme to underwater vehicles

In the previous section, the guidance law which ensure interception with desired impact angle are discussed. This guidance law is based on closed-loop feedback methodology. Since in the underwater situations, information about targets are not available, this guidance law cannot be implemented directly. Hence, in this case, the implementation of guidance laws can be done only in open loop. The robustness of above-derived SMC-based impact angle guidance law against uncertainties can also be utilised here. In this section, different methods to implement the proposed guidance law, in open loop, are proposed. First, the guidance command profile is generated using SMC-based guidance law based on last known information about the targets using closed-loop feedback. This generation of guidance command can be done through simulation with known target's velocity and heading angle information. These guidance command profile is used to further generate the open-loop guidance command using the different sampling techniques described below.

4.4.1 Sample and hold

In this subsection, the implementation of the guidance law, with the guidance commands generated in closed loop, is discussed using usual sample and hold (SH) technique. With this technique, the guidance command are sampled at a fixed rate, that is, with the equal sampling interval denoted by T, and hold the same value during the sampling period as shown in Figure 4.3(a). Hence, the guidance commands have constant value during the sampling period. At a time instant $t = nT + \Delta t$ lies between the interval $[nT, (n+1)T]$, the guidance command can be given by

$$a_T[nT + \Delta t] = a_T[nT], \tag{4.19}$$

where $a_T[nT]$ is the guidance command at nth sampling instant.

4.4.2 Linear interpolation

In the previous subsection, the guidance law is proposed to be implemented in open loop with sample and hold technique. With this technique, the performance of the guidance law may degrade when the sampling period T becomes large. In order to avoid these situations while keeping the number of storage point same, the guidance law can also be implemented using a linear interpolation method. In this method, the guidance commands at any time during the sampling interval is no more constant but has a value which is determined by linearly interpolating the two successive samples taken just before and after that time as shown in Figure 4.3(b). Consider a time instant

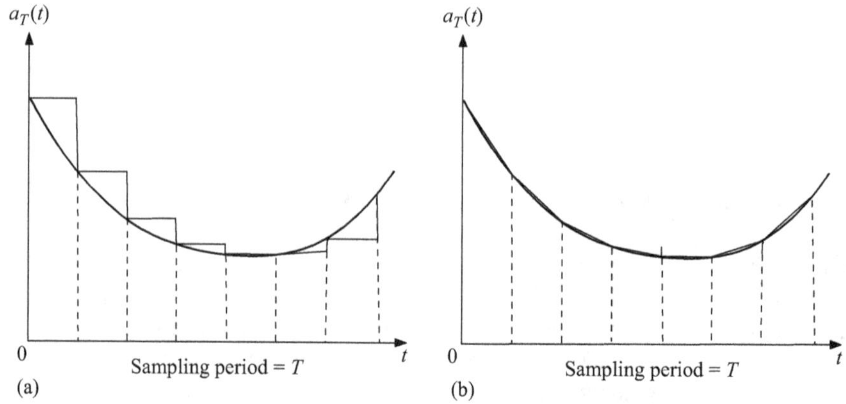

Figure 4.3 Implementation of guidance command with: (a) sample and hold (b) linear interpolation

$t = nT + \Delta t$ lies between the interval $[nT, (n+1)T]$ at which we want to determine the guidance command. At this time instant, the guidance command is given by

$$a_T[nT + \Delta t] = a_T[nT] + \left(\frac{a_T[(n+1)T] - a_T[nT]}{T} \right) \Delta t, \tag{4.20}$$

where $a_T[nT]$ and $a_T[(n+1)T]$ are the guidance commands at nth and $(n+1)$th sampling instant.

4.4.3 Improved sample and hold

The constant guidance commands are more easy to implement in a practical system than the linearly time-varying guidance commands. Although the implementation of guidance command with linear interpolation method are expected to give the better performance compared to the usual sample and hold technique, it is always preferred to implement guidance strategy with sample and hold technique. In order to achieve satisfactory performance with sample and hold technique, an improved way of implementing it is proposed as shown in Figure 4.4(a). In the previous sample and hold technique, the guidance command were sampled at the sampling instants and kept the same value till the next sampling instants. With the proposed strategy, the value of guidance command again remains constant during sampling period. But its value is determined based on the average value of guidance commands at current and next sampling instant. Now consider a time instant $t = nT + \Delta t$ at which we want to determine the guidance command. At this time instant, the guidance command is given by

$$a_T[nT + \Delta t] = \frac{a_T[nT] + a_T[(n+1)T]}{2}, \tag{4.21}$$

where $a_T[nT]$ and $a_T[(n+1)T]$ are the guidance commands at nth and $(n+1)$th sampling instant.

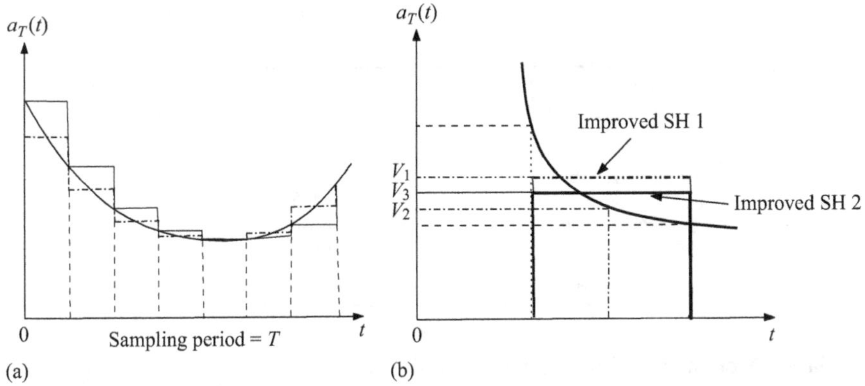

Figure 4.4 Implementation of guidance command with: (a) improved sample and hold 1 and (b) improved sample and hold 2

In order to achieve further improvement in performance of the guidance law using sample and hold technique, another variant of implementing sample and hold, based on double average, is introduced. In this method, we do the same process of taking sample and holding it for the given time interval. But, the sample value is taken as the average of the two averages, one taken at the average of the function value denoted as V_1 in the Figure 4.4(a), and other taken at the mid point of the interval, denoted as V_2 in the Figure 4.4(a). The first value V_1 is the same as which is considered as the sample value in improved sampling and hold method earlier discussed and given by (4.21). These details are demonstrated in the Figure 4.4(a). In this figure, V_3 is considered as the sample value for this improved sampling technique, and mathematically, it is given by

$$V_3 = \frac{V_1 + V_2}{2}, \tag{4.22}$$

where

$$V_1 = \frac{a_T[nT] + a_T[(n+1)T]}{2} \tag{4.23a}$$

$$V_2 = a_T\left[\left(nT + \frac{T}{2}\right)\right] \tag{4.23b}$$

$$V_3 = a_T[nT + \Delta t] \tag{4.23c}$$

4.5 Simulation results

In this section, the numerical simulation results are presented to evaluate the performance of the proposed sampling methods in the implementation of impact angle constrained guidance laws in both closed as well as open-loop sense. These results

Table 4.1 Simulation parameters

r_0	150 m
θ_0	100°
γ_{T0}	60°
γ_{S0}	0°
V_T	16.5 m/s
V_S	6 m/s
θ_F	90°
$a_{T\max}$	200 m/s^2

validate the claim of achieving the desired impact angle criterion under different engagement situations. To reduce chattering, the signum function in a_T^{disc} is replaced by sigmoid function with a boundary layer of 0.2. The sigmoid function is given by

$$\text{Sgmf}(s) = 2\left(\frac{1}{1+\exp^{-as}} - \frac{1}{2}\right), \quad a > 0, \tag{4.24}$$

where the constant a is inversely proportional to the width of the boundary layer ε. The maximum lateral accelerations, that the AUV can provide in both the directions, are assumed to be limited as follows:

$$a_T = \begin{cases} a_T^{\max}\,\text{sign}(a_T) & \text{if } |a_T| \geq a_T^{\max} \\ a_T & \text{if } |a_T| < a_T^{\max} \end{cases} \tag{4.25}$$

where $a_T^{\max}$ is the maximum allowable AUV lateral acceleration. The AUV's initial position is assumed to be the origin of the inertial frame of reference. The initial engagement geometry and speed of AUV and ship, which remains same for all the simulation unless otherwise specified, are listed in Table 4.1. Other various constants used in the simulations are $\varepsilon_1 = 0.001, \varepsilon_2 = 0.15, \varepsilon_\theta = 0.0001, \tilde{M} = 0.25, C = 1$ and $\alpha = 0.5$. These values of the constants are obtained after performing several simulations. In all the subsequent results shown here, the lateral acceleration of AUV is specified in terms of m/s^2 which can be easily interpreted in °/s with conversion of 1 m/s^2 = 3.4725°/s. The AUV lateral acceleration is assumed to be bounded by a maximum lateral acceleration $a_{T\max}$, which in turn imposes a limit on turn rate of AUV. Note that the ship and the AUV are denoted as target and interceptor in subsequent figures, respectively.

4.5.1 Implementation of guidance law with closed-loop feedback

In this subsection, the guidance law is implemented on AUV with closed-loop feedback in order to test its performance. First the impact angle of −90° is selected. All other initial conditions are same as listed in Table 4.1. Simulation results for this engagement scenario is presented in Figure 4.5 which shows the trajectories of AUV and ship, achieved impact angle, lateral acceleration profile and the deviation of sliding variable. From Figure 4.6(a) and (b), it can be easily seen that the AUV is able to intercept the ship with a desired impact angle. The miss distance and achieved impact angle for this case are 0.0427 m, and −90°, respectively. From Figure 4.6(d), it can

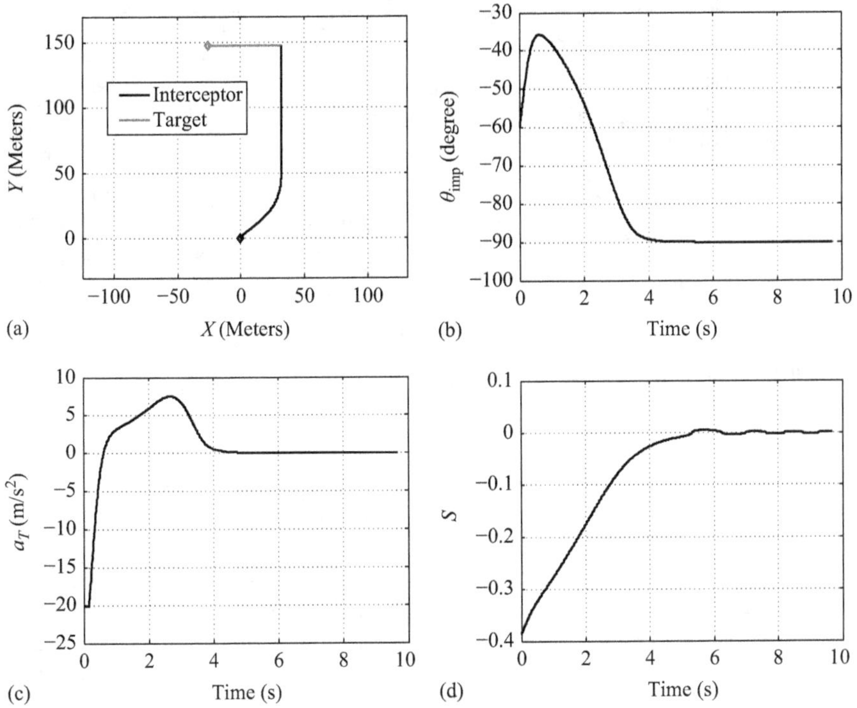

Figure 4.5 Constant velocity target: (a) trajectories in (X, Y)-space, (b) impact angle, (c) AUV lateral acceleration a_T and (d) occurrence of sliding mode

also be observed that sliding mode has occurred much before the interception with the lateral acceleration demand shown in Figure 4.6(c). The AUV demands higher lateral acceleration initially and decreases to very low towards the interception, which is desirable.

In order to evaluate the performance of guidance law under large heading angle error, simulation has been done with four different heading angles of $30°, 120°, 210°$ and $300°$ but with the same desired impact angle of $-90°$. Simulation results are shown in Figure 4.6. For all these engagement scenarios, the AUV is able to intercept the ship, and conclusion similar to previous case can also be drawn. In these cases, the miss distances and the achieved impact angles are given in Table 4.2. To evaluate the performance of guidance law against different desired impact angles, simulation has been done with $\theta_{\text{imp}} = -60°, -90°$ and $-120°$ and results are shown in Figure 4.7. For these engagement scenarios, the miss distances and achieved impact angles are given in Table 4.3.

In order to validate the claim of improvement in transient performance of sliding mode dynamics, the simulations has also been done with the discontinuous control component given by (4.7) and (4.13), and the results are compared. The simulation

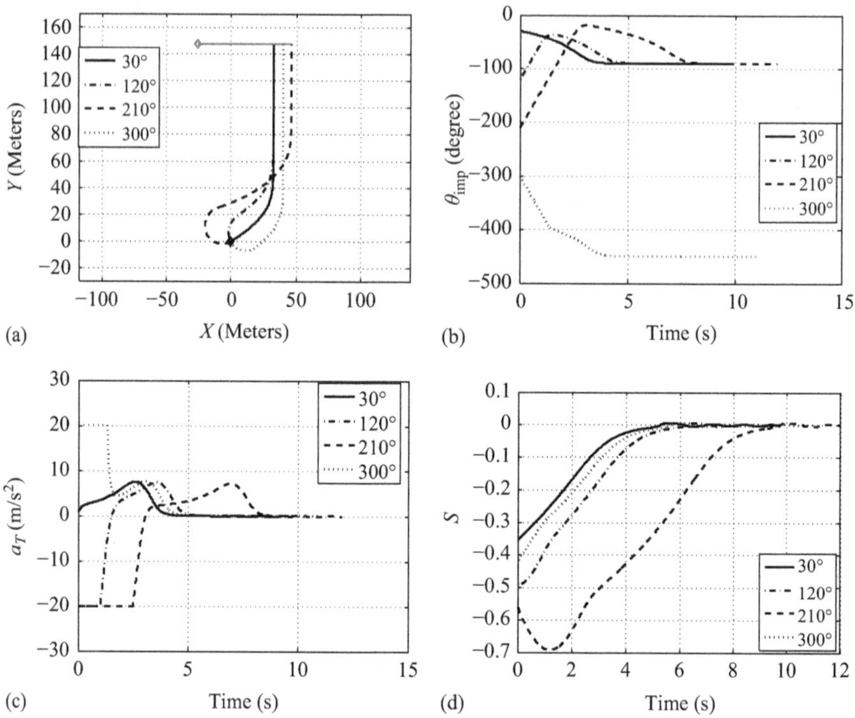

Figure 4.6 Constant velocity target at different launch angles: (a) trajectories in (X, Y)-space, (b) impact angle, (c) AUV lateral acceleration a_T and (d) occurrence of sliding mode

Table 4.2 Constant velocity target at different launch angles

γ_{T0}	30°	120°	210°	300°
r_{miss}(m)	0.0262	0.004	0.0510	0.01
θ_{imp}	−90°	−89.99°	−90°	−90°

results are shown in Figure 4.8. From Figure 4.8(d), it can be seen that guidance law with discontinuous component given by (4.13), which has r in the numerator, gives fast sliding mode dynamics, but demand on missile lateral acceleration is higher as compared to the other one. For both cases, the miss distance and achieved impact angle are listed in Table 4.4.

4.5.2 Implementation of guidance law in open loop

In the previous subsection, the evaluation of proposed guidance strategy with closed-loop feedback is performed. In this subsection, the effectiveness of proposed sampling

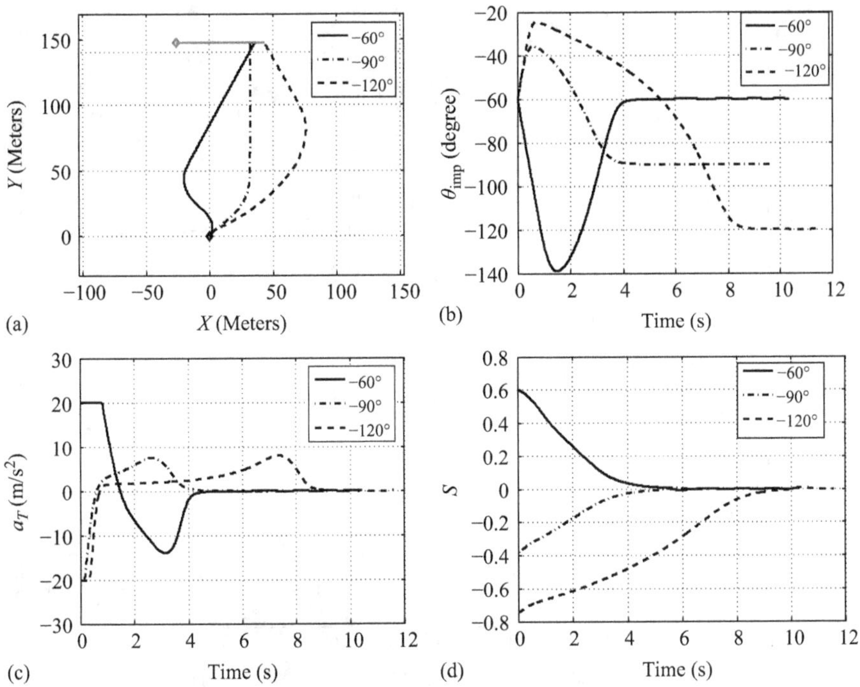

Figure 4.7 Constant velocity target at different impact angles: (a) trajectories in (X, Y)-space, (b) impact angle, (c) AUV lateral acceleration a_T and (d) occurrence of sliding mode

Table 4.3 Constant velocity target at different impact angles

θ_{imp}	−60°	−90°	−120°
r_{miss}(m)	0.01	0.04	0.03
θ_{imp}	−60°	−90°	−119.99°

techniques will be investigated. First the case of impact angle of 90° is considered with the initial conditions same as listed in Table 4.1 and simulation results are shown in Figure 4.9. In this figure, trajectories of AUV and target, impact angle, closed-loop AUV lateral acceleration and sampled AUV lateral acceleration are shown. In this figure, simulation results are shown for closed-loop feedback and two different sampling intervals of 0.05 and 0.1 s. From Figure 4.10(b), it can be seen that the AUV demands higher lateral acceleration initially and decreases to very low towards the interception which is desirable. The miss distances and impact angles for these engagements are listed in Table 4.5. It can be noted from Table 4.5 that both the miss distance and the impact angle error increases with increase in the sampling time, and

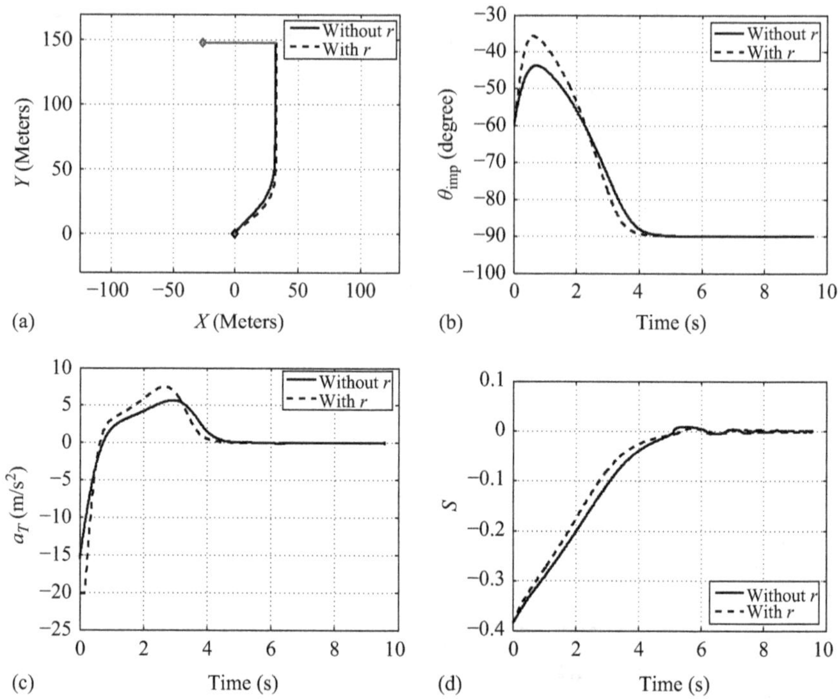

Figure 4.8 Constant velocity target with and without r in a_T^{disc}: (a) trajectories in (X, Y)-space, (b) impact angle, (c) AUV lateral acceleration a_T and (d) occurrence of sliding mode

Table 4.4 Constant velocity target with and without r in a_T^{disc}

a_T^{disc}	**With r**	**Without r**
r_{miss}(m)	0.0427	0.0499
θ_{imp}	−90°	−90°

hence as expected, performance of the guidance law degrades with sampled guidance command.

In order to improve the performance of the proposed guidance law, the AUV's guidance command is implemented with the linear interpolation technique. In this case, the engagement geometry, as listed in Table 4.1, is considered with the three different sampling times of 0.1, 0.3 and 0.5 s. The simulation results are presented in Figure 4.10, and the miss distances and achieved impact angles are listed in Table 4.6. It can be observed that the miss distances and impact angle errors are relatively smaller

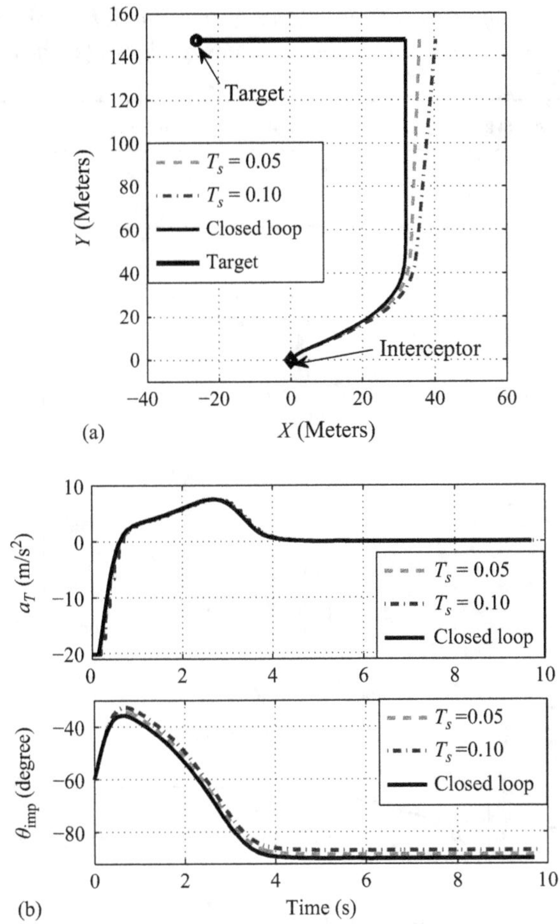

Figure 4.9 Constant velocity targets using sample and hold with different sampling intervals: (a) trajectories and (b) interceptor's lateral acceleration and achieved impact angles

than those for sample and hold technique. With this linear interpolation technique, the errors in miss distance and impact angles for sampling intervals of 0.5 s are also acceptable.

Although the performance of the guidance law in open loop is improved using the improved sample and hold technique, it would not be much appreciable if the sampling interval becomes larger. In order to further improve the results, another sample and hold technique, introduced in the earlier subsection, is implemented. The guidance law is implemented for the −90° case with the same initial condition listed in Table 4.1 and the results are shown in Figure 4.12. In this figure, results are shown for the different sampling intervals. The miss distances and impact angle errors are

Table 4.5 Comparison of miss distance and achieved impact angles using sample and hold for a desired impact angle of −90°

Sampling intervals	Closed loop	0.05	0.1
r(m)	0.0427	3.56	7.56
θ_{imp}	−90°	−88.563°	−86.93°

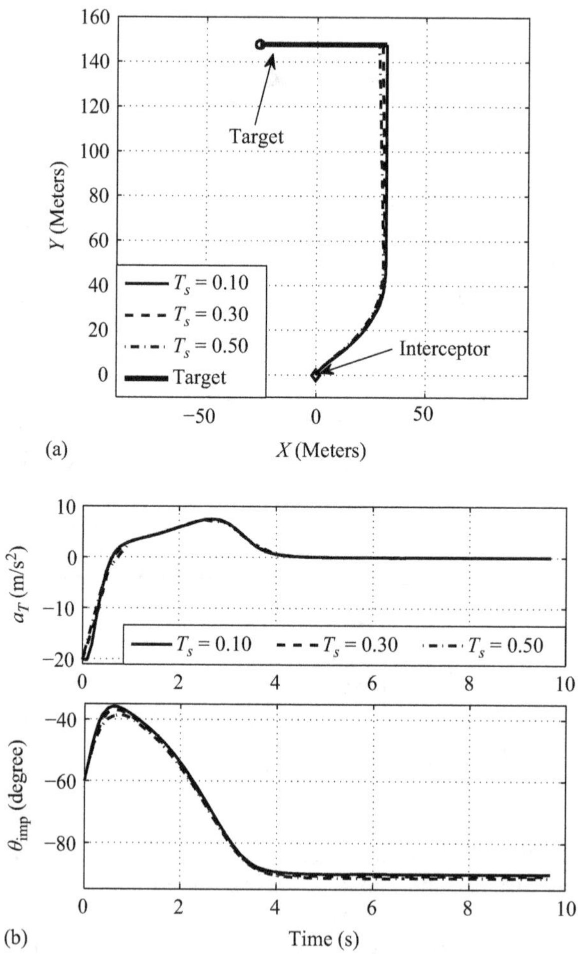

Figure 4.10 Constant velocity targets using linear interpolation with different sampling intervals: (a) trajectories and (b) interceptor's lateral acceleration and achieved impact angles

Table 4.6 Comparison of miss distance and achieved impact angles using linear interpolation for a desired impact angle of −90°

Sampling interval T_s	**0.1**	**0.3**	**0.5**
MD(m)	0.1845	0.2388	1.8132
θ_{imp}	−90.0833°	−90.10°	−90.78°

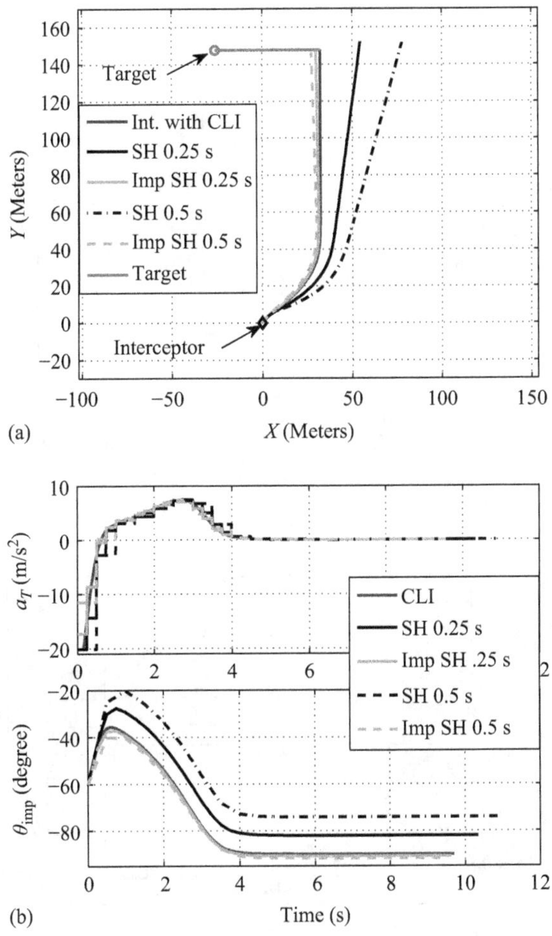

Figure 4.11 Constant velocity targets using closed-loop implementation, SH techniques with different sampling times (a) Trajectories and (b) lateral acceleration and achieved impact angles

Table 4.7 Comparison of miss distances and achieved impact angles using sample and hold techniques for a desired impact angle of −90°

	Closed		SH		SH1	
T_s	*MD*	θ_{imp}	*MD*	θ_{imp}	*MD*	θ_{imp}
0.25	0.0427	−90°	19.51	−82.07°	2.03	−90.82°
0.50	0.04	−90°	38.93	−74.24°	4.28	−91.74°

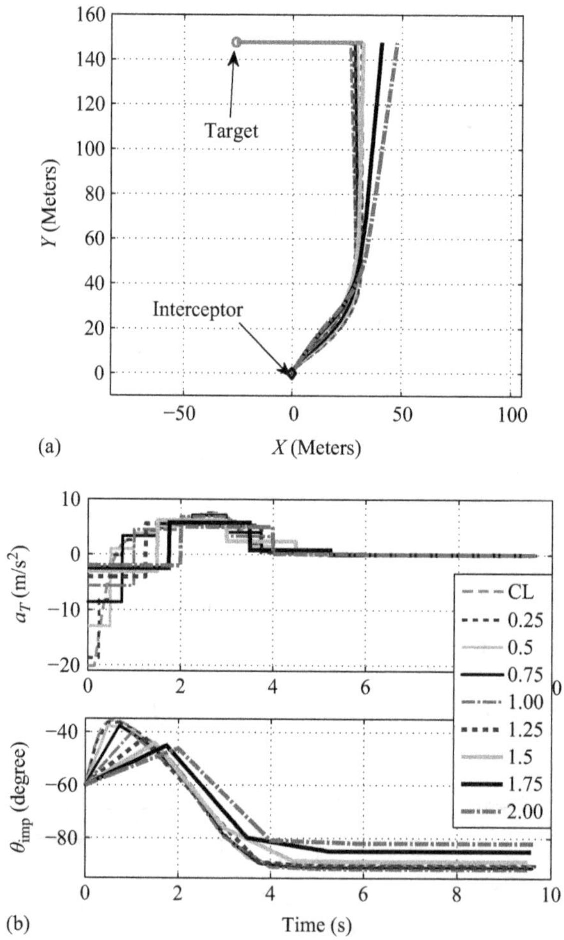

Figure 4.12 Constant velocity targets using sample and hold SH2 with different sampling intervals: (a) trajectories and (b) interceptor's lateral acceleration and achieved impact angles

Table 4.8 Comparison of miss distances and achieved impact angles using sample and hold techniques for a desired impact angle of −90°

Data	**Miss distance (m)**			**Achieved impact angle (degree)**		
Sampling intervals	**Normal**	**Improved 1**	**Improved 2**	**Normal**	**Improved 1**	**Improved 2**
Closed loop	0.04			−90.00°		
0.1	7.22	1.01	0.75	−86.93°	−90.43°	−90.31°
0.25	19.51	2.03	0.54	−82.07°	−90.82°	−90.21°
0.50	38.93	4.28	2.10	−74.24°	−91.74°	−90.82°
0.75	69.37	2.50	2.97	−62.56°	−88.78°	−91.13°
1.0	NI	18.29	4.64	NI	−81.90°	−91.75°
1.25	NI	43.47	2.92	NI	−70.33°	−90.89°
1.50	NI	55.72	1.50	NI	−66.33°	−88.78°
1.75	NI	97.48	9.44	NI	−47.29°	−84.92°
2.00	NI	NI	15.93	NI	NI	−81.93°

Table 4.9 Comparison of miss distances and achieved impact angles using sample and hold techniques for different launch angles

Heading angle	**Closed**		**SH $T_s = .05$ s**		**SH1 $T_s = .25$ s**		**SH2 $T_s = 0.5$ s**	
γ_{T0}	*MD*	θ_{imp}	*MD*	θ_{imp}	*MD*	θ_{imp}	*MD*	θ_{imp}
30°	0.0262	−90°	0.3437	−90.08°	0.59	−89.75°	0.63	−89.75°
120°	0.004	−89.99°	3.11	−88.61°	0.056	−89.97°	0.11	−89.97°
210°	0.0510	−90°	2.74	−88.6139°	0.01	−89.99°	0.01	−89.98°
300°	0.01	−90°	3.24	−91.37°	2.55	−91.03°	2.49	−91.03°

compared for the cases of the closed loop implementation, normal SH and improved SH technique and listed in Table 4.8. Note that NI in this table denotes the cases where the interception could not achieved with the corresponding sampling methods. It can be easily seen that the performance for the second improved version of sample and hold is better than other sample and hold techniques.

In order to evaluate the performance of guidance law under large heading angle error, simulation has been done with four different heading angles of 30°, 120°, 210° and 300° but with the same desired impact angle of −90° and with different sampling methods. Note that the sampling time is considered to be larger for improved sample and hold techniques than normal sampling techniques. The miss distances and achieved impact angles are listed in Table 4.9. From these data, it can be concluded that even with larger sampling intervals, the improved sample and hold methods perform better than the normal SH method. Simulation results for sample and hold $SH1$ with $t_s = 0.25$ s are shown in Figure 4.13. Similarly, simulations have also been

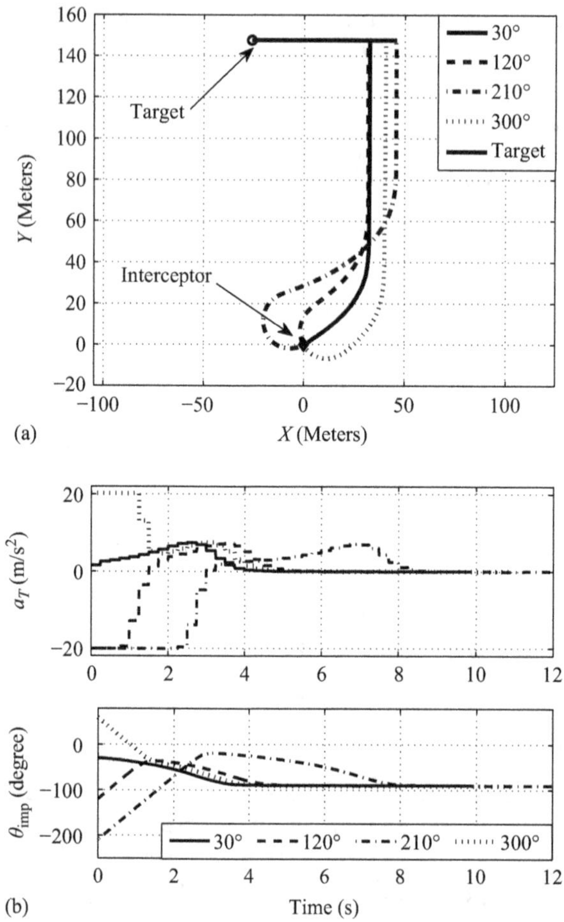

Figure 4.13 Constant velocity targets using sample and hold SH1 with different heading angles: (a) trajectories and (b) interceptor's lateral acceleration and achieved impact angles

carried out for different impact angles of $-60°, -90°$ and $-120°$ but same launch angle with different proposed sampling techniques. The miss distances and achieved impact angles for these simulations are listed in Table 4.10 which again prove the superiority of improved sample and hold techniques. Simulation results for sample and hold $SH2$ with $t_s = 1$ s are shown in Figure 4.14. From Figure 4.14, it can be seen that with only few guidance commands, generated using improved sampling method, the target can be intercepted at the desired impact angle.

In order to demonstrate the effect of the gain of discontinuous control component, simulations has also been carried out for different values of gain, and results are shown in Figure 4.15. The corresponding miss distance and impact angle errors are listed in

Table 4.10 Comparison of miss distances and achieved impact angles using sample and hold techniques for different impact angles

Impact angle	**Closed**		**SH T_s = .05 s**		**SH1 T_s = .25 s**		**SH2 T_s = 1 s**	
θ_{imp}	***MD***	θ_{imp}	***MD***	θ_{imp}	***MD***	θ_{imp}	***MD***	θ_{imp}
−60°	0.01	−60°	3.71	−61.39°	1.40	−59.49°	4.59	−58.65°
−90°	0.04	−90°	3.56	−88.56°	2.03	−90.82°	4.62	−91.75°
−120°	0.03	−119.99°	2.88	−118.58°	1.32	−120.81°	1.32	−121.41°

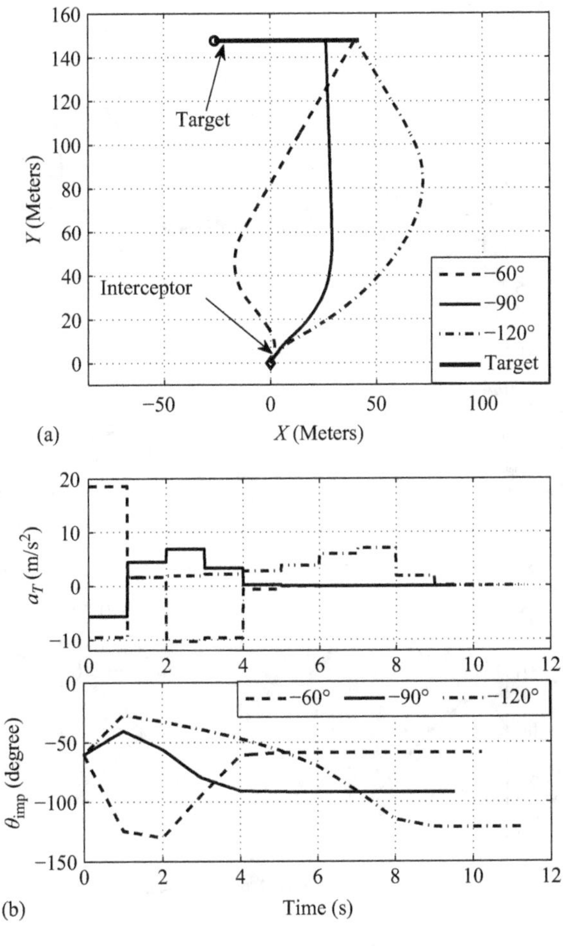

Figure 4.14 Constant velocity targets using sample and hold SH2 with different impact angles: (a) trajectories and (b) interceptor's lateral acceleration and achieved impact angles

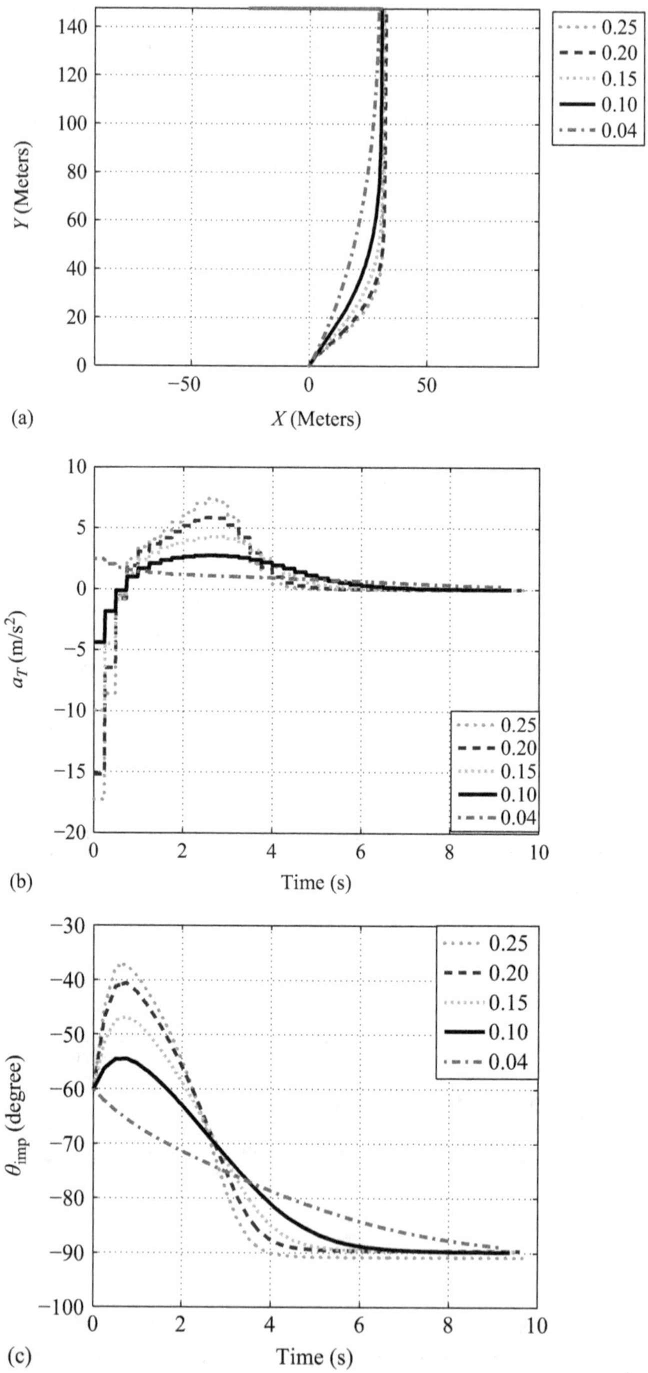

Figure 4.15 Constant velocity target with different gain and 0.25 s sampling time: (a) trajectories in (X, Y)-space, (b) interceptor latax a_T and (c) impact angle

Table 4.11 Comparison of the miss distances and achieved impact angle for different gain and 0.25 s sampling time but heading angle of 60°

M	0.04	0.10	0.15	0.20	0.25
MD(m)	0.01	0.01	0.03	0.07	0.57
θ_{imp}	−88.93°	−89.97°	−89.98°	−90.07°	−90.23°

Table 4.11. From Figure 4.15 and Table 4.11, it can be seen that with all these values of gain, the interceptor is able to intercept the target with negligible miss distance and impact angle errors. The different values of gain allow the designer to control the trajectory and hence gives the flexibility to control the time of interceptor depending on the requirement. This is also an advantage for the proposed guidance law.

It was observed that the results with linear interpolation is relatively better than that with the sample and hold technique. But, from implementation point of view, the sample and hold technique is preferable than the linear interpolation technique. In order to evaluate the performance of the guidance law under improved sampled and hold technique, simulations have been carried out for the same engagement scenario as that of Figure 4.9 and results are compared as shown in Figure 4.11. Here, the sampling time of 0.25 and 0.5 s are considered. From Figure 4.9, it be clearly seen that the interceptor are able to intercept the targets with improved sampling method even with a higher sampling time. The improvement in the miss distances and achieved impact angles can also be observed from the Table 4.7 for the sampling interval of 0.25 and 0.5 s, respectively. Note that the miss distances in the case of normal sample and hold are very high.

4.6 Conclusions and future work

In this chapter, different sampling methods to implement impact angle guidance law, based on sliding mode control, are proposed. Extensive simulation results are shown for different situations, having different sampling times, desired impact angles, etc. It has been shown that with the proposed sampling technique, the guidance update time can be increased to a large extent which is desirable for easier implementation of guidance command. The miss distances and impact angle errors have been analysed for different sampling time and sampling methods. The sampling intervals are assumed to be equally spaced. The exploration of using unequal sampling intervals, which can reduce the number of storage points, could be a possible future work.

References

[1] L.E. Dubins, "On curves of minimal length with a constraint on average curvature, and with prescribed initial and terminal positions and tangents", *American Journal of Mathematics*, Vol. 79, 1957, pp. 497–516.

[2] Y. Meyer, P. Isaiah, and T. Shima, “On Dubins paths to intercept a moving target”, *Automatica*, Vol. 53, 2015, pp. 256–263.

[3] A. Gopalan, A. Ratnoo, and D. Ghose, “Time-optimal guidance for lateral interception of moving targets”, *Journal of Guidance, Control, and Dynamics*, Vol. 39, No. 3, 2016, pp. 510–525.

[4] N. Indig, J.Z. Ben-Asher, and E. Sigal, “Near-optimal minimum-time guidance under spatial angular constraint in atmospheric flight”, *Journal of Guidance, Control, and Dynamics*, Vol. 39, No. 7, 2016, pp. 1563–1577.

[5] S.R. Kumar, S. Rao, and D. Ghose, “Sliding-mode guidance and control for all-aspect interceptor with terminal angle constraints”, *Journal of Guidance, Control, and Dynamics*, Vol. 35, No. 4, 2012, pp. 1230–1246.

[6] S.R. Kumar, S. Rao, and D. Ghose, “Non-singular terminal sliding mode guidance and control with terminal angle constraints for non-maneuvering targets”, *Proceedings of the 12th International Workshop on Variable Structure Systems*, Mumbai, India, January 2012, pp. 291–296.

[7] S.R. Kumar, S. Rao, and D. Ghose, “Nonsingular terminal sliding mode guidance with impact angle constraints”, *Journal of Guidance, Control, and Dynamics*, Vol. 37, No. 4, 2014, pp. 1114–1130.

[8] C.H. Lee, T.H. Kim, and M.J. Tahk, “Design of impact angle control guidance laws via high-performance sliding mode control”, *Proceedings of the Institution of Mechanical Engineers, Part G: Journal of Aerospace Engineering*, Vol. 227, No. 2, 2013, pp. 235–253.

[9] A. Ratnoo and D. Ghose, “Impact angle constrained interception of stationary targets”, *Journal of Guidance, Control, and Dynamics*, Vol. 31, No. 6, 2008, pp. 1816–1821.

[10] A. Ratnoo and D. Ghose, “Impact angle constrained guidance against nonstationary nonmaneuvering targets”, *Journal of Guidance, Control, and Dynamics*, Vol. 33, No. 1, 2010, pp. 269–275.

[11] B.S. Kim, J.G. Lee, and H.S. Han, “Biased PNG law for impact with angular constraint”, *IEEE Transactions on Aerospace and Electronic Systems*, Vol. 34, No. 1, 1998, pp. 277–288.

[12] S.K. Jeong, S.J. Cho, and E.G. Kim, “Angle constraint biased PNG”, *Proceedings of the 5th Asian Control Conference*, IEEE, Piscataway, NJ, 2004, pp. 1849–1854.

[13] K.S. Erer and O. Merttopçuoğlu, “Indirect control of impact angle against stationary targets using biased pure proportional navigation”, *Journal of Guidance, Control, and Dynamics*, Vol. 35, No. 2, 2012, pp. 700–703.

[14] T.H. Kim, B.G. Park, and M.J. Tahk, “Bias-shaping method for biased proportional navigation with terminal-angle constraint”, *Journal of Guidance, Control, and Dynamics*, Vol. 36, No. 6, 2013, pp. 1810–1816.

[15] Y.A. Zhang, G.X. Ma, and H.L. Wu, “A biased proportional navigation guidance law with large impact angle constraint and the time-to-go estimation”, *Proceedings of the Institution of Mechanical Engineers, Part G: Journal of Aerospace Engineering*, Vol. 228, No. 10, 2014, pp. 1725–1734.

[16] K.S. Erer, R. Tekin, and M.K. Ozgoren, "Look angle constrained impact angle control based on proportional navigation", *Proceedings of Guidance, Navigation, and Control Conference*, Kissimmee, FL, USA, 2015, AIAA 2015-0091.
[17] M. Kim and K.V. Grider, "Terminal guidance for impact attitude angle constrained flight trajectories", *IEEE Transactions on Aerospace and Electronic Systems*, Vol. 9, No. 6, 1973, pp. 852–859.
[18] C.K. Ryoo, H. Cho, and M.J. Tahk, "Optimal guidance laws with terminal impact angle constraint", *Journal of Guidance, Control, and Dynamics*, Vol. 28, No. 4, 2005, pp. 724–732.
[19] C.K. Ryoo, H. Cho, and M.J. Tahk, "Time-to-go weighted optimal guidance with impact angle constraints", *IEEE Transactions on Control Systems Technology*, Vol. 14, No. 3, 2006, pp. 483–492.
[20] X.Y. Xu and Y.L. Cai, "Optimal guidance law and control of impact angle for the kinetic kill vehicle", *Proceedings of the Institution of Mechanical Engineers, Part G: Journal of Aerospace Engineering*, Vol. 225, 2011, pp. 1027–1036.
[21] Y.I. Lee, S.H. Kim, and M.J. Tahk, "Optimality of linear time-varying guidance for impact angle control", *IEEE Transactions on Aerospace and Electronic Systems*, Vol. 48, No. 4, 2012, pp. 2802–2817.
[22] B.G. Park, T.H. Kim, and M.J. Tahk, "Optimal impact angle control guidance law considering the seeker's field-of-view limits", *Proceedings of the Institution of Mechanical Engineers, Part G: Journal of Aerospace Engineering*, Vol. 227, No. 8, 2013, pp. 1347–1364.
[23] I. Taub and T. Shima, "Intercept angle missile guidance under time varying acceleration bounds", *Journal of Guidance, Control, and Dynamics*, Vol. 36, No. 3, 2013, pp. 686–699.
[24] Y.I. Lee, S.H. Kim, and M.J. Tahk, "Analytic solutions of optimal angularly constrained guidance for first-order lag system", *Proceedings of the Institution of Mechanical Engineers, Part G: Journal of Aerospace Engineering*, Vol. 227, No. 5, 2013, pp. 827–837.
[25] Y.I. Lee, S.H. Kim, J.I. Lee, and M.J. Tahk, "Analytic solutions of generalized impact-angle-control guidance law for first-order lag system", *Journal of Guidance, Control, and Dynamics*, Vol. 36, No. 1, 2013, pp. 96–112.
[26] C.H. Lee, T.H. Kim, M.J. Tahk, and I.H. Whang, "Polynomial guidance laws considering terminal impact angle and acceleration constraints", *IEEE Transactions on Aerospace and Electronic Systems*, Vol. 49, No. 1, 2013, pp. 74–92.
[27] T.H. Kim, C.H. Lee and M.J. Tahk, "Time-to-go polynomial guidance laws with terminal impact angle/acceleration constraints", *Proceedings of 18th IFAC World Congress*, Elsevier Science Ltd., Kidlington, Oxford, UK, 2011, pp. 3915–3919.
[28] T. Shima, "Intercept-angle guidance", *Journal of Guidance, Control, and Dynamics*, Vol. 34, No. 2, 2011, pp. 484–492.

[29] I.R. Manchester and A.V. Savkin, "Circular navigation guidance law for precision missile/target engagements", *Journal of Guidance, Control, and Dynamics*, Vol. 29, No. 2, 2006, pp. 314–320.
[30] M.G. Yoon, "Relative circular navigation guidance for the impact angle control problem", *IEEE Transactions on Aerospace and Electronic Systems,* Vol. 44, No. 4, 2008, pp. 1449–1463.
[31] R. Tsalik and T. Shima, "Inscribed angle guidance", *Journal of Guidance, Control, and Dynamics*, Vol. 38, No. 1, 2015, pp. 30–40.
[32] V. Utkin, J. Guldner, and J. Shi, *Sliding Mode Control in Electro-Mechanical Systems*, 2nd ed. Automation and Control Engineering Series, New York: CRC Press, 2009.

Chapter 5

Experimentally based analysis of low altitude terrain following by autonomous underwater vehicles

Sophia M. Schillai[1]*, Alexander B. Phillips*[2]*, Eric Rogers*[3]*, and Stephen R. Turnock*[1]

Abstract

Operating a flight style autonomous underwater vehicle in close proximity to a terrain is very often completely reliant on the vehicle sensors for terrain detection. This challenges the manoeuvrability of such vehicles, which are also required to be energy efficient. In this chapter, recent experimentally based results on the performance of such vehicles are given. These results use the fully understood environment of a lake and are for the critical tasks of repeatability, obstacle detection and the actuation strategy used.

5.1 Introduction

Many applications exist where photographic surveys taken by autonomous underwater vehicles (AUV) are required, e.g. monitoring the adverse effects of climate change [1], managing food supply [2,3] or understanding the impact of mining activity at the seafloor [4]. In addition to enhancing existing approaches, AUV-based photographic surveys can also replace alternatives that are damaging the survey area or that put human divers at risk [2].

Surveys at low attitudes with AUVs encounter challenges relating to the sensors, planning and control employed. Information about the terrain ahead must be obtained, the vehicle state relative to the terrain ahead must be determined, and the correct control actions for achieving the goal altitude or avoiding an obstacle whilst maintaining a constant forwards speed must be designed and applied.

[1] Civil, Maritime and Environmental Engineering and Science, School of Engineering Sciences, University of Southampton, UK
[2] Marine Autonomous and Robotic Systems, National Oceanography Centre, UK
[3] Department of Electronics and Computer Science, University of Southampton, UK

Figure 5.1 Delphin2 AUV

In the next section, background on low altitude terrain-tracking surveys is given, together with the relevant background on the vehicle, known as Delphin2, see Figure 5.1. This is followed by descriptions of current terrain following strategies, terrain detection, horizon tracking. Next the set-up for the lake-based experiments, on which the results of this chapter are based, is given. This includes details on the experiment parameter variation and how the data is analysed for performance. The penultimate section gives the results from the experiments, and the last section summarises the progress made and discusses possible follow on research.

5.2 Background

Typically, AUVs performing photographic missions are optimised for either range or manoeuvrability. Long endurance flight style vehicles use stern planes to control the vehicle heading and pitch and a main propeller at the rear to maintain forwards speed. They are optimised for range and endurance, but are limited in their capability to avoid obstacles, see, e.g. [5].

A reduction in operating speed comes at the cost of reducing the maximum achievable pitch angle and increasing the required pitch to keep a constant depth [6]. The more manoeuvrable hover capable AUVs are typically fitted with thrusters, providing actuation with more degrees of freedom. An AUV so equipped is capable of performing surveys in complex terrains but at reduced operational speeds, see, e.g. [2,7]. This feature, combined with the power consumption of the thrusters, increases the time and energy costs for covering the same area [2]. Attempting to compromise between the two options, several hover-enhanced flight style vehicles have been developed, e.g. [8].

Low altitude AUV surveys vary widely in water depth, survey altitude and size of the survey area, from grids over a 25 m × 25 m area at 15 m depth and less [2] to large scale grids over a 10 km × 10 km area at 5,000 m depth [9]. Limited by the location-specific light attenuation and scattering of light in water, see, e.g. [10], coupled with the available means of lighting and required image resolution, the permissible survey altitude is always very low.

In practice, altitude demands vary between 0.4 and 12 m, see, e.g. [7,9,11,12]. Generally, the vehicle path planning is split between the vertical and horizontal planes. In the horizontal plane, the waypoints are typically evenly spaced in a grid, as required by sampling strategies or for a large area reconstruction [2,9]. Recent work has investigated more complex path planning in the horizontal plane [13], but still a separation of horizontal waypoints and vertical altitude goals remains. This allows the simplification of obstacle avoidance to a two-dimensional problem. Typically, the engineering data and the photo analysis are treated independently. In particular, a report is produced if a collision occurred and how tall a step was successfully navigated. Separately, the total number of photos or a total area that was successfully mapped is reported without investigating the detailed vehicle state for the less successful photographs.

5.3 Current terrain following strategies

Pre-deployment planning based on existing maps can improve the altitude tracking results and vehicle safety [12] but applicability is limited since for most of the ocean, no high-resolution maps are available [14,15]. Most vehicles rely on in-situ measurements for terrain detection in the vertical plane. If the water quality permits, camera systems can be used for obstacle detection [7], but sonar sensors are more widely used for their longer ranges and lower sensitivity to water quality.

Sensor configurations include downwards altimeter only [16], mechanical scanning sonar [5] or multibeam sonar [12]. The range measurement is performed in discrete time steps termed 'bins'. Based on the speed of sound in water, a distance can be assigned to each bin. If a single beam or mechanical scanning sonar is used, the device sends a single acoustic pulse through the water. The acoustic pulse is approximately cone shaped, with a given vertical and horizontal opening angle.

For the mechanical scanning sonar, the angle position of the sonar head can be configured within one plane of rotation. Detecting an obstacle based on the returns of a single beam or mechanical scanning sonar involves analysing the numbers representing the return intensity in each bin to establish the distance at which an obstacle is reflecting sound. Algorithms for this task include using a fixed, constant threshold above which a return is considered an obstacle, automated threshold detection, e.g. using Otsu's method [17] or through a first-order infinite impulse response digital filter [5]. More complex analysis algorithms model the sound propagation within the sonar beam to calculate a probability of occupancy for a grid map [18].

5.4 Terrain following with Delphin2

Delphin2 is an AUV developed at the University of Southampton [19]. For increased manoeuvrability, Delphin2 is additionally equipped with four through-body thrusters, which allows hovering and operation at slower speeds. It has been used for altitude tracking, in Lower Lough Erne, Northern Ireland, [16] using a downwards facing echo sounder (altimeter) mounted in the front section of the vehicle. Delphin2 was

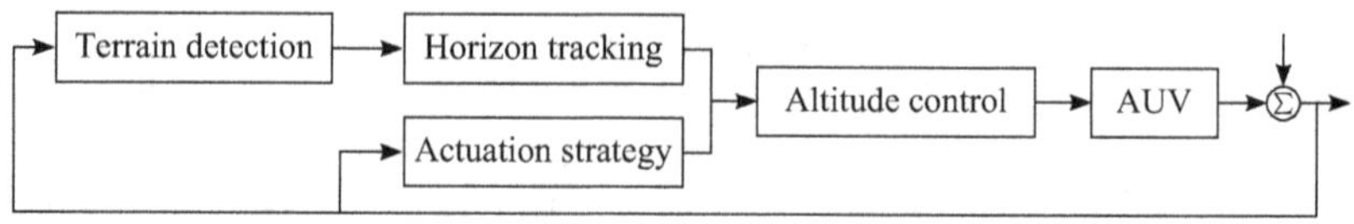

Figure 5.2 Block diagram illustrating the components of terrain following on Delphin2

flown at an altitude of 0.75 m with a surge velocity of 0.5 ms^{-1}. The absence of forwards looking situational awareness resulted in a crash into the lake floor during these trials.

To further investigate the horizon-tracking challenges, terrain following was added to the existing AUV control system on Delphin2. The added system consists of three components: (i) terrain detection with the mechanical scanning sonar, (ii) horizon tracking based on the terrain detection data and (iii) use of this data for altitude control. The choice of actuation method available on Delphin2 adds an additional component of the altitude tracking which allows comparison of hover-to-flight-style actuation as well as evaluating if thrusters at higher speeds, albeit less efficient, can improve the survey success.

The actuation force allocation is based on the surge velocity estimate. Moreover, the surge-dependent weight function for the actuators (further described in Section 5.4.4) is then applied to the actuator demands of the controller (Section 5.4.3). An overview of the system components is shown in Figure 5.2 and the remainder of this section describes their implementation on Delphin2.

5.4.1 Terrain detection

Sensors for terrain detection on Delphin2 are an altimeter, mounted downwards facing in the front section of the vehicle and the mechanical scanning sonar, orientated to scan the vertical plane ahead. The sensors and their measuring principle are illustrated in Figure 5.3(a). In this case, the altimeter is a Tritech Micron Echosounder, the sonar frequency is 500 kHz and the beam shape is a 6° conical beam. The altimeter has a minimum range of 0.5 m, a digital resolution of 1 mm [20] and the update frequency achieved with the software framework on Delphin2 is 2.5 Hz (Figure 5.4).

If the altitude is less than the minimum sensor range, the altimeter returns larger readings, resulting in a terrain estimate that is less than the actual terrain. An alternative is to use only an altimeter. In this case, the vehicle risk is significantly increased and in the worst case, the AUV will try to reduce its altitude further, based on the high altimeter readings.

After choosing the sonar scanning pattern, terrain detection methods were tested on sonar data collected at Testwood lake, see later in this section, where the experimental results in this chapter were obtained. In the collected data, no side lobe returns were observed but, due to the overall low reflectivity of the lake bed, problems with water surface reflections emerged as discussed next.

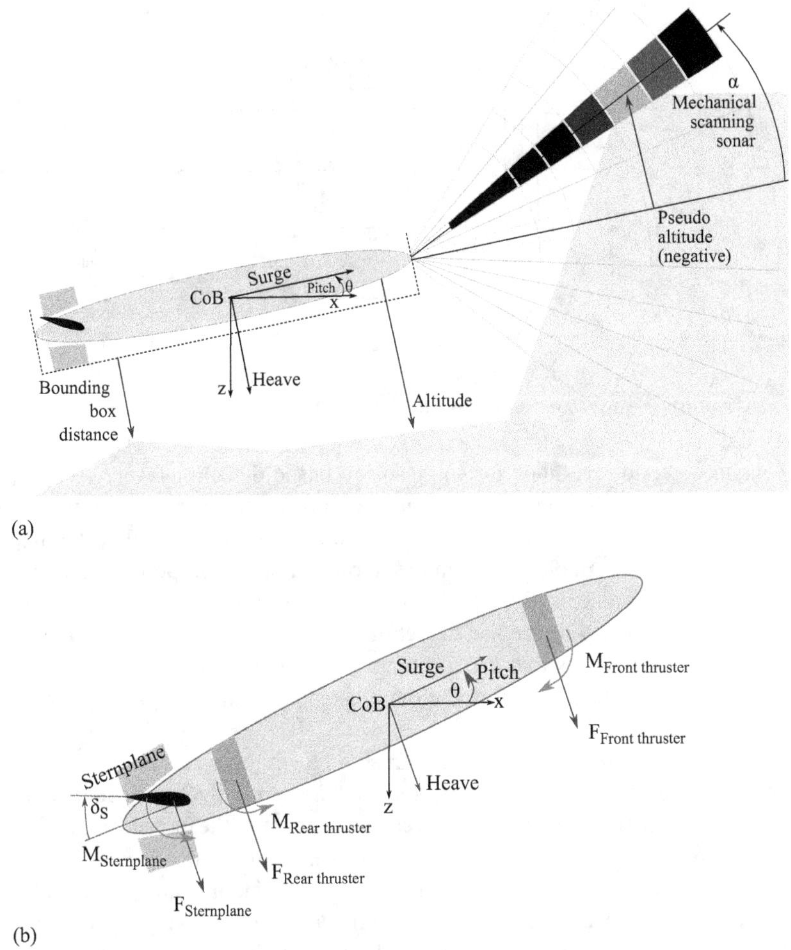

Figure 5.3 Vertical plane navigation and obstacle avoidance of the hover capable, flight style Delphin2 AUV. (a) Terrain detection sensors and illustration of the bounding box altitude estimate method. (b) Vertical plane actuation, forces, moments and reference frame

In addition to the detection of water surface reflections when pointing the sonar beam towards the surface, a second type of water surface reflections was found to cause problems. Due to reflections within the vehicle, a portion of each sonar measurement takes a path to the water surface, causing false detection at a range roughly corresponding to the vehicle depth (see Figure 5.5). Combined with low reflectivity of the lake bed, the two are barely distinguishable in the returns of one measurement. Since in the experiment set-up the typical detection range for the terrain and the required vehicle depth are very similar, the option of filtering in a band around the current depth was not used.

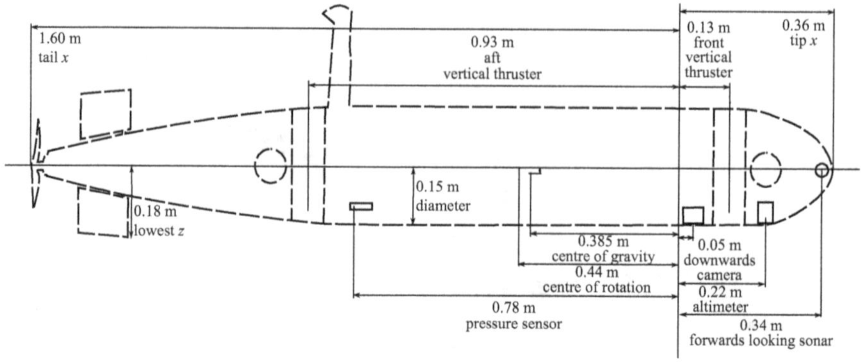

Figure 5.4 Sensor and actuator positions on the Delphin2 AUV

Several detection methods for single-beam based detections were compared to establish the most suitable method. Otsu thresholding, both with and without a reduced window size, was discarded since it was difficult to tune and hence detect a majority of correct returns. A variable threshold based on an impulse response filter could be tuned to detect terrain correctly, but the tuning was very sensitive and small changes led to missing the main return, or detection of increased noise levels. Figure 5.5 illustrates detection ranges that are too long, where the main terrain is missed and also too short with detection before the main lake bed return (the tuning parameters used are also given).

A manually selected constant sonar threshold S_{th} was also tested and found to be more reliable in detection, but with an associated range of only 2–4 m. To increase the range, the threshold was reduced with increasing distance to the transducer, with a total reduction of S_r per metre. Figure 5.5(a) shows an example from an experiment using a too low, constant threshold, resulting in the vehicle surfacing due to the detection of the surface reflections as an obstacle. The same data was used to produce examples of other thresholding methods, as shown in Figure 5.5. For experimental evaluation, several fixed and range-dependent thresholds were chosen, with most experiments performed with a threshold of $S_{th} = 95$, resulting in low range but reliable detection.

As in other AUVs, the vertical and horizontal planes are assumed to be independent and only actuation and control of the former are considered. Figure 5.3(b) shows the nomenclature of the vertical plane actuation and reference frame. Assuming positive surge velocity, a positive pitch angle leads to the AUV's surfacing. Moments that contribute to a positive pitch angle can be obtained by setting a positive stern plane angle or by configuring the thruster forces. Hence the difference between the aft and front thruster and hence the sum of their moments is greater than zero ($0 \leq M_{aft} + M_{front} = F_{aft} \cdot l_{aft} - F_{front} \cdot l_{front}$).

5.4.2 Horizon tracking

Detection range limits in the test environment makes the scan speed a more significant factor in obstacle detection. Figures 5.6 and 5.7 illustrate the impact of scanning speed,

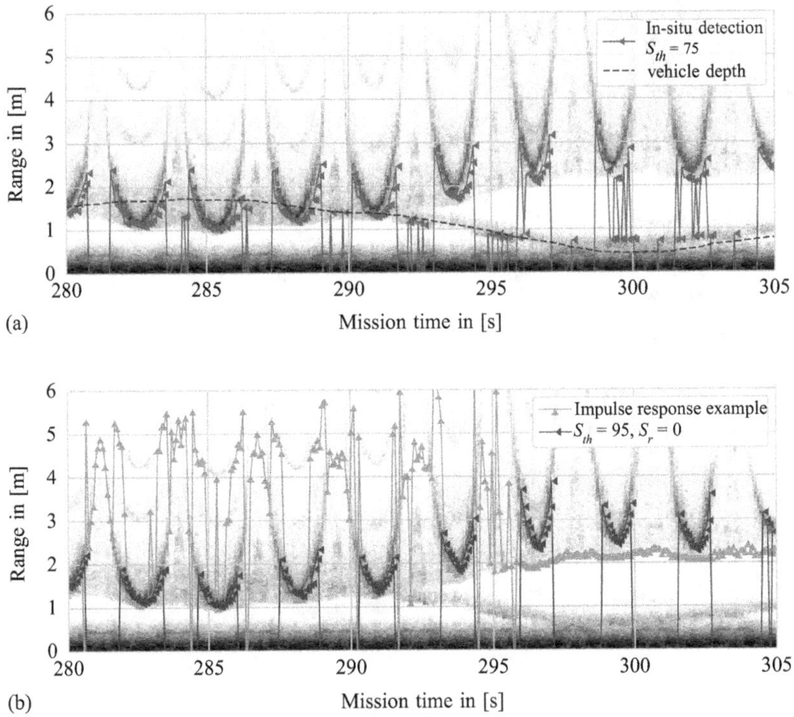

Figure 5.5 *Terrain detection example: the sonar is continuously scanning the angle sector over time, as the vehicle moves forwards over the terrain, each individual vertical line represents return intensity at a new scan angle. During the mission, a constant threshold of 75 was used, and returns from the water surface were incorrectly detected as an obstacle, resulting in the vehicle surfacing. Detection results using a higher threshold and an impulse response filter over the same sonar returns are shown in (b)*

vehicle speed, sonar range, vehicle altitude and step height using a fixed sonar beam or a continuous sector scan. A step of given height was positioned at $x = 2$ m and the plots show the AUV's horizon estimate at each position.

It was found that the time taken to update the scan angle for individual measurements was too long, and hence a fixed scan sector was used. In addition to increasing scan speed and avoiding detections of the water surface, the regular scanning pattern has the advantages that (i) it is easier to visually inspect after the experiment, (ii) all measurements are updated regularly and (iii) false detections cannot influence the scanning angle. However, the scan sector limits need reconsideration for different missions.

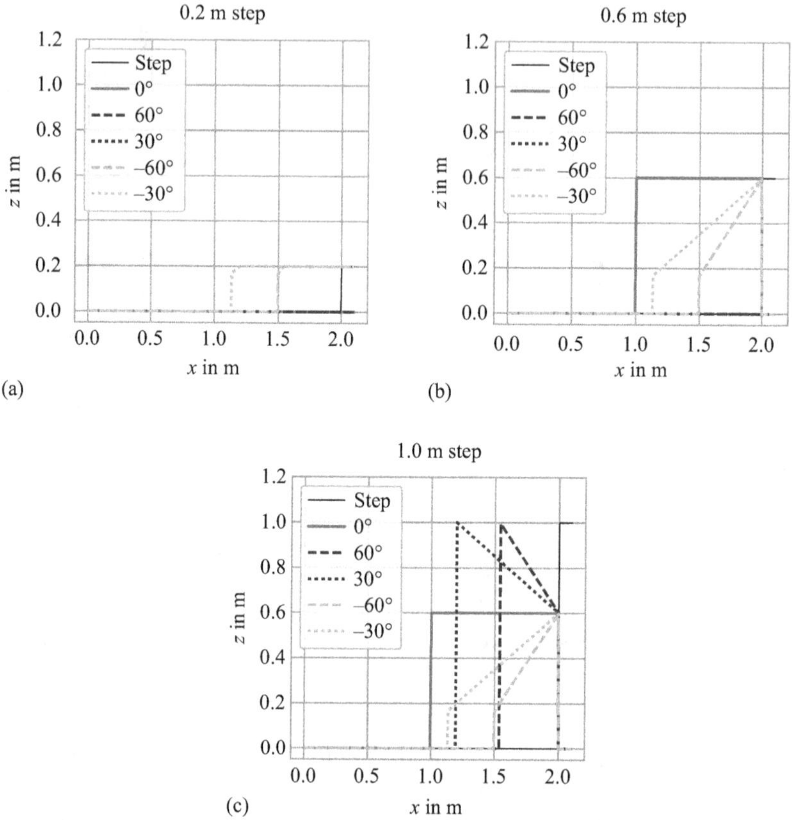

Figure 5.6 Detection height and range of a terrain step using fixed sonar angles. Parameters used for the simulation: AUV altitude 0.6, detection range 1 m. (a) Detection only by 60 degree downwards facing beam, (b) detection by downwards and forwards facing beam, (c) detection by all beams, reduced height estimate with increasing proximity.

For each angle position of the scanning sonar, two values are stored: the detection range when the angle was last visited (-1 for no detection) and the pitch angle of the vehicle at the point of detection. For every new measurement, the highest angle with a valid detection is used in the estimation of the height of the horizon ahead.

Figure 5.6 illustrates that the scanning angle limits determine how well the size of an obstacle can be determined and also how obstacles that are higher than the vehicle altitude tend to be underestimated. Expressing the obstacle height h as a pseudo altitude $a_{pseudo} = a - h$, the range Δx at which an obstacle can first be detected by a scanning sonar with range r is $\Delta x = \sqrt{r^2 - (a-h)^2}$. This assumes that the full height can actually be detected by the lowest (for small obstacles) or highest (for high obstacles) beam angle α. The detection limits for obstacle heights are $h_{\min} = r \cdot$

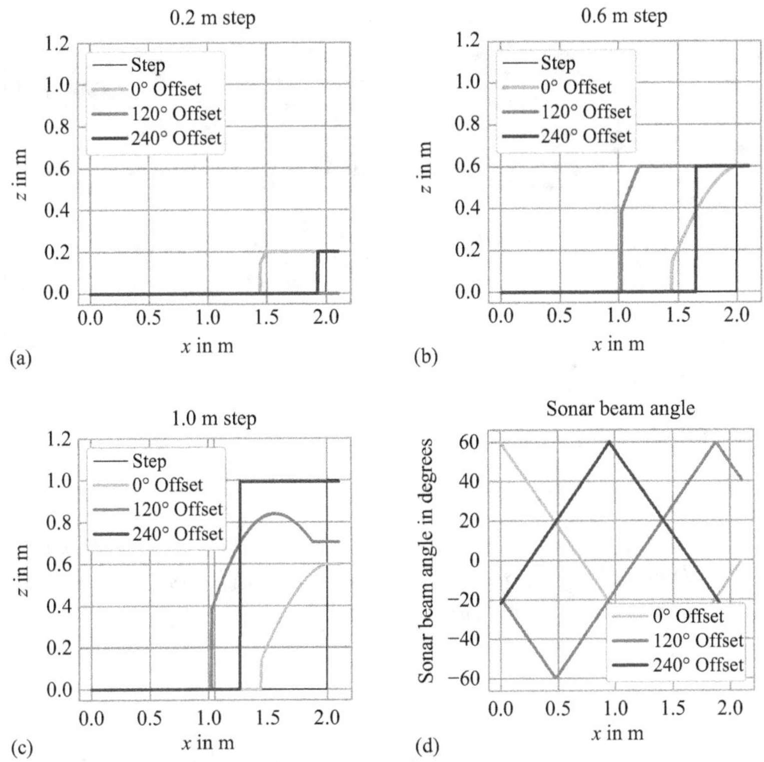

Figure 5.7 Detection height and range of a terrain step using a scanning sonar, scanning a sector between −60° and 60°, with a scan period of 2.8 s at 1 m/s surge speed and varied offsets of the initial scan angle. AUV altitude 0.6 m, detection range 1 m. (a) With the correct offset, a shallow step can be missed completely. (b) With increasing step height, the chances of missing the step reduce. The correct step height is approached. (c) Step is detected reliably, detected height and distance varies. (d) Sonar scan angle vs. x-position.

$\sin(\alpha_{min}) + a$ and $h_{max} = r \cdot \sin(\alpha_{max}) + a$. As the vehicle approaches, all obstacles that are not exactly at the same height will not be detectable at full height. The range at which obstacles with $a - h > 0$ will be overlooked and obstacles with $a - h < 0$ will be underestimated is $(|a - h|)/(\sin(\alpha_{min/max}))$.

Moving from a fixed to a slow scanning sonar, Figure 5.7 illustrates that it is detrimental for the obstacle detection if the scan period is shorter than the obstacle detection range. With a shorter scan period (increased speed of a complete sector scan or reduced vehicle speed), the offset in terrain detection becomes smaller.

A scanning sector of −60° to 0° and 3.6° angle step size were chosen for the experiments to reduce the risk of falsely detecting the water surface as terrain but still detecting all steps of a similar height to the goal altitude. To maximise the speed

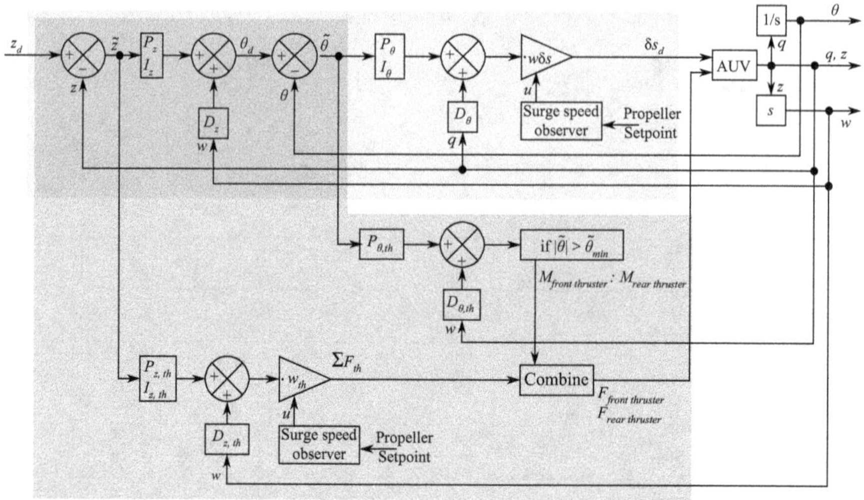

Figure 5.8 PID control scheme unifying flight style and hover capable actuation of Delphin2. The flight style (top) and hover style (top left, bottom) components are indicated.

of one full sector scan, a step angle of 3.6° was chosen (above the recommended maximum step size of 1.8°). At this step size, the speed per degree is $0.023\frac{s}{1^\circ}$ (4.1 s for a 90° sector scan period). The resulting time for a full scan with 3.6° steps is 2.82. In the same time, a horizon scanning beam pattern would complete 3.5 measurements, covering a maximum section of 12°.

5.4.3 Altitude controller

Previous studies into the depth tracking performance of Delphin2 have used model predictive control [16] and a cascaded proportional plus integral plus derivative (PID) controller [21]. This controller is a combination of hover capable and flight style control, unified to work over a range of operation speeds. The contribution of the thrusters and horizontal stern planes is allocated by applying a weighting function based on the forwards speed. The full control design is shown in Figure 5.8, with the split of flight style control (top) and hover capable control marked.

For the altitude tracking experiments, the altimeter was used as the only input to the depth controller. The depth demand was calculated from the current depth, the current pitch, the current altitude and the altitude demand. During the horizon tracking experiments, the altitude from the altimeter was compared to the pseudo altitude calculated from the forward-looking sonar and the lowest of the two values was used to calculate the depth demand.

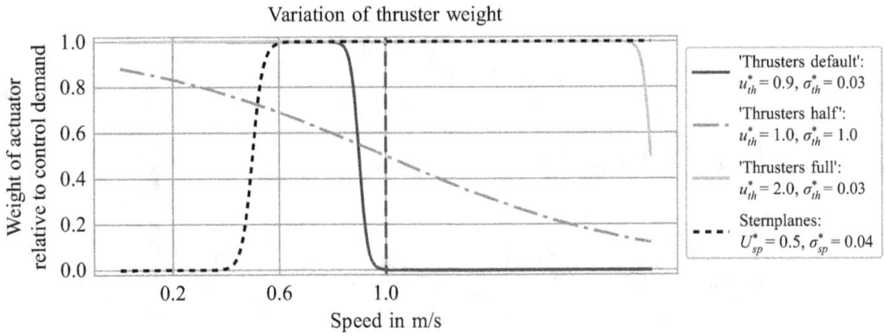

Figure 5.9 Thruster and stern plane weight functions: the stern plane weight function remains unchanged, the default thruster weight function is used for all three surge speeds, and the 'thrusters half' and 'thrusters full' variation is tested in cases 4 and 5, at 1 m/s surge speed

5.4.4 Actuation strategy

The weight functions w_{th} for the thrusters and w_{sp} for the horizontal stern planes are, respectively, given in (5.1) and (5.2). During the experiments, the weight function for the thrusters was modified for the fastest speed, to test if the altitude tracking performance could be improved by using thrusters outside their optimal speed. The parameters used for the mid-transition speeds u^*_{th} and u^*_{sp} and the width of the transition zone are given in Figure 5.9, which also shows the resulting weight functions.

$$w_{th} = 1 - \frac{1}{2}\left(\tan h\left(\frac{u - u^*_{th}}{\sigma^*_{th}}\right) + 1\right) \tag{5.1}$$

$$w_{sp} = \frac{1}{2}\left(\tan h\left(\frac{u - u^*_{sp}}{\sigma^*_{sp}}\right) + 1\right) \tag{5.2}$$

5.5 Testwood lake experiment set-up

To test terrain following with flight style AUVs, a body of water that allows a long enough path for the vehicle to reach a steady state is required. Testwood lake, a 4 m deep reservoir north-west of Southampton, with an area of approximately 200 m by 500 m was chosen. This lake has a high step feature, which is 300 m wide and has a height of 0.6, which is very suitable for the tests performed, see Figure 5.10. Other advantages of the lake environment over the use of towing tanks are the absence of sonar reflections from side walls and the low magnetic disturbance to the compass.

Before the tests started, a depth map of the entire lake and a detailed map of the area of interest was acquired from the water surface using GPS positioning and

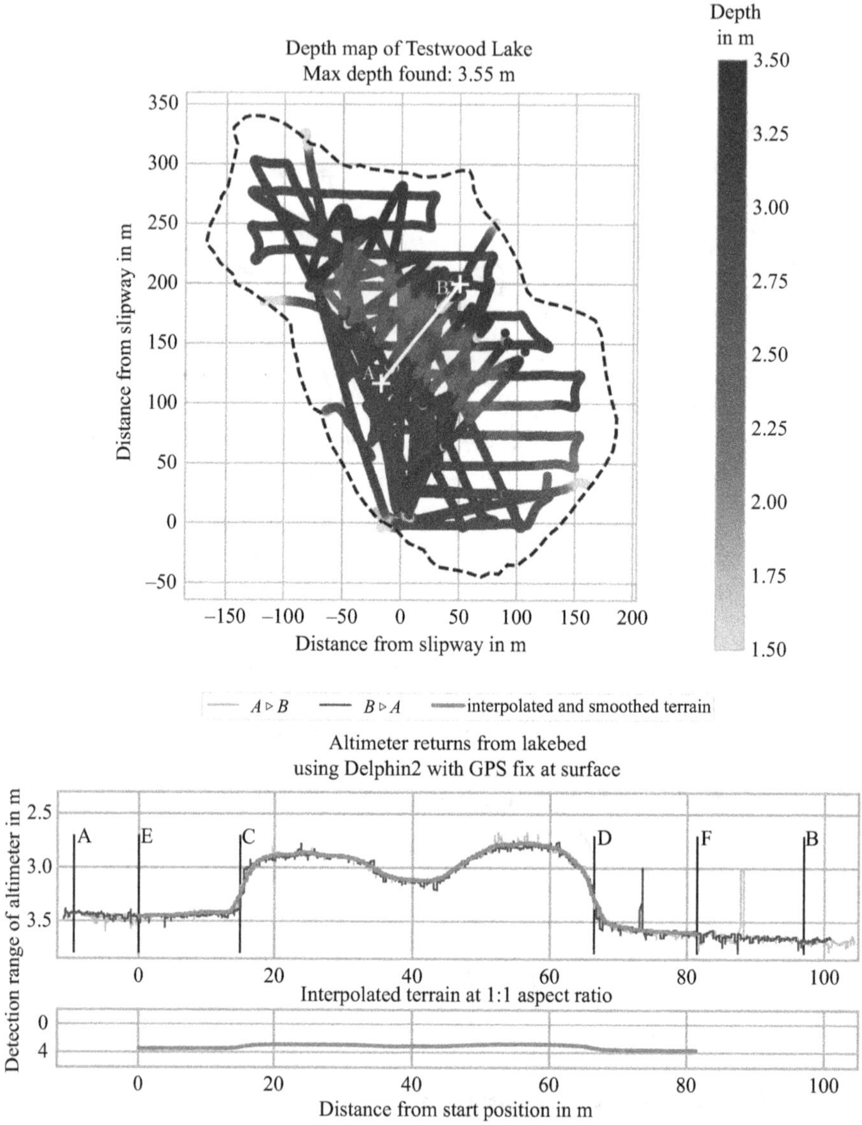

Figure 5.10 Testwood lake depth map and profile of the experiment path, obtained from altimeter measurements made with a GPS fix at the water surface. The end points of the experiment path are A and B. It was structured with further points for analysis purposes

altimeter readings. A transect for crossing the step feature at a 90° angle was chosen and a distance of 25 m was added before and after the step to ensure the vehicle operated in a steady state when it reached the step (see Figure 5.10 (bottom plot)). During the lake experiments, the vehicle crossed the 0.7 m high step that was identified in the lake, performing altitude tracking from point A to point B ($A \triangleright B$) and in the reverse direction ($B \triangleright A$). All experiments were then executed by reaching the start position at the surface with a GPS fix, diving to the goal altitude with the thrusters and then moving forwards under a constant heading and the chosen altitude tracking method for each of the experiments.

5.5.1 Experiment parameter variation

To attempt to get within an altitude region close to the terrain, the vehicle operator can influence the quality of the results by modifying the goal altitude and the vehicle speed. Also (if required) the vehicle configuration, the thruster weight and the terrain detection threshold can be used to modify the altitude tracking behaviour. The parameter variations were chosen to investigate the impact of the choices made, and 1.0 and 0.6 m were chosen as the goal altitudes for the experiments.

The lower altitude of 0.6 m was chosen to make the step a real obstacle in the path that required avoidance. Also the experiment paths were crossed at 0.3, 0.6, and 1.0 m/s surge velocity. These represent low, medium and high speeds for Delphin2 and are well understood from previous experiments on diving [21] and heading [22] control at the surface.

During the horizon tracking experiments, the control allocation at the 1 m/s speed was varied to increase the use of thrusters from the default thruster weight of 0 to a factor of 0.5 ("half") and 1.0 ("full") (see Figure 5.9). All combinations of actuator settings, goal altitude and obstacle detection settings are given in the experiment matrix (Table 5.1).

Two cases were identified as references and to ensure repeatability they were repeated several times on separate days. All other experiments were only repeated twice, though sometimes these numbers vary since missions with indications of a sensor failure or an early mission abort at the lake side were repeated as a precaution.

5.5.2 Performance analysis

To ensure comparability of the collected data, several steps of post-processing were applied to account for changes in the water level between experiment days and restrict all data to a well-defined experiment area. An estimate of the terrain shape was obtained by crossing the area of interest at the surface, using GPS for more accurate positioning.

The area of interest was crossed twice, once from each direction. Based on the GPS position, the distance to the start point A was calculated. Both depth maps show spikes which stem from altimeter noise. A running average was used to combine the two terrains into an estimate of the real terrain. Figure 5.10 (bottom plot) shows the depth measurement from the two crossings, and the interpolated terrain estimate. To enable comparison of the experiments in both directions, symmetry was achieved by including a fixed length of 15 m before and after the step, resulting in a 81.5 m long path (A to E in Figure 5.10). The distance between the start positions (A, B) and the considered experiment path (E, F) allows for the vehicle to reach a steady state.

Table 5.1 Experiment matrix for Testwood lake experiments. An empty detection threshold S_{th} in the obstacle detection indicates that only the altimeter was used for obstacle avoidance. Repeats vary due to time limitations at the lake not allowing for all missions with a mission abort to be rerun

Case No.	Altitude		Speed			Thrusters		Detection		Repeats	
	$0.6m$	$1.0m$	$0.3\frac{m}{s}$	$0.6\frac{m}{s}$	$1.0\frac{m}{s}$	original	varied	S_{th}	S_r	$A \triangleright B$	$B \triangleright A$
1)		■	■			■				2	1
2)		■		■		■				2	2
3)		■			■	■				2	2
4)		■			■		half			1	1
5)		■			■		full			1	1
6)		■	■			■		95		2	3
7)		■		■		■		95		3	2
8)		■			■	■		95		10	11
9)		■			■	■		85		2	2
10)		■			■	■		75		2	1
11)		■			■	■		95	1	1	1
12)	■		■			■		95		3	2
13)	■			■		■		95		2	2
14)	■				■	■		95		4	4
15)	■				■	■		75		1	1
16)	■				■	■		95	1	1	1

The recorded data was restricted to the area between the points E and F by matching the terrain estimate, obtained from the depth, pitch and altitude during the mission, to the terrain profile obtained from the surface. This method was chosen since it was more reliable compared to acceleration-based positioning. The x-position gives the distance from the start point, along the goal heading (see Figure 5.10).

The various sensors on Delphin2 record data asynchronously at rates between 5 Hz and 20 Hz. To combine data for calculating values derived from the sensor data, the recorded measurements were first interpolated based on the time between subsequent measurements. The terrain as seen during a mission is combined from the depth of the vehicle, the altitude and compensating the altitude based on the vehicle pitch angle.

To better align the missions, the time of crossing the upwards and downwards step were read manually from a plot of the terrain as seen during a mission. The error of this reading is estimated at ± 0.5 s. Given a constant propeller setting and the maximum pitch angle at $-20°$, the speed along the experiment path x-axis is assumed constant. Using the step length read as 51.5 m and the time between crossing the beginning and end of the step, the speed of Delphin2 along the global x-direction was estimated, cross referenced with existing speeds estimates of this vehicle and used to estimate the x-position along the experiment path. Hence, experiments at different speeds can be compared based on the x-position.

It was found that despite attempts of normalising the depth sensor at the water surface, it had a constant error that varied by several centimetres on the same day. To compensate for this offset, the mean water depth of the first section of each experiment was calculated and compared to that of the re-constructed terrain. For comparison between different experiments, all data sets were finally interpolated and re-sampled at a resolution of 1 cm, starting at $x = 0$ m. All data post-processing used the python pandas library [23] for interpolation, re-sampling, combination and plotting of data.

The altimeter minimum range can cause an overestimate of the altitude and therefore a second altitude estimate was calculated by subtracting the depth measurement from the water depth at the estimated position. Unlike the altitude measurement from the altimeter, which is parallel to the vehicle-fixed z-axis, this altitude measurement is parallel to the global z-axis. The measurement zero is also at the bottom of the hull, and its x-position is that of the centre of buoyancy.

A third measure for the terrain distance, first introduced in [24], was implemented by fitting a bounding box around the Delphin2 vehicle (see Figure 5.3(a) for illustration and Figure 5.4 for the length and lowest z-coordinate of Delphin2). Using the estimate of the x-position and the terrain estimate and considering the current pitch angle, the shortest distance of the bounding box to the terrain was used to better describe the collision risk of the entire vehicle rather than that of just a selected point. Whilst the control goal may be a constant altitude, when performing altitude tracking surveys in complex terrain, the real goal is often to keep the survey vehicle within a suitable range of the terrain rather than at a fixed altitude. To account for this, the analysis method of risk and success zones introduced in [24] was applied to the lake experimental results.

5.6 Results

In general terms, the results obtained are as expected with almost perfect altitude tracking for hover capable mode and less accurate results when at higher speeds under flight style control. For all cases, the return to the goal altitude after the downwards step takes between 1.5 and 2 times longer than when navigating the step in the reverse direction. This is due to the positive buoyancy of Delphin2, and when using the forward-looking avoidance, this effect is further increased due to overestimating the horizon ahead.

Closer examination of the sensor measurements show how the terrain step and false detections propagate through the measured values as the AUV passes over the terrain. In particular, the vehicle path looks smooth for all cases and the altitude line only shows sensor noise on the altimeter; the impacts of sensor noise, false detections and controller oscillation can be recognised in the altitude results for the terrain detection sensors, the pitch angle and the stern plane setpoints. Figure 5.11 is annotated to highlight this feature for the slowest case with hover actuation at 0.3 m/s, using the forward-looking obstacle avoidance (case 6), in addition to the fastest 1.0 m/s case with flight style actuation both using the altimeter only (case 3) and using altimeter plus the forward-looking obstacle avoidance (case 8).

5.6.1 *Repeatability and obstacle detection*

Overall, the experimental results are very similar in one case, whilst showing distinction between the different test cases, indicating a good repeatability and clear separation between different configurations. To further test for the repeatability of the results, cases 8 and 14 were selected as references. For each reference case, all

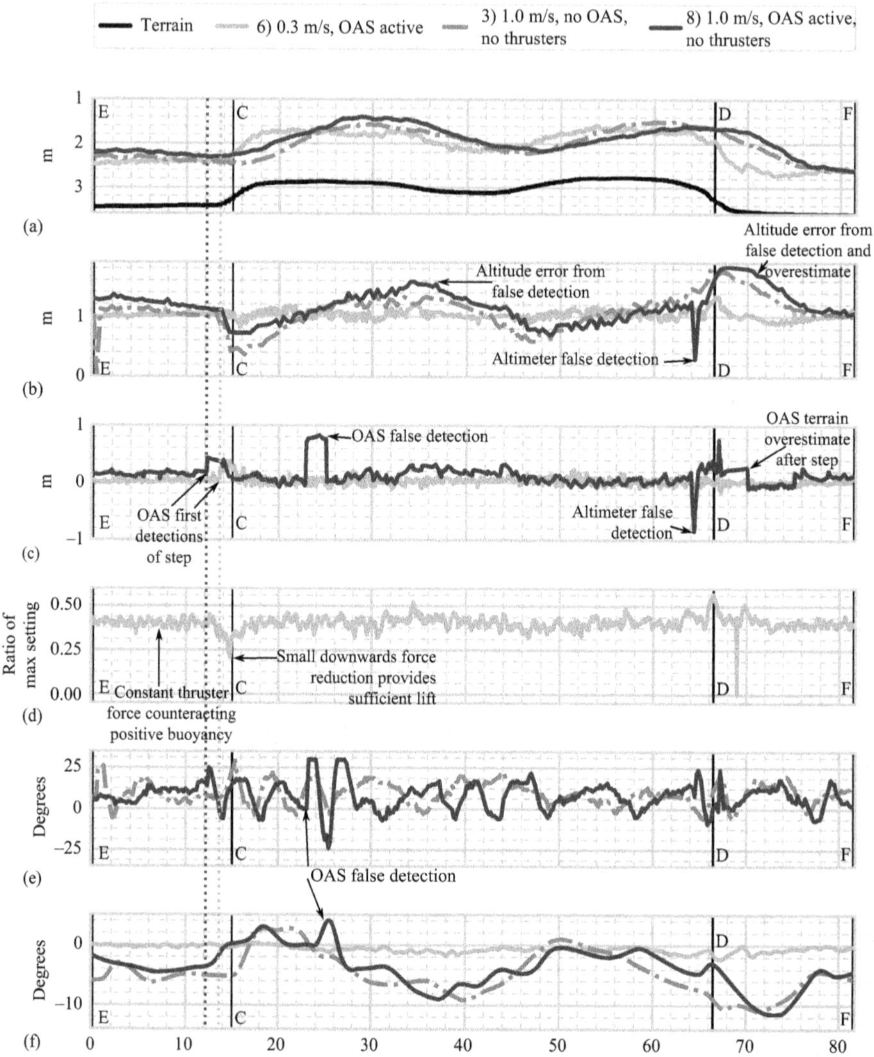

Figure 5.11 Comparison of terrain following performance over full terrain section for key cases, in the direction A ▷ B: at slowest speed with forwards looking obstacle avoidance (case 6), and at highest speed with and without forward-looking obstacle avoidance (cases 8 and 3). (a) Vehicle depth, (b) altitude, (c) difference: pseudo altitude – altitude (positive=pseudo altitude is used), (d) thruster setpoints, (e) sternplane setpoints, (f) pitch angle

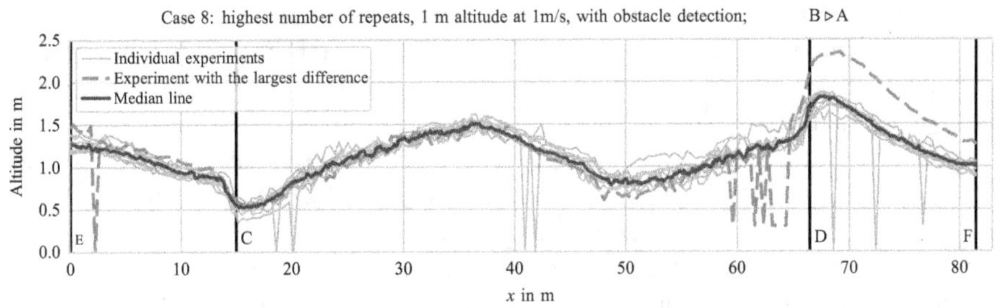

Figure 5.12 Comparison of the altitude median line of reference case 8, to the actual experiments of case 8

repeats of one heading ($A \triangleright B$ or $B \triangleright A$) were combined in a median line. To determine a median line for a measurement, the median of the value from each individual repeat was calculated for each point along the x-direction. The median lines are used as a reference when comparing experiments for the same case, or to compare the reference cases amongst themselves.

The mean difference of the reference cases to their own altitude median line is less than 0.1 m, with the exception of one experiment for case 8, $B \triangleright A$. This reference case is shown in Figure 5.12 with the one exception case highlighted. In contrast to most of the altitude measurements, which only have brief individual false measurements, this case has a significant number of repeated false measurements. This, in turn, led to the vehicle increasing its altitude significantly and thus differing greatly in the path-generated relative to other paths.

The impact of the obstacle-detection parameters on the repeatability can be seen in the altitude, photographic success and risk analysis of cases 10, 11, 15 and 16, which all only differ from one of the reference cases in their sonar setting but have a greater variation in altitude, mission success and vehicle risk. Other experiment cases vary in their difference to the median line of the reference cases. However, for most cases, the repeat of the same case leads to similar results. This supports the assumption that provided the terrain detections are correct, the experiments are very repeatable and that a single experiment is representative of the expected performance, despite timing variation in terrain detection and unsteady control.

On extrapolating the altitude tracking results for a full battery charge and estimating the total path length that was successfully photographed, the effects of the false terrain detection on the photographic success are evident. This can reduce the expected photographed area to even below that of hover capable operation at 0.3 m/s speed for the obstacle configurations of cases 10 and 11, rather than doubling the covered length in the case of correct detection at 1 m/s, case 8.

To estimate the overall obstacle-detection success, the detection of each experiment was manually compared to the sonar scan and categorised into three groups, i.e. no false detection, occasional false detections and significant false detections. No false detections were found for most of the experiments with a fixed threshold S_{th} set at 95. Reducing the threshold to 85 or varying it proportionally to the distance resulted in occasional false detections for each of the experiments and a further reduction to $S_{th} = 75$ resulted in significant false detections for all experiments. It was expected that, due to the larger depth, detection would be less impacted by the surface reflections at lower altitude but the improvement found is not significant.

At 1 m altitude, the detection range of the first step was on average 3 m. This distance varied by up to ± 1.4 m, as expected with a 2.8 s sector scan time. The sonar settings of cases 10 and 11 increased this distance to over 4 m. For a 0.6 m altitude goal, the average range is drastically reduced to between 1.5 and 2 m. With the variation over the scan period, this led to detection ranges as low as 0.1 m. However, this is still a significant safety improvement as it compensates for false altimeter readings. Figure 5.13 compares examples of both very late and very early detection to the median line for case 8. Whilst the experiment with a late detection measures a significantly higher terrain altitude, the detection range of just above 1 m does not leave enough time for the vehicle to achieve a sufficient depth change rate, resulting in one of the highest risk results for case 8. This illustrates the importance of considering repeatability and variation in sensor performance in understanding the vehicle risk.

5.6.2 Actuation strategy

The impact of the transition between flight style and hover capable actuation is measurable in the distance made good until the goal altitude is reached again (without obstacle detection). It is below 3 m for 0.3 m/s hover capable control and increases to over 7 m for flight style control. Using flight style actuation only at 1 m/s speed, a significant portion of the experiments show an oscillation in the pitch angle (see Figures 5.11 and 5.13) that is not observed when tracking a depth with the same controller (see [21]). Due to the dynamics of the AUV acting as a low-pass filter, this oscillation is barely noticeable in the altitude and depth measurements.

In the standard control for Delphin2, the use of the thrusters is phased out between surge velocities of 0.8 and 1.0 m/s (see Figure 5.9) and the thruster weight was varied at maximum speed to determine if the thrusters can support the flight style control for obstacle avoidance. Figure 5.14 compares the three 1 m/s cases with stern plane only actuation and variations of the thruster usage. The selected cases use only the altimeter for obstacle avoidance to remove timing effects from the obstacle detection.

The flight style control with added thruster actuation overall maintains the goal altitude better, but an impact of the thrusters on the efficiency on the pitch control can clearly be seen. In particular, despite stern planes set for pitching the vehicle upwards, and the expected thruster moment supporting this, the vehicle error indicates that the vehicle tends to pitch downwards more than demanded.

A fast pitch change rate is still achieved at the upwards step, see the altitude and pitch error plots in Figure 5.14, but the goal altitude is approached much slower since the vehicle soon acquires a downwards pitching error again. This contributes to a lower overshoot for the current control parameters and reduces the time it takes to return to the goal height after a downwards step. Thus the overall results are still better for these actuation strategies, but the loss of pitch control makes them unsuitable.

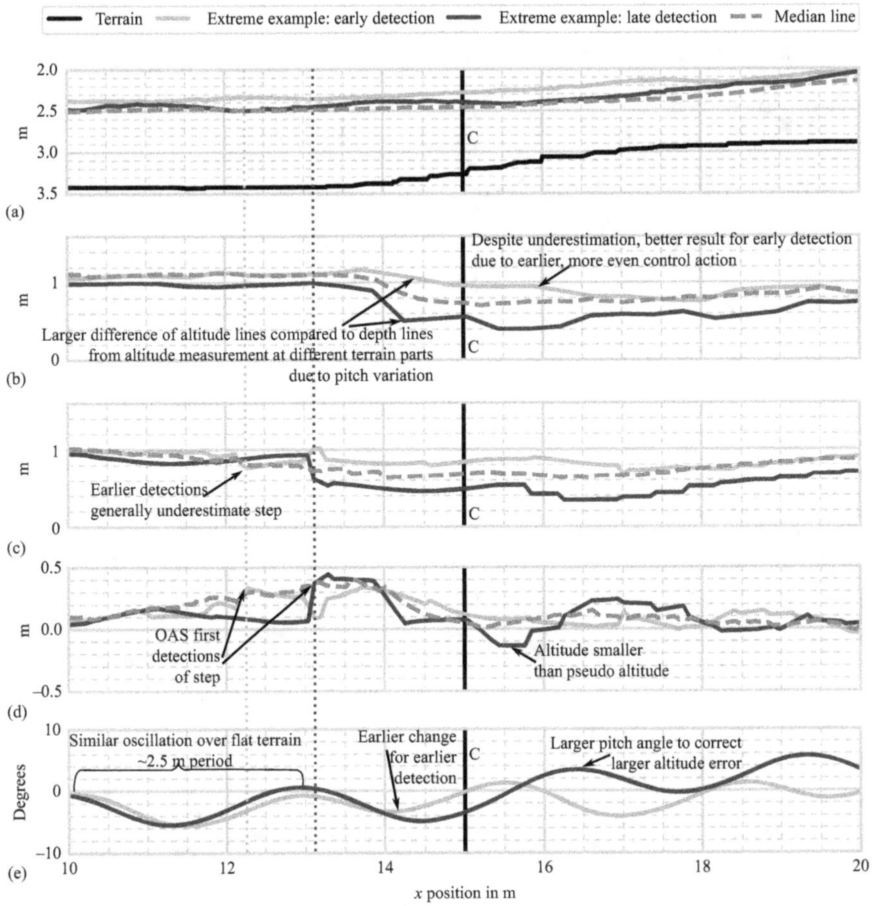

Figure 5.13 Extreme examples of detection variation compared to median line of reference case 8, direction A ⊳ B, at 1 m/s. The section shows includes 5 m before and after the upwards step at C. (a) Depth, (b) altitude, (c) pseudo altitude, (d) difference: pseudo altitude − altitude (positive=pseudo altitude is used), (e) pitch angle

The reasons behind the difficulties tracking a given pitch angle when combining the thrusters at high speed is most likely a combination of (i) the stern planes being less efficient due to the turbulence cause by the thrusters and (ii) the thruster performance degrading with surge speed at different rates as observed in [25]. Furthermore, ground effects from operating at low distances to the terrain, similar to those observed at the free surface, may play a role, though the thruster tests at up to 0.8 m/s altitude only show variations in differential pressure at larger depths (lower altitudes) for the highest thruster speeds [26].

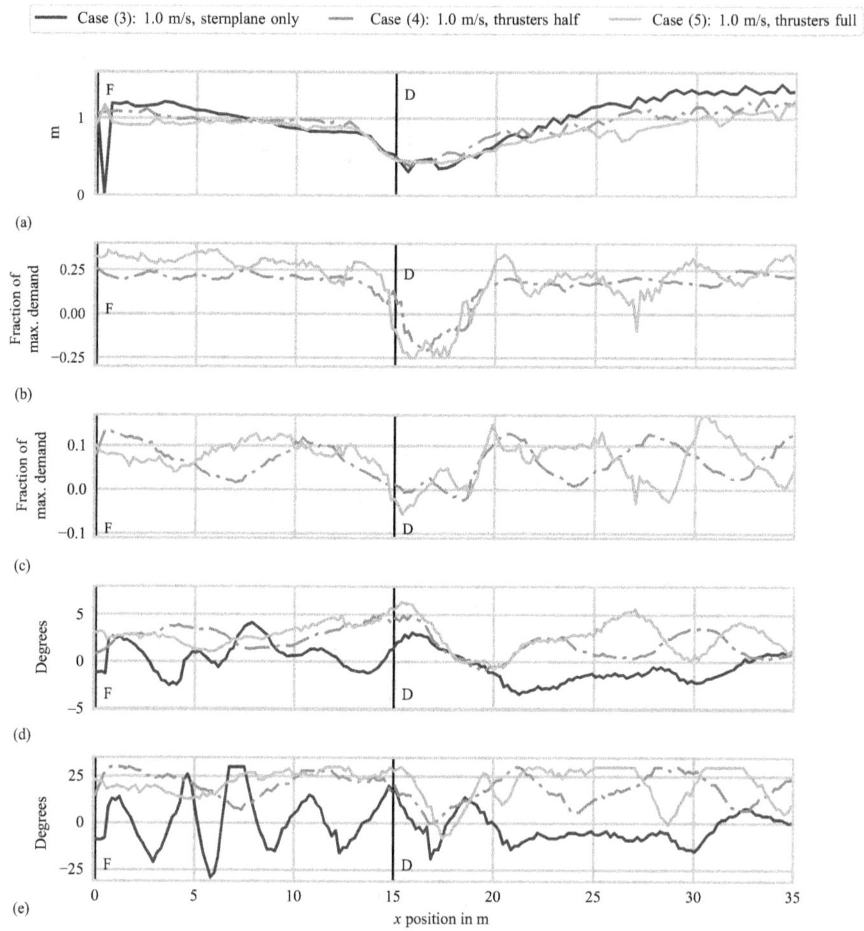

Figure 5.14 Using the altimeter only for tracking, and a constant surge velocity of 1 m/s, the thruster weight is varied (cases 3, 4, 5, variation as in 5.9). The B ⊳ A direction is shown, since for this direction none of the three cases had altimeter spike measurements. To estimate the pitching moment generated by the thrusters, the thruster setpoints are weighted to consider their distance to the centre of rotation and subtracted to get the differential forces that contribute to the pitching moment. (a) Altitude, (b) Aft thruster setpoint, (c) Weighted difference of aft- and front thruster to estimate rotational moment (pos. = to surface), (d) Pitch error: $\tilde{\theta} = \theta_d - \theta$ (pos. = needs to pitch towards surface), (e) sternplane angle (positive = to surface)

5.7 Conclusion

This chapter has undertaken a detailed analysis of AUV terrain following, using data from repeated experiments over the same terrain to compare altitude tracking success and repeatability for different configurations. For reliable terrain detection, the altitude tracking was shown to be repeatable, with a distinction in the results for different actuation and control methods.

As expected, the best altitude-tracking performance with minimum variation in pitch angle was observed for hover capable actuation at slow speed. With the transition to more energy efficient flight style actuation, the slower reactions of the vehicle reduce the altitude-tracking capability. At this point, the forward-looking obstacle avoidance system becomes crucial for minimising collision risk.

Increasing the forwards speed of the vehicle increases the area that can be surveyed. The mission success rates show that this effect decreases with increased terrain complexity and unreliable terrain detection. The error in altitude increases especially on a downwards slope or step. On further increasing the risk and reducing the photo quality, no steady state pitch angle was reached with the current controller. It was found that the addition of thrusters at full speed does not improve the altitude tracking capabilities but reduces controllability of the AUV.

The analysis of the experiment results demonstrates how the introduction of a risk zone in the altitude-tracking analysis can reveal vehicle risks that may not be visible from altitude plots and statistical analysis directly and that the mission success zone can help translate the results into more meaningful information for mission planners. Rather than choosing a flight style or hover capable vehicle, a choice is made between increased or minimal vehicle risk, larger or smaller area covered, lowered photo density or almost complete area coverage and lower vs. higher operational cost.

Future work on combining hover and flight style actuation for altitude tracking should include optimising energy consumption by varying the surge velocity based on the altitude error. The most significant improvements in survey results are expected from further research to improve the terrain detection and also the altitude control for a downwards slope. Limited by software protocol and mechanical factors rather than the speed of sound in water, significant speed improvements may be possible by choosing a multibeam sonar over the mechanical scanning device. This should also increase the repeatability of experiments since it removes the impact of the relative timing of the scan angle. Improving the ability to better compare the altitude tracking performance over complex terrain not only directly impacts the choices of vehicle operators, but will also support a complete automation of the process for adaptive missions planning.

References

[1] Ling SD, Mahon I, Marzloff MP, *et al.* Stereo-imaging AUV detects trends in sea urchin abundance on deep overgrazed reefs. Limnology and Oceanography: Methods. 2016;14(5):293–304. Available from: http://doi.wiley.com/10.1002/lom3.10089.

[2] Smale DA, Kendrick GA, Harvey ES, *et al.* Regional-scale benthic monitoring for ecosystem-based fisheries management (EBFM) using an autonomous underwater vehicle (AUV). ICES Journal of Marine Science. 2012;69: 1108–1118.

[3] Nishida Y, Tamaki U, Hamatsu T, *et al.* Investigation method for the biomass of kichiji rockfish by hovering type AUV. In: OCEANS 2014 – TAIPEI; 2014. p. 1–4.

[4] Nakajima R, Yamamoto H, Kawagucci S, *et al.* Post-drilling changes in seabed landscape and megabenthos in a deep-sea hydrothermal system, the Iheya North Field, Okinawa Trough. PLoS One. 2015;10(4). https://doi.org/10.1371/journal.pone.0123095.

[5] McPhail SD, Furlong M, and Pebody M. Low-altitude terrain following and collision avoidance in a flight-class autonomous underwater vehicle. Proceedings of the Institution of Mechanical Engineers, Part M: Journal of Engineering for the Maritime Environment. 2010;224(4):279–292. Available from: http://eprints.soton.ac.uk/169447/.

[6] Eng TH, and Chitre M. Minimum speed seeking control for nonhovering autonomous underwater vehicles. Journal of Field Robotics. 2015;33(5): 706–733.

[7] Marouchos A, Muir B, Babcock R, *et al.* A shallow water AUV for benthic and water column observations. In: MTS/IEEE OCEANS 2015 – Genova: Discovering Sustainable Ocean Energy for a New World; 2015.

[8] Wynn RB, Huvenne VAI, Le Bas TP, *et al.* Autonomous Underwater Vehicles (AUVs): their past, present and future contributions to the advancement of marine geoscience. Marine Geology. 2014;352:451–468. Available from: http://dx.doi.org/10.1016/j.margeo.2014.03.012.

[9] Morris KJ, Bett BJ, Durden JM, *et al.* A new method for ecological surveying of the abyss using autonomous underwater vehicle photography. Limnology and Oceanography: Methods. 2014;12(11):795–809.

[10] Akkaynak D, Treibitz T, Shlesinger T, *et al.* What is the space of attenuation coefficients in underwater computer vision? In: 2017 IEEE Conference on Computer Vision and Pattern Recognition (CVPR). IEEE; 2017. p. 568–577. Available from: http://ieeexplore.ieee.org/document/8099551/.

[11] Bodenmann A, Thornton B, Nakajima R, *et al.* Wide area 3D seafloor reconstruction and its application to sea fauna density mapping. In: Proceedings of MTS/IEEE Oceans. vol. 13; 2013.

[12] Houts SE, and Rock SM. Trajectory planning for motion-constrained AUVs in uncertain environments. In: 2014 Oceans – St John's, OCEANS 2014; 2015.

[13] Otsuki Y, Thornton B, Maki T, *et al.* Real-time autonomous multi resolution visual surveys based on seafloor scene complexity. Autonomous Underwater Vehicles (AUV), IEEE/OES; 2016. p. 330–335.

[14] Becker JJ, Sandwell DT, Smith WHF, *et al.* Global bathymetry and elevation data at 30 arc seconds resolution: SRTM30_PLUS: Marine Geodesy. 2009;32(4):355–391.

[15] Harris PT, Macmillan-Lawler M, Rupp J, *et al.* Geomorphology of the oceans. Marine Geology. 2014;352:4–24. Available from: http://dx.doi.org/10.1016/j.margeo.2014.01.011.

[16] Steenson LV, Turnock SR, Phillips AB, *et al.* Model predictive control of a hybrid autonomous underwater vehicle with experimental verification. Proceedings of the Institution of Mechanical Engineers, Part M: Journal of Engineering for the Maritime Environment. 2014;228(2):166–179.

[17] Chew JL, and Chitre M. Object detection with sector scanning sonar. In: Oceans-San Diego. San Diego: IEEE; 2013. p. 1–8.

[18] Zhou M, Bachmayer R, and Deyoung B. Mapping for control in an underwater environment using a dynamic inverse-sonar model. In: OCEANS 2016 MTS/IEEE Monterey, OCE 2016; 2016.

[19] Phillips AB, Steenson LV, Rogers E, *et al.* Delphin2: An over actuated autonomous underwater vehicle for manoeuvring research. Transactions of the Royal Institution of Naval Architects Part A: International Journal of Maritime Engineering. 2013;155(PART A4):171–180.

[20] Tritech International Ltd. Micron Echosounder Ultra Compact Underwater Altimeter. Westhill, UK. Available from: http://www.tritech.co.uk/media/products/tritech-micron-echo-sounder-ultra-compact-altimeter.pdf.

[21] Tanakitkorn K, Wilson PA, Turnock SR, *et al.* Depth control for an over-actuated, hover-capable autonomous underwater vehicle with experimental verification. Mechatronics. 2016;41:67–81.

[22] Tanakitkorn K, Wilson PA, Turnock SR, *et al.* Sliding mode heading control of an over-actuated , hover-capable autonomous underwater vehicle with experimental verification. Journal of Field Robotics. 2017;35(3):396–415. Available from: https://onlinelibrary.wiley.com/doi/abs/10.1002/rob.21766.

[23] McKinney W. Data structures for statistical computing in Python. In: Proceedings of the 9th Python in Science Conference (Scipy 2010); 2010. p. 51–56. 1697900. Available from: http://conference.scipy.org/proceedings/scipy2010/mckinney.html.

[24] Schillai SM, Turnock SR, Rogers E, *et al.* Evaluation of terrain collision risks for flight style autonomous underwater vehicles. In: IEEE/OES Autonomous Underwater Vehicles (AUV). Tokyo; 2016. p. 311–318. Available from: http://eprints.soton.ac.uk/400679/.

[25] Palmer A, Hearn GE, and Stevenson P. Experimental testing of an autonomous underwater vehicle with tunnel thrusters. In: First International Symposium on Marine Propulsion; June 2009. p. 1–6.

[26] Steenson LV, Phillips AB, Furlong ME, *et al.* The performance of vertical tunnel thrusters on an autonomous underwater vehicle operating near the free surface in waves. In: Second International Symposium on Marine Propulsors; June 2011. p. 1–8. Available from: http://eprints.ecs.soton.ac.uk/22165/.

Chapter 6
Nonlinear H_∞ control of autonomous underwater vehicles

Subhasish Mahapatra[1] and Bidyadhar Subudhi[2]

Abstract

This chapter concerns about the design of nonlinear state feedback H_∞ control algorithm for an autonomous underwater vehicle (AUV) for both vertical and horizontal planes which will lead to design a three-dimensional path following control. A three degree-of-freedom nonlinear model of an AUV has been considered in both horizontal and vertical planes for developing the diving and steering control laws, respectively. In this, the energy dissipative theory is used which leads to form a Hamilton–Jacobi–Isaac's (HJI) inequality. The nonlinear H_∞ control algorithm has been developed by solving HJI equation such that the AUV tracks the desired depth and the desired yaw angle in diving and steering planes, respectively. Furthermore, the nonlinear control strategy has been extended for path following control algorithm in the steering. Simulation studies have been carried out using MATLAB®/Simulink® environment to verify the efficacies of the proposed control algorithm for AUV. From the results obtained, it is concluded that the proposed robust control algorithms exhibit a good tracking performance ensuring internal stability and significant disturbance attenuation.

6.1 Introduction

In the recent years, autonomous marine vehicles (AMVs) played an important role to provide solutions for many oceanographic applications such as defense organizations for underwater surveillance, petroleum industries for detection of oil wells, inspection of underwater pipelines and study of aqua life. The oceanographic solutions are mostly the guidance and control of these AMVs in oceanic environment in presences of many challenges. These challenges arose due to the disturbances that

[1]Department of Electrical and Electronic Engineering, Madanapalle Institute of Technology and Science, Andhra Pradesh, India
[2]Department of Electrical Engineering, National Institute of Technology Rourkela, Odisha, India

occur during the maneuver of autonomous underwater vehicle (AUV) oceanic environment as ocean current, wave disturbances, etc. Besides, other parameters such as variation in hydrodynamic parameters and environmental disturbances (such as temperature, pressure, viscosity) play an important role during the deployment of AUV. This chapter focuses on the development of guidance control algorithms on AUVs which are one of the subgroups of marine vehicles. In general, the control and guidance problems associated with AUVs are path following, trajectory tracking [1], and way point tracking [2], etc. Hence, the following section provides the relevant information on control and guidance of AUVs developed in the past few decades.

Many linear and nonlinear control strategies have been reported in literature [3] for the control and guidance of marine vehicles. As linear control algorithms need linearization of the nonlinear plant of AUV, the nonlinear control strategies have been developed to avoid these linearization issues. In [4,5], a nonlinear gain scheduling is developed by considering a set of linear finite static output feedback controllers. The control problem is designed using linear-matrix-inequality-based techniques in order to find out steering and diving control laws for a flat fish type AUV. Some of the other nonlinear control strategies such as backstepping control, sliding mode control (SMC) and nonlinear model-based control have been addressed for path following and trajectory tracking of AUVs. In [6], a backstepping control with Lyapunov direct method was described for underactuated vehicles for path following control in presence of environmental disturbances such as wave, wind and ocean current. A robust nonlinear adaptive control algorithm was proposed in [7] using backstepping and Lyapunov-based techniques to achieve the path following of the AUV. Subsequently, a nonlinear trajectory tracking of marine vehicles is addressed in [1,8] using nonlinear SMC and Lyapunov-based tracking control respectively. These control strategies render good tracking capability by possessing global stability. However, the work highlighted in the aforementioned literature has focused less attention toward robustness and internal stability. In view of this, a nonlinear control strategy using nonlinear H_∞ control [9] is developed in this chapter in which the authors have addressed the problems associated with robustness and internal stability.

The H_∞ control techniques have been proved to be an important robust control tool to address nonlinear problems. In [10], Doyle–Glover–Khargonekar–Francis (DGKF) have reported various state space formulations for different H_∞ control problems addressed by several authors. The development of nonlinear H_∞ control was initiated by Ball–Helton [11,12], Isidori–Astolfi [13,14] and van der Schaft [15,16]. This describes the problems associated with internal stability and robustness by formulating the control problem in terms of Hamilton–Jacobi–Isaac's (HJI) inequality by using the energy dissipativity theory [17]. The energy dissipative theory suggests that the input energy of any system is always greater than or equal to the output energy. In [15,16], L_2-induced norm between the disturbance input and the penalty variable (considered as output) is optimized to obtain the nonlinear control law by solving the Hamilton–Jacobi inequality. This article illustrates that if an H_∞ control problem is solved for a linear system, then locally one can also obtain a solution to the nonlinear H_∞ control problem. In [13,14], disturbance attenuation problems are addressed that ensures internal stability as well as robustness by using state and measurement feedback techniques for nonlinear affine systems. In this approach, two

Hamiltonian functions are formulated for both state feedback and output feedback systems. These nonlinear H_∞ control algorithms are used in many applications such as tracking of rigid body motion, flexible link manipulators and other benchmark problems in [18–21]. However, the design of H_∞ control algorithm is associated with the selection of suitable weighting functions for the penalty variable that includes control variable, states, and errors. The penalty variable can be weighted with both static [22] and dynamic [23] weighting functions. The static weighing functions are designed using constants, whereas dynamic weighting function are designed using static weights, frequency weighting functions (such as filters) as well as nonlinear weighting functions.

In this chapter, the nonlinear H_∞ control algorithm is designed for both diving and steering planes. Furthermore, a three-dimensional (3D) path following control algorithm is designed using both diving and steering control algorithms. The path following control algorithm is designed using Serret–Frenet (S–F) frame [24]. To design the 3D path following control of an AUV, the dynamics of AUV is transformed into a nonlinear structure. In this nonlinear controller technique, the nonlinear dynamics of AUV is formulated in terms of an infinite sum of Taylor series in order to transform into a model through which the diving and steering control laws can be obtained. In this approach, the AUV dynamics has been transformed into a second-order Taylor series form to obtain a second-order nonlinear H_∞ state feedback controller for 3D path following of an AUV. The increment of system order causes the improvement of robustness as well as internal stability. However, the complexity of the controller structure will also increase. This design helps to improve the internal stability and robustness, which had not been typically addressed by other control algorithms.

This chapter is organized as follows: Section 6.2 describes the preliminary idea about the HJI formulation. Subsequently, Section 6.3 exploits the modeling of AUV both in diving and steering planes. This also describes the formulation of kinematic model in terms of S–F frame to design the 3D path to achieve the path following control. The design of the nonlinear H_∞ control scheme has been explored in Section 6.4. Section 6.5 describes the synthesis of both diving and steering algorithms numerically as well as using simulation. The analysis of path following control strategy has been highlighted in Section 6.6. Section 6.7 concludes the chapter.

6.2 Preliminaries

Consider a generalized nonlinear plant $G_p(x, w, u)$ with a state vector $x \in \Re^{n\times 1}$ is considered which is comprising two input vectors: an exogenous disturbance input vector $w \in \Re^{r\times 1}$ and a control input vector $u \in \Re^{m\times 1}$, and two output vectors: a penalty vector $z \in \Re^{s\times 1}$ and a measurement vector $y \in \Re^{p\times 1}$ and is represented as

$$G_p(x, w, u) : \begin{cases} \dot{x} = f_a(x) + g_w(x)w + g_u(x)u \\ z = h_z(x) + l_{z1}(x)w + l_{z2}(x)u \\ y = h_y(x) + l_{y1}(x)w \end{cases} \tag{6.1}$$

where $f_a(x)$, $g_w(x)$, $g_u(x)$, $h_z(x)$, $h_y(x)$, $l_{z1}(x)$, $l_{z2}(x)$ and $l_{y1}(x)$ are smooth mappings of class C^k, k should be as large as possible defined in a neighborhood of the origin in $\Re^n$. Assuming $f_a(0) = 0, h_z(0) = 0$ and $h_y(0) = 0$, an HJI inequality is derived from the energy dissipative theory [17] which is described in the following definition.

Definition 6.1. *([16]) For a positive real value function $\gamma \geq 0$, the above nonlinear system $G_p(x, w, u)$ has an L_2-gain$\leq \gamma$ if it satisfies*

$$\int_0^{T_s} \|z(t)\|^2 dt \leq \gamma^2 \int_0^{T_s} \|w(t)\|^2 dt, \quad 0 < \gamma \leq 1 \tag{6.2}$$

for all values of $T_s \geq 0$ and $w : [0, T_s]$.

The nonlinear system adopts the concept of dissipativity, if there exists a nonnegative storage function $\xi(x) \in \Re^+$, $\xi(0) = 0$ for a supply rate of $S(w, z)$ such that for any T_s and w, the dissipation inequality can be written as

$$\xi(x(T_s)) - \xi(0) \leq \int_0^{T_s} S(w, z) dt \tag{6.3}$$

The above dissipative system has L_2-gain from w to z, if the supply rate is considered as

$$S(w, z) = \gamma^2 \|w\|^2 - \|z\|^2, \quad 0 < \gamma \leq 1$$

The (6.3) is rewritten as

$$\xi_x(x)^T \dot{x} \leq \gamma^2 \|w\|^2 - \|z\|^2 \tag{6.4}$$

Hence, the inequality (6.4) is transformed to HJI inequality as follows:

$$\xi_x(x)^T \dot{x} + z^T z - \gamma^2 w^T w \leq 0 \tag{6.5}$$

where $\xi_x(x)^T = \partial \xi(x)/\partial x$.

Hence, a solution of the HJI inequality (6.5) is derived by choosing a proper nonnegative storage function $\xi(x)$. The left side of (6.5) is treated as Hamiltonian function $H[x, \xi_x(x), w, u]$ for simplification of controller design. The nonlinear H_∞ control algorithms which are developed in the subsequent chapters are intended to ensure the internal stability along with the attenuation of external disturbance. The following assumptions are considered for the control design as reported in [10].

Assumption 6.1. *The following assumptions are considered for controller design*

1. $l_{z1}(x) = 0$
2. $h_z^T(x) l_{z2}(x) = 0$
3. $l_{z2}^T(x) l_{z2}(x) = \chi$

Remark 6.1. *Assumption 6.1 is considered based on the fact that the penalty variable is not directly fed to the external disturbance input. Subsequently, Assumption 6.2 ensures that there is no existence of any cross product terms. In Assumption 6.3, χ is a constant diagonal matrix to penalize the control input.*

6.3 Modeling of AUV

The dynamics of AUV involves six-degrees-of-freedom equations of motion associated with coupled and nonlinear terms. The nonlinear terms are generally hydrodynamic damping, added mass coefficients along with environmental disturbances. The AUV body frame with respect to NED (North-East-Down) frame is shown in Figure 6.1. The nonlinear structures of AUV both in diving and steering plane are designed by considering the notations as stated in Tables 6.1 and 6.2, respectively. The corresponding states of diving as well as steering planes are shown in Figures 6.2 and 6.3. Various modeling techniques are described in [25,26] for different structures of AUVs. However, this work has considered the mathematical modeling of an INFANTE AUV as described in [26]. The coupling of diving and steering motion is not considered in this work. However, separate diving and steering control laws are derived to achieve the 3D path following task. The nonlinear state space models of both the planes are formulated by considering the AUV parameters from [26]. Apart from that, a kinematic model is formulated for AUV in terms of S–F frame for path following task in 3D space.

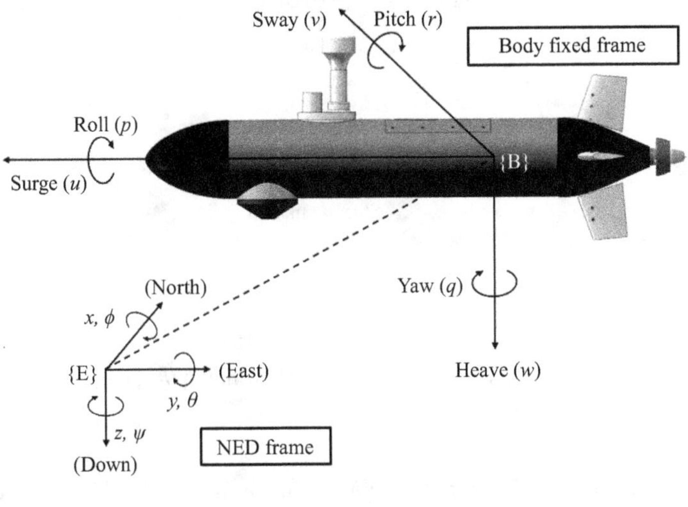

Figure 6.1 General AUV structure with reference frames

Table 6.1 Notations used for AUV in diving plane

Degree of freedom	Direction/Orientation	Forces/ Moments	Linear/Angular velocities	Positions/Euler angles
1	In x-direction (surge motion)	X_v	u_v	x_v
2	In z-direction (heave motion)	Z_v	w_v	z_v
3	About y-axis (pitch motion)	M_v	q_v	θ_v

Table 6.2 Notations used for AUV in steering plane

Degree of freedom	Direction/Orientation	Forces/ Moments	Linear/Angular velocities	Positions/ Euler angles
1	In x-direction (surge motion)	X_h	u_h	x_h
2	In y-direction (sway motion)	Y_h	v_h	y_h
3	About z-axis (yaw motion)	N_h	r_h	ψ_h

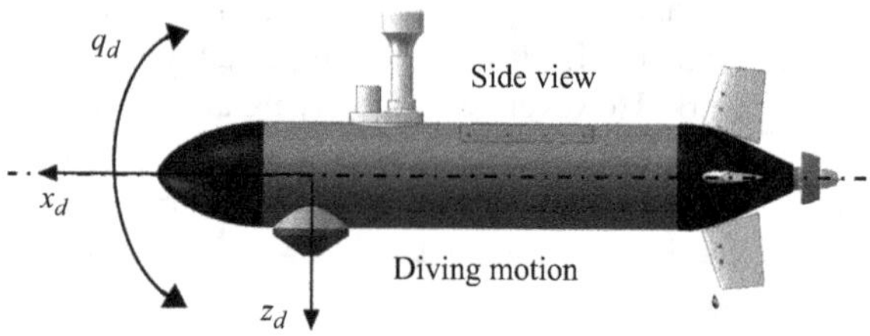

Figure 6.2 Diving motion

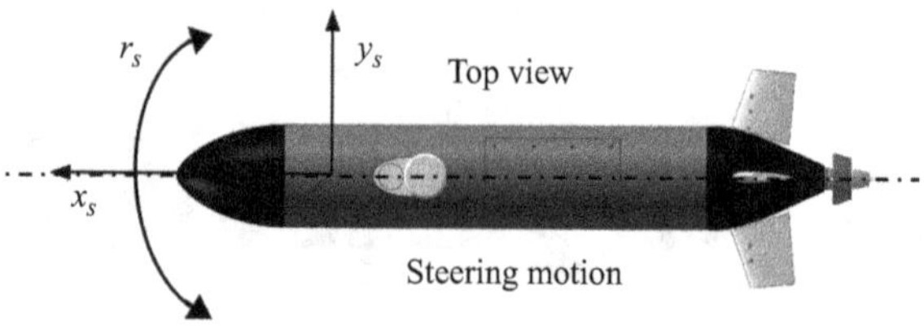

Figure 6.3 Steering motion

Assumption 6.2. *The surge velocity is kept constant during the synthesis of the control algorithm.*

Remark 6.2. *In view of designing the path-following control which is not time restrictive, the surge velocity is kept constant. However, a separate control algorithm can be implemented to regulate the forward speed of the AUV.*

Assumption 6.3. *The roll motion of the AUV is neglected.*

Remark 6.3. *The roll effect is significant during 3D motion of AUV where diving and steering dynamics are coupled. Here, the controller is designed for the diving and heading control separately where the effect of roll motion is not significant. However, the batteries are kept at the bottom of the AUV so that the buoyancy and gravity forces will contribute the roll motion toward zero which will passively stabilize the roll effect.*

6.3.1 AUV modeling: diving plane

The AUV dynamics in diving plane is described by neglecting the roll motion, which involves the surge, heave and pitch motion. It is represented as follows:

Surge motion:

$$m\dot{u}_v = C_{X_{uu}}u_v^2 + C_{X_{ww}}w_v^2 + C_{X_{qq}}q_v^2 + u_v^2 C_{X_{\delta_s\delta_s}}\delta_s^2 + C_{X_{\dot{u}}}\dot{u}_v + T_v, \tag{6.6}$$

$$\dot{x}_v = u_v\cos(\theta_v) + w_v\sin(\theta_v), \tag{6.7}$$

Heave motion:

$$m(\dot{w}_v - u_v q_v) = (W - B)\cos(\theta_v) + C_{Z_w}u_v w_v + C_{Z_q}u_v q_v + C_{Z_{\delta_s}}u_v^2\delta_s + C_{Z_{\dot{w}}}\dot{w}_v, \tag{6.8}$$

$$\dot{z}_v = -u_v\sin(\theta_v) + w_v cos(\theta_v), \tag{6.9}$$

Pitch motion:

$$I_y\dot{q}_v = z_B B\sin(\theta_v) + C_{M_w}u_v w_v + C_{M_q}u_v q_v + C_{M_{\delta_s}}u_v^2\delta_s + C_{M_{\dot{q}}}\dot{q}_v, \tag{6.10}$$

$$\dot{\theta}_v = q_v, \tag{6.11}$$

where C_* is the representation of the coefficients of the AUV in diving plane considered from [26] and T_v is the thrust force applied during diving motion. To develop a controller for diving plane, the forward speed is assumed to be constant (i.e. $u_v = u_0$). Hence, one can neglect the surge motion to design control algorithms.

Referring to Figure 6.2, a state vector is chosen as

$$x_{v,i}(t) = [w_v, q_v, z_v, \theta_v]^T = [x_{v,1}, x_{v,2}, x_{v,3}, x_{v,4}]^T,$$

and a disturbance vector

$$w_{v,i}(t) = [w_{v,1}(t), w_{v,2}(t), w_{v,3}(t), w_{v,4}(t)]^T,$$

where $w_{v,1}(t)$ is the external disturbance associated with the AUV dynamics and $w_{v,2}(t)$, $w_{v,2}(t)$ and $w_{v,2}(t)$ are the measurement noises, the nonlinear structure of AUV in the vertical plane can be written as

$$\dot{x}_{v,i}(t) = f_a(x_{v,i}) + g_w(x_{v,i})w_{v,i}(t) + g_u(x_{v,i})u_{v,i}(t) \tag{6.12}$$

where

$$\begin{aligned}
f_a(x_{v,i}) &= \left[f_{a_1}(x_{v,i}), f_{a_2}(x_{v,i}), f_{a_3}(x_{v,i}), f_{a_4}(x_{v,i})\right]^T,\\
f_{a_1}(x_{v,i}) &= C_w[(W - B)\cos x_{v,4} + C_{Z_w}u_v x_{v,1} + C_{Z_q}u_v x_{v,2} + mu_v x_{v,2}],\\
f_{a_2}(x_{v,i}) &= C_q[Z_B B\sin x_{v,4} + C_{M_w}u_v x_{v,1} + C_{M_q}u_v x_{v,2}],\\
f_{a_3}(x_{v,i}) &= -u_v\sin x_{v,4} + x_{v,1}\cos x_{v,4},\\
f_{a_4}(x_{v,i}) &= x_{v,2},
\end{aligned}$$

$$g_w(x_{v,i}) = \begin{bmatrix} C_w C_{v1} & 0 & 0 & 0 \\ C_q C_{v2} & \rho & 0 & 0 \\ 0 & 0 & \rho & 0 \\ 0 & 0 & 0 & \rho \end{bmatrix}, \quad g_u(x_{v,i}) = \begin{bmatrix} C_w C_{Z_{\delta_s}} \\ C_q C_{M_{\delta_s}} \\ 0 \\ 0 \end{bmatrix},$$

$$u_{v,i}(t) = \delta_s, \quad C_w = (m - C_{Z_{\dot{w}}})^{-1}, \quad C_q = (m - C_{M_{\dot{q}}})^{-1},$$

where ρ is the coefficient of sensor noises during depth measurement and $C_v = [C_{v1}, C_{v2}]$ involves the coefficients of internal disturbances incorporated with heave and pitch dynamics of AUV, respectively.

The corresponding penalty vector is given by

$$z_{v,i}(t) = h_z(x_{v,i}) + l_{z2}(x_{v,i})u_{v,i}(t) \tag{6.13}$$

where

$$h_z(x_{v,i}) = \begin{bmatrix} W_{v1} & 0 & 0 & 0 \\ 0 & W_{v2} & 0 & 0 \\ 0 & 0 & W_{v3} & 0 \\ 0 & 0 & 0 & W_{v4} \\ 0 & 0 & 0 & 0 \end{bmatrix}, \quad l_{z2}(x_{v,i}) = \begin{bmatrix} 0 \\ 0 \\ 0 \\ 0 \\ W_{v,u} \end{bmatrix}$$

and the measurement vector is considered as

$$y_{v,i}(t) = h_y(x_{v,i}) + l_{y1}(x_{v,i})w_{v,i}(t) \tag{6.14}$$

where

$$h_y(x_{v,i}) = \begin{bmatrix} 0 & 1 & 0 & 0 \\ 0 & 0 & 1 & 0 \\ 0 & 0 & 0 & 1 \end{bmatrix}, \quad l_{y1}(x_{v,i}) = \begin{bmatrix} 0 & \rho & 0 & 0 \\ 0 & 0 & \rho & 0 \\ 0 & 0 & 0 & \rho \end{bmatrix}$$

The penalty variable $z_{v,i}(t)$ is weighted by the static weighting functions $W_{v1}, W_{v2}, W_{v3}, W_{v4}$ and $W_{v,u}$ with respect to state variables $x_{v,1}, x_{v,2}, x_{v,3}, x_{v,4}$ and control input $u_{v,i}(t)$, respectively.

6.3.2 AUV modeling: steering plane

Similarly, the steering plane dynamics is considered by neglecting the roll effect. The states associated with this are surge, sway and yaw that are defined as [26]

Surge motion:

$$m\dot{u}_h = C_{X_{uu}} u_h^2 + C_{X_{vv}} v_h^2 + C_{X_{rr}} r_h^2 + u_h^2 C_{X_{\delta_r \delta_r}} \delta_r^2 + C_{X_{\dot{u}}} \dot{u}_h + T_h, \tag{6.15}$$

$$\dot{x}_h = u_h \cos(\psi_h) - v_h \sin(\psi_h), \tag{6.16}$$

Sway motion:

$$
\begin{aligned}
m(\dot{v}_h + u_h r_h) &= C_{Y_{\delta r}} u_h^2 \delta_r + C_{Y_{r|r|}} r_h |r_h| + C_{Y_{v|v|}} v_h |v_h| + C_{Y_r} u_h r_h \\
&\quad + C_{Y_v} u_h v_h + C_{Y_{\dot{v}}} \dot{v}_h, \qquad (6.17)\\
\dot{y}_h &= -u_h \sin(\psi_h) + v_h \cos(\psi_h), \qquad (6.18)
\end{aligned}
$$

Yaw motion:

$$
\begin{aligned}
I_z \dot{r}_h &= C_{N_v} u_h v_h + C_{N_r} u_h r_h + C_{N_{r|r|}} r_h |r_h| + C_{N_{v|v|}} v_h |v_h| \\
&\quad + C_{N_{\dot{r}}} \dot{r}_h + C_{N_{\delta r}} u_h^2 \delta_r, \qquad (6.19)\\
\dot{\psi}_h &= r_h, \qquad (6.20)
\end{aligned}
$$

where C_* is the representation of the coefficients of AUV in steering plane considered from [26] and T_h is the thrust force for the steering plane dynamics. Similarly, by keeping the forward speed constant (i.e. $u_h = u_0$), the surge motion can be neglected in the steering plane. If the yaw angle for steering control is to be controlled, the surge and sway velocities in earth-fixed frame can also be neglected to obtain a simplified model for controller design.

Consider a state vector

$$x_{h,i}(t) = [v_h, r_h, \psi_h]^T = [x_{h,1}, x_{h,2}, x_{h,3}]^T,$$

and disturbance vector $w_{h,i}(t)$

$$w_{h,i}(t) = \left[w_{h,1}(t), w_{h,2}(t)\right]^T,$$

where $w_{h,1}(t)$ is the disturbance added with in the dynamics and $w_{h,2}(t)$ is the sensor noise. Then, the nonlinear state space model of AUV in the horizontal plane is represented as follows

$$\dot{x}_{h,i}(t) = f_a(x_{h,i}) + g_w(x_{h,i}) w_{h,i}(t) + g_u(x_{h,i}) u_{h,i}(t) \qquad (6.21)$$

where

$$
\begin{aligned}
f_a(x_{h,i}) &= \left[f_{a_1}(x_{h,i}),\ \ f_{a_2}(x_{h,i}),\ \ f_{a_3}(x_{h,i})\right]^T,\\
f_{a_1}(x_{h,i}) &= C_v \left[C_{Y_r} u_s x_{h,2} + C_{Y_v} u_s x_{h,1} + C_{Y_{r|r|}} x_{h,2} \left|x_{h,2}\right| + C_{Y_{v|v|}} x_{h,1} \left|x_{h,1}\right|\right],\\
f_{a_2}(x_{h,i}) &= C_r \left[C_{N_v} u_s x_{h,1} + C_{N_r} u_s x_{h,2} + C_{N_{r|r|}} x_{h,2} \left|x_{h,2}\right| + C_{N_{v|v|}} x_{h,1} \left|x_{h,1}\right|\right],\\
f_{a_3}(x_{h,i}) &= x_{h,2},\\
g_w(x_{h,i}) &= \begin{bmatrix} C_v C_{h1} & 0 \\ C_r C_{h2} & 0 \\ 0 & 0 \end{bmatrix}, \quad g_u(x_{h,i}) = \begin{bmatrix} C_v u_h^2 C_{Y_{\delta r}} \\ C_r u_h^2 C_{N_{\delta r}} \\ 0 \end{bmatrix},\\
u_{h,i}(t) &= \delta_r, \quad C_v = (m - C_{Y_{\dot{v}}})^{-1}, \quad C_r = (I_z - C_{N_{\dot{r}}})^{-1},
\end{aligned}
$$

where $C_h = [C_{h1}, C_{h2}]$ includes the coefficients of internal disturbances involved in sway and yaw dynamics of AUV, respectively.

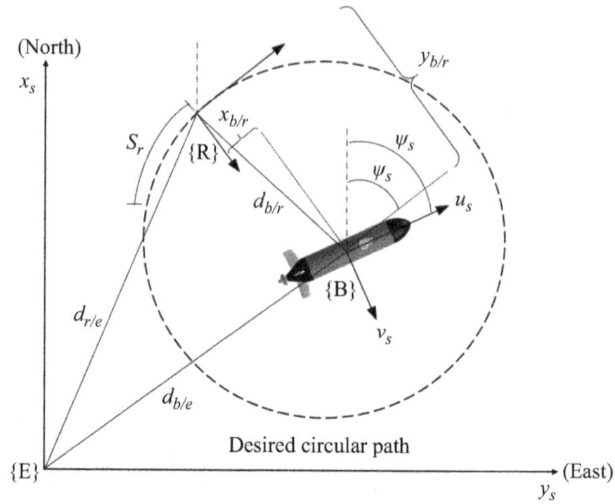

Figure 6.4 Description of S–F frame in steering plane

The corresponding penalty vector can be written as

$$z_{h,i}(t) = h_z(x_{h,i})x_{h,i}(t) + l_{z2}(x_{h,i})u_{h,i}(t) \tag{6.22}$$

where

$$h_z(x_{h,i}) = \begin{bmatrix} W_{h1} & 0 & 0 \\ 0 & W_{h2} & 0 \\ 0 & 0 & W_{h3} \\ 0 & 0 & 0 \end{bmatrix}, \quad l_{z2} = \begin{bmatrix} 0 \\ 0 \\ 0 \\ W_{h,u} \end{bmatrix}$$

and the corresponding measurement vector is given by

$$y_{h,i}(t) = h_y(x_{h,i})x_{h,i}(t) + l_{y1}(x_{h,i})w_{h,i}(t) \tag{6.23}$$

where

$$h_y(x_{h,i}) = \begin{bmatrix} 0 & 0 & 1 \end{bmatrix}, \quad l_{y1}(x_{h,i}) = \begin{bmatrix} 0 & C_{h3} \end{bmatrix}$$

and C_{h3} is the coefficient of sensor noise in yaw angle measurement. Similarly, the penalty variable $z_{h,i}(t)$ is weighted with the static weighting functions W_{h1}, W_{h2}, W_{h3} and $W_{h,u}$ with respect to state variables $x_{h,1}, x_{h,2}, x_{h,3}$ and control input $u_{h,i}(t)$, respectively.

6.3.3 Path kinematics: Serret–Frenet frame

The 3D path following of an AUV is achieved by designing a kinematic model for the AUV in terms of S–F frame. The time is not a constraint for achieving the path following task. The objective is to converge the AUV into a specified path. Here a circular path is considered which is placed at a certain depth and the S–F frame travels along a circular path as shown in Figure 6.4. As the kinematic model of AUV in the horizontal plane is three degree-of-freedom structure in which the circular path is considered, thus the differential kinematic model in S–F frame [27] is given by

$$\begin{bmatrix} \dot{x}_{b/r} \\ \dot{y}_{b/r} \end{bmatrix} = \begin{bmatrix} \cos(\psi_{rb}) & -\sin(\psi_{rb}) \\ \sin(\psi_{rb}) & \cos(\psi_{rb}) \end{bmatrix} \begin{bmatrix} u_h \\ v_h \end{bmatrix} - \begin{bmatrix} \dot{s}_r \\ 0 \end{bmatrix}$$
$$-\dot{s}_r \begin{bmatrix} 0 & -p_c(s_r) \\ p_c(s_r) & 0 \end{bmatrix} \begin{bmatrix} x_{b/r} \\ y_{b/r} \end{bmatrix} \tag{6.24}$$

where $\psi_{rb} = \psi_h - \psi_r$, is the steering angle of the body relative to the steering angle of S–F frame, $p_c(s_r)$ is the path curvature along the circular path and $[x_{b/r}, y_{b/r}]^T$ is the error space vector between the body and S–F frame along x and y axes.

6.4 Development of nonlinear control algorithm

This section exploits the design of nonlinear control algorithms using the H_∞ technique. The control law is formulated using the dissipative theory. As discussed in the preliminary section, the HJI inequality is formulated in order to find out the control laws for both diving and steering planes. The HJI inequality is transformed into an infinite sum of Taylor series expansion. The control structures of both diving and steering planes are shown in Figure 6.5(a) and (b), respectively.

6.4.1 Nonlinear state feedback H_∞ controller

The control law is formulated by solving HJI inequality as described in preliminary section and is rewritten as

$$H[x, \xi_x(x), w, u] = \xi_x(x)^T \dot{x} + z^T z - \gamma^2 w^T w \leq 0 \tag{6.25}$$

where $H[x, \xi_x(x), w, u]$ is called as the Hamiltonian function. Furthermore, this Hamiltonian is again transformed into an infinite sum of Taylor series in order to find out the nonlinear H_∞ control laws of appropriate order (Taylor series order) for both diving and steering planes.

The solution of the HJI inequality (6.25) results in the control law. An analytical approach is pursued to find out a saddle point condition, i.e. $(\breve{w}, \widehat{u})$ for the Hamiltonian function such that

$$H[x, \xi_x(x), w, \widehat{u}] \leq H[x, \xi_x(x), \breve{w}, \widehat{u}] \leq H[x, \xi_x(x), \breve{w}, u] \tag{6.26}$$

where $\breve{w}$ represents the worst-case disturbance which maximizes $H[\bullet]$ and $\widehat{u}$ represents the control law that minimizes $H[\bullet]$.

Referring Assumption 6.1, (6.25) is rewritten as

$$\xi_x(x)^T [f_a(x) + g_w(x)d + g_u(x)u] + [h_z(x) + l_{z2}(x)u]^T [h_z(x) + l_{z2}(x)u]$$
$$-\gamma^2 w^T w \leq 0 \tag{6.27}$$

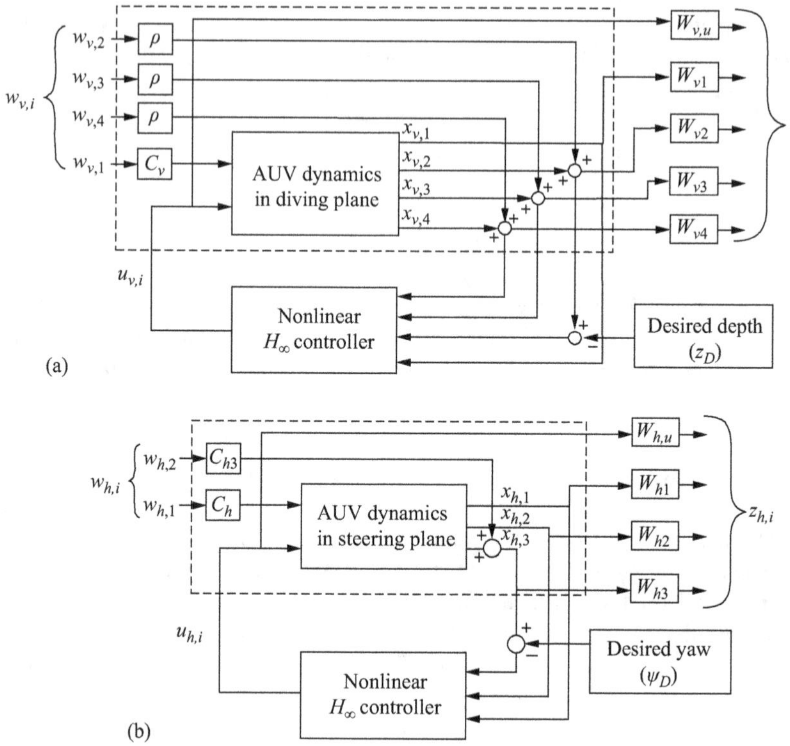

Figure 6.5 Structures of nonlinear $\mathcal{H}_\infty$ controller: (a) diving control and (b) steering control

The control input and worst-case disturbance (i.e. saddle point condition $(\breve{w}, \hat{u})$) are obtained by considering the partial derivative of the Hamiltonian function (6.27). Therefore, the saddle point $(\breve{w}, \hat{u})$ is given as follows:

$$\hat{u} = \hat{u}[x, \xi_x(x)] = -\frac{1}{2}\chi^{-1}g_u(x)^T\xi_x(x) \tag{6.28}$$

$$\breve{w} = \breve{w}[x, \xi_x(x)] = \frac{1}{2}\gamma^{-2}g_w(x)^T\xi_x(x) \tag{6.29}$$

Hence, the storage function $\xi(x)$ will satisfy the above saddle point condition and it is obtained by applying $(\breve{w}, \hat{u})$ in (6.27) that results

$$H[x, \xi_x(x)] = \xi_x(x)^T f_a(x) - \hat{u}^T\chi\hat{u} + h_z^T(x)h_z(x) + \gamma^2\breve{w}^T\breve{w} \leq 0 \tag{6.30}$$

The inequality (6.30) is transformed to an infinite sum of Taylor's series of order k to find out an analytical solution for (6.30) to obtain the control law and worst-case disturbance. Hence, (6.30) can be transformed as

$$\xi_x(x)^{[2]T} f_a(x)^{[k]} + \cdots + \xi_x(x)^{[k+1]T} F_a x^{[1]} - \left[\widehat{u}^{[1]T} \chi \widehat{u}^{[k]} + \cdots + \widehat{u}^{[k]T} \chi \widehat{u}^{[1]}\right]$$
$$+ h_z^T(x) h_z(x) + \gamma^2 \left[\breve{w}^{[1]T} \breve{w}^{[k]} + \cdots + \breve{w}^{[k]T} \breve{w}^{[1]}\right] \leq 0 \tag{6.31}$$

where

$$\widehat{u}^{[k]} = -\frac{1}{2}\chi^{-1}\left[g_u(x)^{[1]T}\xi_x(x)^{[k+1]} + \cdots + g_u(x)^{[k]T}\xi_x(x)^{[2]}\right] \tag{6.32a}$$

$$\breve{w}^{[k]} = \frac{1}{2}\gamma^{-2}\left[g_w(x)^{[1]T}\xi_x(x)^{[k+1]} + \cdots + g_w(x)^{[k]T}\xi_x(x)^{[2]}\right] \tag{6.32b}$$

$$\xi(x) = \xi(x)^{[2]} + \xi(x)^{[3]} + \cdots + \xi(x)^{[k+1]} \tag{6.32c}$$

where $\{F_a x^{[1]}, g_w(x)^{[1]}, g_u(x)^{[1]}, \xi(x)^{[2]}, \breve{w}^{[1]}, \widehat{u}^{[1]}\}$, $\{f_a(x)^{[2]}, g_w(x)^{[2]}, g_u(x)^{[2]}, \xi(x)^{[3]}, \breve{w}^{[2]}, \widehat{u}^{[2]}\}, \ldots, \{f_a(x)^{[k]}, g_w(x)^{[k]}, g_u(x)^{[k]}, \xi(x)^{[k+1]}, \breve{w}^{[k]}, \widehat{u}^{[k]}\}$ are the first, second, $\cdots$ kth terms of Taylor's series expansion of $\{f_a(x), g_w(x), g_u(x), \xi(x), \breve{w}, \widehat{u}\}$, respectively. Higher order nonlinear H_∞ control algorithms can be formulated by using universal expressions of this power series. The algorithms for designing the first- and second-order nonlinear H_∞ control laws by choosing proper value of k are as follows:

- **First-order controller:** Applying the Taylor's series expansion of each term, the state equation (6.1) and considering first order, i.e. $k = 1$, the first-order controller can be obtained. The control input and worst-case disturbance are presented as $\widehat{u} = \widehat{u}^{[1]}$ and $\breve{w} = \breve{w}^{[1]}$, respectively. The HJI inequality is obtained as

$$\xi_x(x)^{[2]T} F_a x^{[1]} - 2\widehat{u}^{[1]T} \chi \widehat{u}^{[1]} + 2\gamma^2 \breve{w}^{[1]T} \breve{w}^{[1]} + h_z(x)^T h_z(x) \leq 0 \tag{6.33}$$

  where

$$\widehat{u}^{[1]} = -\frac{1}{2}\chi^{-1} g_u(x)^{[1]T}\xi_x(x)^{[2]} \tag{6.34a}$$

$$\breve{w}^{[1]} = \frac{1}{2}\gamma^{-2} g_w(x)^{[1]T}\xi_x(x)^{[2]} \tag{6.34b}$$

  and $\xi(x)^{[2]}$ is the quadratic polynomial type storage function.
- **Second-order controller:** Similarly, for deriving a second-order control law, apply the Taylor's series expansion of each term of the state equation (6.1) and consider second order, i.e. $k = 2$. Consequently, a second-order controller can be obtained as $\widehat{u} = \widehat{u}^{[1]} + \widehat{u}^{[2]}$ and worst-case disturbance as $\breve{w} = \breve{w}^{[1]} + \breve{w}^{[2]}$. The HJI inequality is written as

$$\xi_x(x)^{[2]T} f_a(x)^{[2]} + \xi_x(x)^{[3]T} F_a x^{[1]} - \widehat{u}^{[1]T} \chi \widehat{u}^{[2]} - \widehat{u}^{[2]T} \chi \widehat{u}^{[1]}$$
$$+\gamma^2 \left(\breve{w}^{[1]T}\breve{w}^{[2]} + \breve{w}^{[2]T}\breve{w}^{[1]}\right) \leq 0 \tag{6.35}$$

where

$$\breve{u}^{[2]} = -\frac{1}{2}\chi^{-1}\left[g_u(x)^{[1]T}\xi_x(x)^{[3]} + g_u(x)^{[2]T}\xi_x(x)^{[2]}\right] \tag{6.36a}$$

$$\breve{w}^{[2]} = \frac{1}{2}\gamma^{-2}\left[g_w(x)^{[1]T}\xi_x(x)^{[3]} + g_w(x)^{[2]T}\xi_x(x)^{[2]}\right] \tag{6.36b}$$

and $\xi(x)^{[3]}$ is the cubic polynomial type storage function.

The first- and second-order nonlinear H_∞ controllers are derived for both diving and steering control laws, respectively.

6.5 Analysis of nonlinear $\mathcal{H}_\infty$ controller

This section explores the synthesis of the nonlinear H_∞ control algorithms numerically which are developed for both diving and steering planes. The realization for both first and second-order nonlinear H_∞ control algorithms is pursued by using MATLAB/Simulink environment. In this section, the simulation results for both diving and steering are also analyzed which are used for designing the path following control in the subsequent section.

6.5.1 Diving control

The nonlinear H_∞ control algorithm using state feedback technique is obtained by considering a constant forward velocity $u_d = 1\ \mathrm{ms}^{-1}$. The initial states for the diving motion is considered as $[w_{v0}, q_{v0}, z_{v0}, \theta_{v0}]^T = [0, 0, 3, 0]^T$, $\delta_s = 0$ and $\rho = 0.001$. Hence, the Taylor's series for each term of state equation (6.12) considering up to second order is given as follows:

$$f_{a_v}(x_{v,i}) = F_{a_v}x_{v,i}^{[1]} + f_a(x_{v,i})^{[2]},$$

$$g_{w_v}(x_{v,i}) = g_{w_v}^{[1]} + g_{w_v}^{[2]},$$

$$g_{u_v}(x_{v,i}) = g_{u_v}^{[1]} + g_{u_v}^{[2]}$$

In this work, a first-order control law is obtained considering the terms F_{a_v}, $g_{w_v}^{[1]}$ and $g_{u_v}^{[1]}$ (the subscript v indicates the vertical plane dynamics). Subsequently, the second-order law is evaluated by considering $F_{a_v}, f_{a_v}(x_{v,i})^{[2]}$, $g_{w_v}^{[1]}$, $g_{w_v}^{[2]}$, $g_{u_v}^{[1]}$, $g_{u_v}^{[2]}$. The higher order terms of Taylor's series are neglected to reduce the complexity in designing the control law. Hence, the following matrices and vectors are computed by considering the parameters from [26] and are given as follows:

$$F_{a_v} = \begin{bmatrix} -0.9468 & 0.1982 & 0 & 0 \\ 2.9493 & -1.7724 & 0 & -8.7532 \\ 1.0000 & 0 & 0 & -1.0000 \\ 0 & 1.0000 & 0 & 0 \end{bmatrix}, \quad f_{a_v}(x_{v,i})^{[2]} = \begin{bmatrix} -0.0143x_{v,4}^2 \\ 0 \\ 0 \\ 0 \end{bmatrix},$$

$$g_{w_v}^{[1]} = \begin{bmatrix} 0.0239 & 0 & 0 & 0 \\ 0.2769 & 0.001 & 0 & 0 \\ 0 & 0 & 0.001 & 0 \\ 0 & 0 & 0 & 0.001 \end{bmatrix}, \quad g_{w_v}^{[2]} = \begin{bmatrix} 0 & 0 & 0 & 0 \\ 0 & 0 & 0 & 0 \\ 0 & 0 & 0 & 0 \\ 0 & 0 & 0 & 0 \end{bmatrix},$$

$$g_{u_v}^{[1]} = \begin{bmatrix} -0.5950 \\ -2.8164 \\ 0 \\ 0 \end{bmatrix}, \quad g_{u_v}^{[2]} = \begin{bmatrix} 0 \\ 0 \\ 0 \\ 0 \end{bmatrix},$$

and $\chi_v = l_{z2}^T l_{z2} = W_{v,u}^2$.

- **First-order controller:** A first-order control law and worst-case disturbance are derived by formulation an HJI inequality as described in (6.33). The control law and worst-case disturbance are then determined using (6.34a) and (6.34b), respectively. Thus, a quadratic storage function $\xi(x_{v,i})^{[2]} = x_{v,i}{}^T \Psi_v x_{v,i}$, where $\Psi_v > 0$, is chosen which can be treated as a Lyapunov function that possesses a local minimum at the equilibrium point for a constant γ value. Subsequently, the HJI inequality (6.33) is formulated which further transformed to a Riccati equation that holds good for all values of Ψ_v and is presented as follows:

$$f_{a_v}^{[1]T}\Psi_v + \Psi_v f_{a_v}^{[1]} + \Psi_v(\gamma^{-2} g_{w_v}^{[1]} g_{w_v}^{[1]T} - g_{u_v}^{[1]} \chi_v^{-1} g_{u_v}^{[1]T})\Psi_v + h_{z_v}^T h_{z_v} = 0 \tag{6.37}$$

The solution of the above Riccati equation is obtained using MATLAB Toolbox by choosing $\gamma = 1$, $C_{v1} = C_{v2} = 1$, $W_v = diag\{W_{v1}, W_{v2}, W_{v3}, W_{v4}\} = I_4$ and $W_{v,u} = 1$ and is given as

$$\Psi_v = \begin{bmatrix} 8.5496 & -1.3172 & 12.1252 & -4.0140 \\ -1.3172 & 0.4720 & -2.1170 & 0.6961 \\ 12.1252 & -2.1170 & 18.8731 & -5.6516 \\ -4.0140 & 0.6961 & -5.6516 & 3.8721 \end{bmatrix}.$$

Therefore, the first-order nonlinear H_∞ control law and worst-case disturbance are determined as follows:

$$\widehat{u}_{v,i}^{[1]} = 0.956x_{v,1} + 0.378x_{v,2} + 0.869x_{v,3} - 0.297x_{v,4} \tag{6.38a}$$

$$\breve{w}_{v,i}^{[1]} = \begin{bmatrix} 0.0993x_{v,1} - 0.161x_{v,2} - 0.297x_{v,3} + 0.097x_{v,4} \\ 0 \\ 0 \\ 0 \end{bmatrix} \tag{6.38b}$$

- **Second-order controller:** Subsequently, a second-order nonlinear H_∞ control law $\widehat{u}_{v,i} = \widehat{u}^{[1]}_{v,i} + \widehat{u}^{[2]}_{v,i}$ and the worst-case disturbance $\breve{w}_{v,i} = \breve{w}^{[1]}_{v,i} + \breve{w}^{[2]}_{v,i}$ are derived by choosing a cubic polynomial function as described in previous section.

$$\xi(x_{v,i})^{[3]} = c_{v,1}x^3_{v,1} + c_{v,2}x^2_{v,1}x_{v,2} + c_{v,3}x^2_{v,1}x_{v,3} + \cdots + c_{v,20}x^3_{v,4},$$

where the unknown coefficients $c_{v,j}, j = 1, 2, \ldots, 20$ are evaluated by formulating a second-order HJI equation as described in (6.35) which can be further transformed into (6.39) considering $\xi(x_{v,i})^{[2]}$, $\widehat{u}^{[1]}_{v,i}$ and $\breve{w}^{[1]}_{v,i}$ from the first-order controller.

$$\begin{aligned} &\xi_{x_{v,i}}(x_{v,i})^{[3]}\left(F_{a_v}x^{[1]}_{v,i} + g^{[1]T}_{w_v}\breve{w}^{[1]}_{v,i} + g^{[1]T}_{u_v}\widehat{u}^{[1]}_{v,i}\right) \\ &+\xi_{x_{v,i}}(x_{v,i})^{[2]}\left(f_{a_v}(x_{v,i})^{[2]} + g^{[2]T}_{w_v}\breve{w}^{[1]}_{v,i} + g^{[2]T}_{u_v}\widehat{u}^{[1]}_{v,i}\right) = 0 \end{aligned} \tag{6.39}$$

Hence, equating the equal power terms from (6.39), a set of 20 equations is obtained. The unknown coefficients $c_{v,j}$ are calculated by solving all the 20 equations using MUPAD Toolbox in MATLAB and then the storage function $\xi(x_{v,i})^{[3]}$ is formulated. Thus, the nonlinear second-order H_∞ control law and the worst-case disturbance are determined as follows:

$$\begin{aligned} \widehat{u}_{v,i} &= \widehat{u}^{[1]}_{v,i} + \widehat{u}^{[2]}_{v,i} \\ &= 0.956x_{v,1} + 0.378x_{v,2} + 0.869x_{v,3} - 0.297x_{v,4} - 0.00614x^2_{v,1} \\ &\quad -0.00798x_{v,1}x_{v,2} - 0.0101x_{v,1}x_{v,3} - 0.025x_{v,1}x_{v,4} \\ &\quad +0.00138x^2_{v,2} - 0.0126x_{v,2}x_{v,3} + 0.00771x_{v,2}x_{v,4} \\ &\quad -0.00196x^2_{v,3} - 0.039x_{v,3}x_{v,4} + 0.00937x^2_{v,4} \end{aligned} \tag{6.40a}$$

$$\begin{aligned} \breve{w}_{v,i} &= \breve{w}^{[1]}_{v,i} + \breve{w}^{[2]}_{v,i} \\ &= \begin{bmatrix} \breve{w}_{v,1} & \breve{w}_{v,2} & \breve{w}_{v,3} & \breve{w}_{v,4} \end{bmatrix}^T \end{aligned} \tag{6.40b}$$

where

$$\begin{aligned} \breve{w}_{v,1} &= 0.0993x_{v,1} - 0.161x_{v,2} - 0.297x_{h,3} + 0.097x_{v,4} + 0.0026x^2_{v,1} \\ &\quad -0.00238x_{v,1}x_{v,2} + 0.00938x_{v,1}x_{v,3} - 0.00663x_{v,1}x_{v,4} \\ &\quad +3.98 \times 10^{-4}x^2_{v,2} - 0.00389x_{v,2}x_{v,3} + 0.00219x_{v,2}x_{v,4} \\ &\quad +0.00827x^2_{v,3} - 0.0106x_{v,3}x_{v,4} + 0.00348x^2_{v,4} \\ \breve{w}_{v,2} &= 0, \quad \breve{w}_{v,3} = 0, \quad \breve{w}_{v,4} = 0 \end{aligned}$$

Thus, the derived nonlinear control law and the worst-case disturbance are implemented for the diving motion of the AUV as shown in Figure 6.5(a) by ensuring the internal stability and robustness.

The diving control algorithm is analyzed by MATLAB/Simulink environment using the parameter of AUV from [26]. The structure of the diving control algorithm is shown in Figure 6.5(a). Considering the initial condition, i.e. $[w_{v0}, q_{v0}, z_{v0}, \theta_{v0}]^T = [0, 0, 3, 0]^T$ and the desired depth

$$z_D = \begin{cases} 6\,\mathrm{m}, & 0 \le t < 200 \\ 8\,\mathrm{m}, & 200 \le t < 400 \\ 5\,\mathrm{m}, & 400 \le t < 600 \end{cases} \tag{6.41}$$

the simulation of diving control analysis is pursued. The control law is obtained by considering a constraint on stern plane, i.e. ± 0.785 radian to achieve the desired depth smoothly. Apart from that, $w_{v,1}$ is subjected with a white noise whose power spectral density (PSD) is ten and $w_{v,2}$, $w_{v,3}$ and $w_{v,4}$ are provided with white noises of PSD 1×10^{-5} each. From Figure 6.6(a), it is observed that the nonlinear control law is effective in tracking the desired depth profile by generating a suitable control input as shown in Figure 6.6(b). Also, it can be observed from Figure 6.6(c) that the error profile diminished to zero with less time. It is observed during the analysis that the settling time can be improved $W_{v,u}$ value. The dynamic response will be improved by considering smaller values of $W_{v,u}$.

For further analysis, an attenuation ratio is defined as

$$\text{Attenuation ratio}(R) = \frac{\int_0^T \|z(t)\|^2 dt}{\int_0^T \|w(t)\|^2 dt} \le \gamma^2. \tag{6.42}$$

According to L_2-gain analysis, if L_2-gain $\le \gamma$, with $\gamma = 1$, the attenuation ratio R must be less than or equal to 1. Hence, the analysis of disturbance attenuation is performed during diving control as shown in Figure 6.6(d). This ensures that the control algorithm is tracking the desired depth profile by achieving internal stability and robustness. The attenuation ratios are evaluated during diving control as follows:

$$\begin{aligned} \text{for} \quad T &= 200, \quad \lim_{T\to 200} R = 0.354 < 1, \\ T &= 400, \quad \lim_{T\to 400} R = 0.5 < 1, \\ T &= 600, \quad \lim_{T\to 600} R = 0.419 < 1 \end{aligned}$$

Furthermore, it is also observed that the disturbance attenuation characteristics can be improved by increasing the order of Taylor's series expansion which will ensure strong internal stability. Referring to (6.40a), the higher order terms present in the control law involved in improving the attenuation ratio of the closed-loop system. However, the increment in system order, i.e. in the form of Taylor's series expansion, will increase the order of the control law (6.40a). Hence, the computational burden for evaluating the control input and worst-case disturbance will also increase. Thus, it is recommended to consider less number of states to design the nonlinear control law.

6.5.2 Steering control

Subsequently, the similar approach as described for diving controller synthesis is followed for deriving the nonlinear H_∞ state feedback controller and the worst-case disturbance for steering motion. Considering a nominal forward speed $u_s = 1\ \mathrm{ms}^{-1}$,

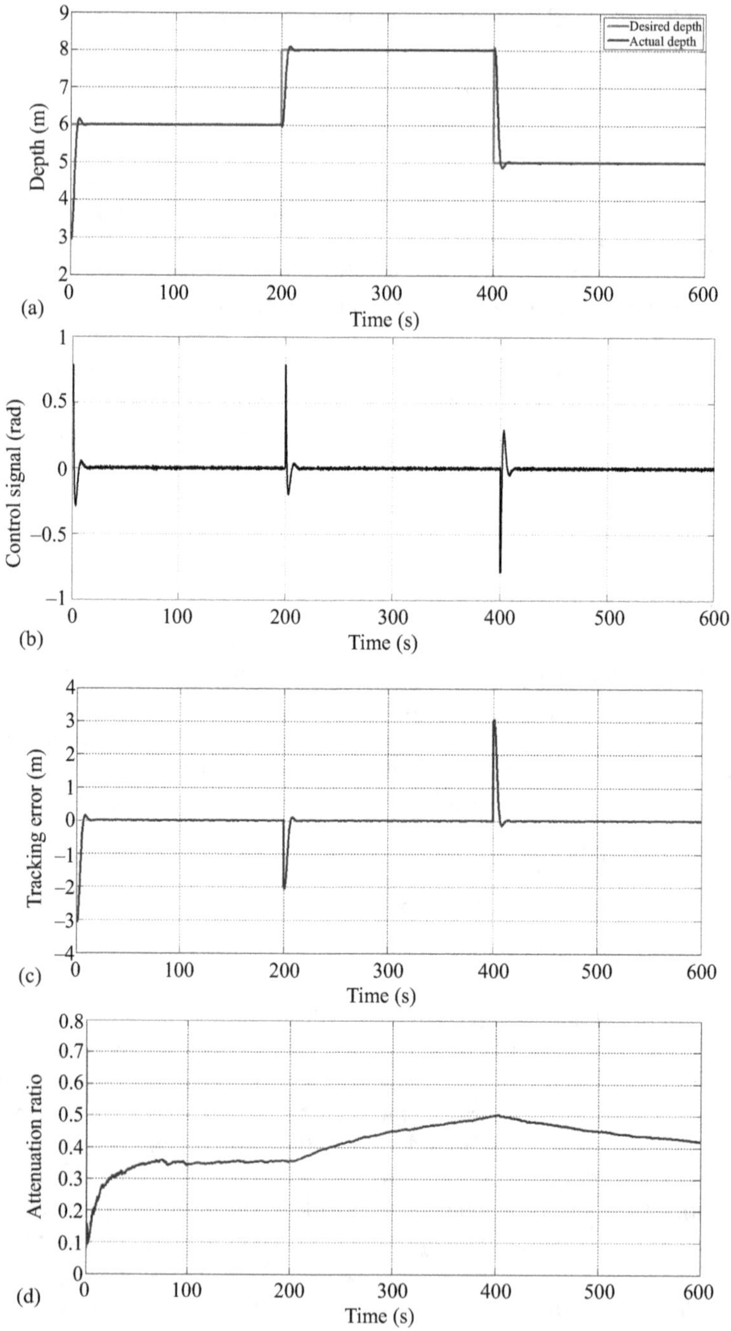

Figure 6.6 (a) Tracking of desired depth, (b) the corresponding controller output, (c) error signal and (d) pitch rate during the tracking of desired depth

an operating point $[v_{h_0}, r_{h_0}, \psi_{h_0}]^T = [0.01, 0.01, 0]^T$ and $\delta_r = 0$, the Taylor's series expansion for the state equation (6.21) up to second order of steering motion is obtained as follows:

$$
\begin{aligned}
f_{a_h}(x_{h,i}) &= F_{a_h} x_{h,i}^{[1]} + f_{a_h}(x_{h,i})^{[2]} \\
g_{w_h}(x_{h,i}) &= g_{w_h}^{[1]} + g_{w_h}^{[2]} \\
g_{u_h}(x_{h,i}) &= g_{u_h}^{[1]} + g_{u_h}^{[2]}
\end{aligned}
$$

Hence, the above Taylor's series terms F_{a_h}, $f_{a_h}(x_{h,i})^{[2]}$, $g_{w_h}^{[1]}$, $g_{w_h}^{[2]}$, $g_{u_h}^{[1]}$, $g_{u_h}^{[2]}$ (the subscript h indicates the horizontal plane dynamics) are evaluated by considering the parameters from [26] and are given as follows:

$$
F_{a_h} = \begin{bmatrix} -0.7471 & -0.3960 & 0 \\ -5.8820 & -3.5391 & 0 \\ 0 & 1.0000 & 0 \end{bmatrix}, \quad f_{a_h}(x_{h,i})^{[2]} = \begin{bmatrix} 0.0236x_{h,2}^2 - 2.168x_{h,1}^2 \\ 0.274x_{h,2}^2 - 0.559x_{h,1}^2 \\ 0 \end{bmatrix},
$$

$$
g_{w_h}^{[1]} = \begin{bmatrix} 0.0239 & 0 \\ 0.2769 & 0 \\ 0 & 0 \end{bmatrix}, \quad g_{w_h}^{[2]} = \begin{bmatrix} 0 & 0 \\ 0 & 0 \\ 0 & 0 \end{bmatrix},
$$

$$
g_{u_h}^{[1]} = \begin{bmatrix} 2.3798 \\ -11.2653 \\ 0 \end{bmatrix}, \quad g_{u_h}^{[2]} = \begin{bmatrix} 0 \\ 0 \\ 0 \end{bmatrix},
$$

and $\chi_h = l_{z2}^T l_{z2} = W_{h,u}^2$.

- **First-order controller:** Similarly, another quadratic storage function is chosen for steering control of AUV in order to achieve a local minimum at the equilibrium point. Assigning $\xi(x_{h,i})^{[2]} = x_{h,i}{}^T \Psi_h x_{h,i}$, where $\Psi_h > 0$, an HJI inequality (6.33) is derived which is transformed to the following Riccati equation:

$$
F_{a_h}^{[1]T}\Psi_h + \Psi_h F_{a_h}^{[1]} + \Psi_h(\gamma^{-2} h_{w_h}^{[1]} g_{w_h}^{[1]T} - g_{u_h}^{[1]} \chi_h^{-1} g_{u_h}^{[1]T})\Psi_h + h_{z_h}^T h_{z_h} = 0 \qquad (6.43)
$$

Thus, the above Riccati equation is simulated using MATLAB/Simulink Toolbox by considering $\gamma = 1$, $C_{h1} = C_{h2} = 1$, $W_h = diag\{W_{h1}, W_{h2}, W_{h3}\} = I_3$, $W_{h,u} = 1$ and $\alpha = 0.001$. Therefore, the solution is given as follows:

$$
\Psi_h = \begin{bmatrix} 0.3155 & 0.0087 & 0.0659 \\ 0.0087 & 0.0746 & 0.1027 \\ 0.0659 & 0.1027 & 1.2086 \end{bmatrix}
$$

Thus, similar approach is used to find out the first-order nonlinear H_∞ control law and worst-case disturbance which are obtained as follows:

$$
\widehat{u}_{h,i}^{[1]} = 0.819x_{h,1} - 0.653x_{h,2} + 1.001x_{h,3} \qquad (6.44a)
$$

$$
\breve{w}_{h,i}^{[1]} = \begin{bmatrix} 0.0099x_{h,1} + 0.021x_{h,2} + 0.030x_{h,3} \\ 0 \\ 0 \end{bmatrix} \qquad (6.44b)
$$

- **Second-order controller:** Another cubic polynomial function is considered for designing a second-order controller $\frown{u}_{h,i} = \frown{u}^{[1]}_{h,i} + \frown{u}^{[2]}_{h,i}$ and the worst-case disturbance $\smile{w}_{h,i} = \smile{w}^{[1]}_{h,i} + \smile{w}^{[2]}_{h,i}$ for steering motion and is presented as follows:

$$\xi(x_{h,i})^{[3]} = c_{h,1}x^3_{h,1} + c_{h,2}x^2_{h,1}x_{h,2} + c_{h,3}x^2_{h,1}x_{h,3} + \cdots + c_{h,10}x^3_{h,3},$$

where the unknown coefficients $c_{h,j}, j = 1, 2, \ldots, 10$ are determined from the HJI inequality (6.35) that is transformed to (6.45) where $\Psi(x_{h,i})^{[2]}$, $\frown{u}^{[1]}_{h,i}$ and $\smile{w}^{[1]}_{h,i}$ are obtained from the first-order controller design.

$$\begin{aligned} &\xi_{x_{h,i}}(x_{h,i})^{[3]}\left(F_{a_h}x^{[1]}_{h,i} + g^{[1]T}_{w_h}\smile{w}^{[1]}_{h,i} + g^{[1]T}_{u_h}\frown{u}^{[1]}_{h,i}\right) \\ &\quad + \xi_{x_{h,i}}(x_{h,i})^{[2]}\left(f_{a_h}(x_{h,i})^{[2]} + g^{[2]T}_{w_h}\smile{w}^{[1]}_{h,i} + g^{[2]T}_{u_h}\frown{u}^{[1]}_{h,i}\right) = 0 \end{aligned} \tag{6.45}$$

Here, the storage function $\xi(x_{h,i})^{[3]}$ is evaluated by obtaining the unknown coefficients $c_{h,j}$ by equating the equal power terms of (6.45). In this, a set of ten equations is solved to obtain the nonlinear second-order H_∞ control law, and the worst-case disturbance is obtained as

$$\begin{aligned} \frown{u}_{h,i} &= \frown{u}^{[1]}_{h,i} + \frown{u}^{[2]}_{h,i} \\ &= 0.819x_{h,1} - 0.653x_{h,2} + 1.001x_{h,3} + 0.315x^2_{h,1} + 0.0676x_{h,1}x_{h,2} \\ &\quad + 0.141x_{h,1}x_{h,3} + 0.021x^2_{h,2} + 0.0219x_{h,2}x_{h,3} \\ &\quad - 1.76 \times 10^{-4}x^2_{h,3} \end{aligned} \tag{6.46a}$$

$$\begin{aligned} \smile{w}_{h,i} &= \smile{w}^{[1]}_{h,i} + \smile{w}^{[2]}_{h,i} \\ &= [\,\smile{w}_{h,1} \quad \smile{w}_{h,2} \quad \smile{w}_{h,3}\,]^T \end{aligned} \tag{6.46b}$$

where

$$\begin{aligned} \smile{w}_{h,1} &= 0.0099x_{h,1} + 0.021x_{h,2} + 0.030x_{h,3} - 0.0191x^2_{h,1} - 0.00506x_{h,1}x_{h,2} \\ &\quad - 0.0172x_{h,1}x_{h,3} + 5.28 \times 10^{-5}x^2_{h,2} - 0.00279x_{h,2}x_{h,3} \\ &\quad - 0.00585x^2_{h,3} \\ \smile{w}_{h,2} &= 0, \quad \smile{w}_{h,3} = 0 \end{aligned}$$

Hence, the steering control of the AUV is achieved by considering the above control law and worst-case disturbance considering the controller structure as shown in Figure 6.5(b).

Similarly, the steering control algorithm is simulated to show the effectiveness of the nonlinear H_∞ control algorithm. The simulation is pursued by using the parameters

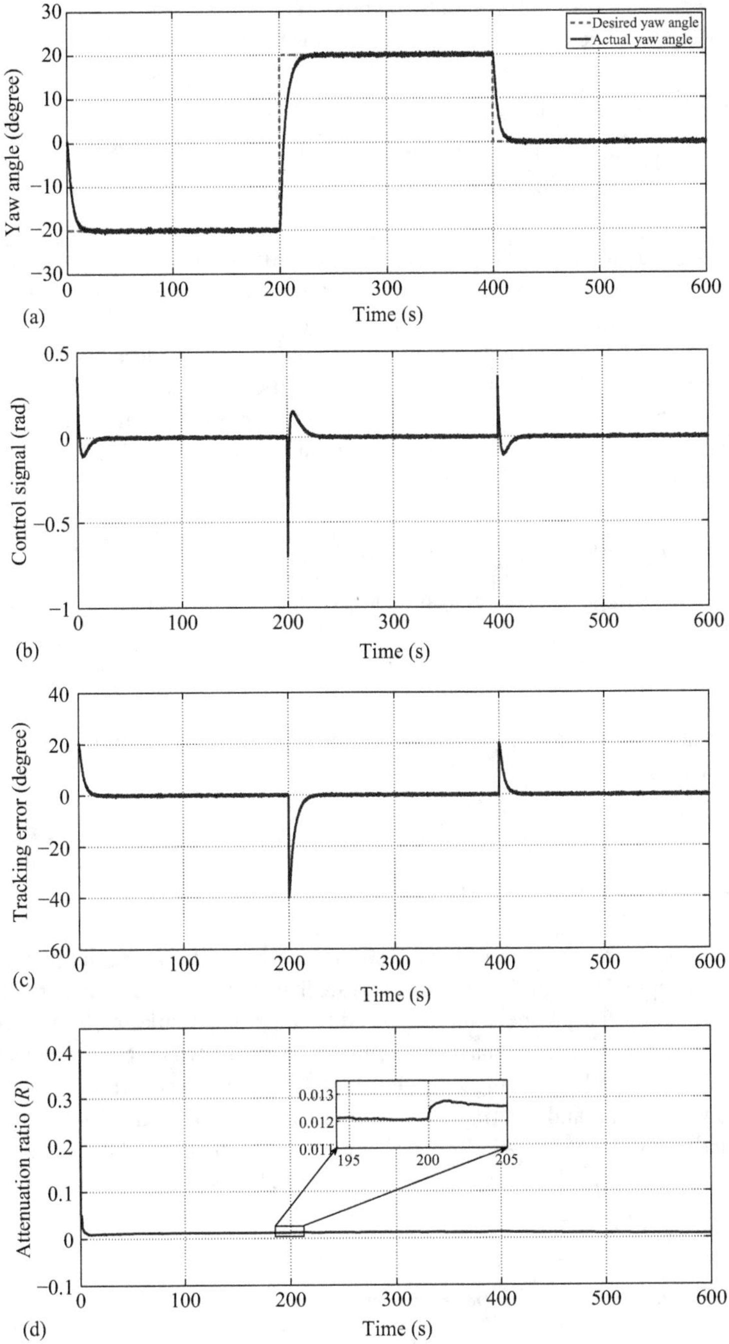

Figure 6.7 (a) Tracking of desired yaw angle, (b) the corresponding controller output, (c) error signal and (d) yaw rate during the tracking of desired yaw angle

from [26] and by considering an initial condition of $[v_{h_0}, r_{h_0}, \psi_{h_0}]^T = [0.01, 0.01, 0]^T$ and $\delta_r = 0$. Here, the desired yaw profile is chosen as

$$\psi_D = \begin{cases} -20^\circ, & 0 \le t < 200 \\ 20^\circ, & 200 \le t < 400 \\ 0^\circ, & 400 \le t < 600 \end{cases} \tag{6.47}$$

During the steering control synthesis, a constraint of $|\delta_r| \le 45^\circ$ is considered. Similarly, the disturbance signals $w_{h,1}$ and $w_{h,2}$ are the white noises with PSD values 10 and 1×10^{-5}, respectively. The tracking of yaw orientation is observed from Figure 6.7(a). The corresponding control input and the error yaw profile are shown in Figure 6.7(b) and 6.7(c). The performance shows the effectiveness of the nonlinear H_∞ control law during steering motion by ensuring the internal stability and robustness which can be observed from Figure 6.7(d). Similarly, the value of R during the steering motion is calculated as

$$\begin{aligned} \text{for} \quad T = 200, \quad & \lim_{T \to 200} R = 0.0120 < 1, \\ T = 400, \quad & \lim_{T \to 400} R = 0.0123 < 1, \\ T = 600, \quad & \lim_{T \to 600} R = 0.0083 < 1 \end{aligned}$$

Thus, the nonlinear control problem in steering plane is solvable for different desired yaw angles.

6.6 Path following control

This section explores the path following control algorithm for a circular path situated at a certain depth. Here, both the diving and steering control laws which are obtained in (6.40a) and (6.46a), respectively, are used to track the circular path. As discussed earlier in introduction section that both diving and steering dynamics are not coupled in this work, the 3D path following is achieved by controlling the depth and yaw profiles separately by diving and steering controller respectively. The simulation realization is pursued using the MATLAB/Simulink environment to ensure the effectiveness of the control algorithm. Further, a guidance law is developed first to achieve the 3D path following task.

6.6.1 Guidance law for path following

The path following control is intended to obtain an appropriate desired yaw angle for AUV in order to track the S–F frame which is moving close to on the circular path. To achieve the 3D path following, the AUV first drive the AUV to the desired depth where the desired circular path is situated. After reaching the desired depth, the AUV converges to the circular path with the error coordinated between body and S–F frame

$$\lim_{t\to\infty} x_{b/r} = 0,$$
$$\lim_{t\to\infty} y_{b/r} = 0.$$

To realize the path following problem, a guidance and an update law were suggested by Børhaug [27] which are given as

$$\psi_D = \psi_r - \tan^{-1}\left(\frac{v_s}{u_s}\right) - \tan^{-1}\left(\frac{y_{b/r}}{\sqrt{\Delta^2 + \left(x_{b/r}\right)^2}}\right) \tag{6.48}$$

$$\dot{s}_r = \sqrt{u_s^2 + v_s^2}\left(\frac{\sqrt{\Delta^2 + \left(x_{b/r}\right)^2} + x_{b/r}}{\sqrt{\Delta^2 + \left(x_{b/r}\right)^2 + \left(y_{b/r}\right)^2}}\right) \tag{6.49}$$

where Δ is a positive design parameter called as look ahead distance. It may be considered as a constant, function of time, error coordinates or any other parameters. In this work, it is assumed as a constant.

6.6.2 Simulation results

A desired circular path is considered at a depth of z_D=30 m for path following. The path coordinates are given by

$$x_D(s_r) = 100\cos(0.01s_r),$$
$$y_D(s_r) = 100\sin(0.01s_r).$$

With an assumption for given s_r, we can evaluate

$$\psi_r(s_r) = \arctan\frac{(y_D)'}{(x_D)'},$$

$$p_c(s_r) = \frac{d\psi_r(s_r)}{ds_r},$$

$$\dot{\psi}_r = p_c(s_r)\dot{s}_r,$$

where $(x_D)' = dx_D/ds_r$ and $(y_D)' = dy_D/ds_r$.

The simulation for path following control is performed by considering $u_s = 2\,\mathrm{ms}^{-1}$, $\Delta = 10$. The performance of the path following controller is shown in Figure 6.8(a). Here, the depth controller is responsible for tracking the desired depth by decreasing the error in z-axis as shown in Figure 6.9(b). From Figure 6.8(c), it is observed that the AUV smoothly dives to achieve the desired depth profile with less time consumption. Subsequently, the steering control is responsible for tracking the desired circular path by generating the desired yaw profile. The tracking of steering dynamics can be observed from the tracking error along x and y axes which are shown in Figures 6.8(d) and 6.9(a), respectively. Also, it can be observed from

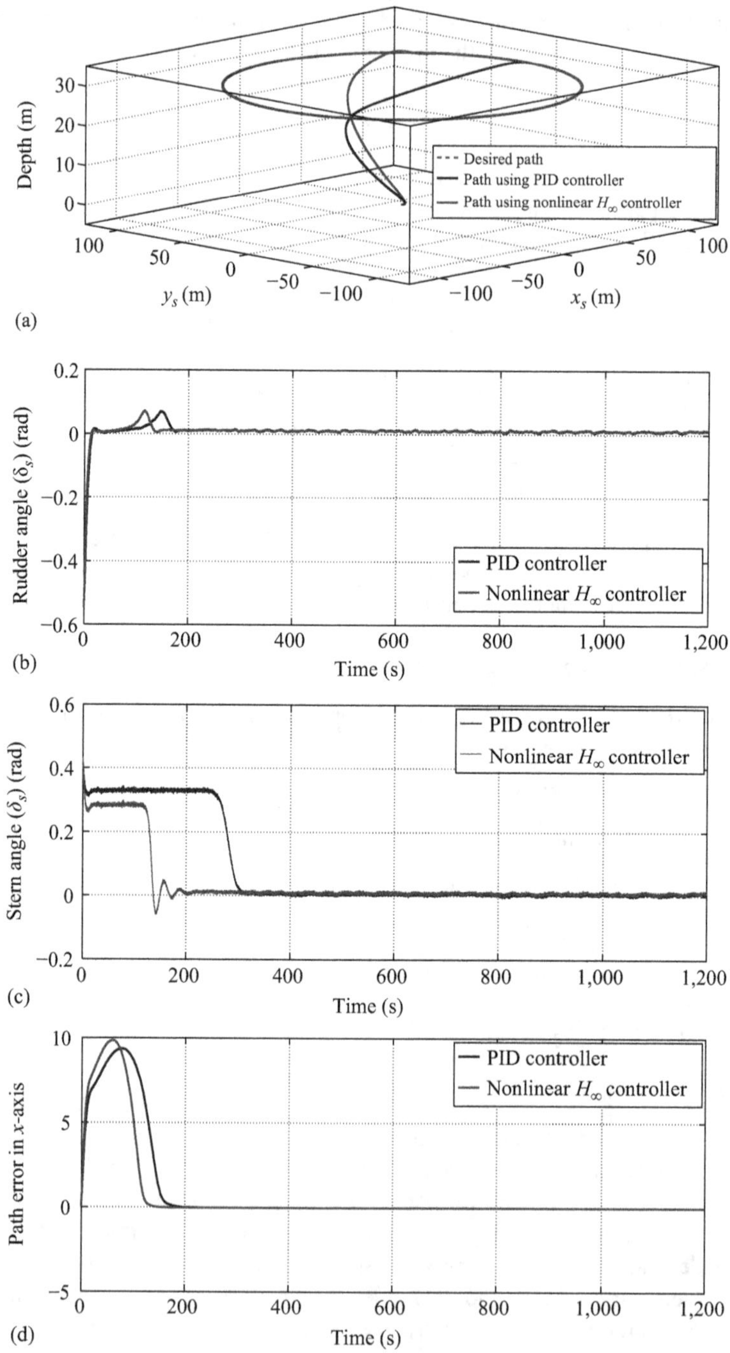

Figure 6.8 *(a) Path following in 3D plane, (b) rudder angle, (c) stern angle and (d) path error in x-axis during path following*

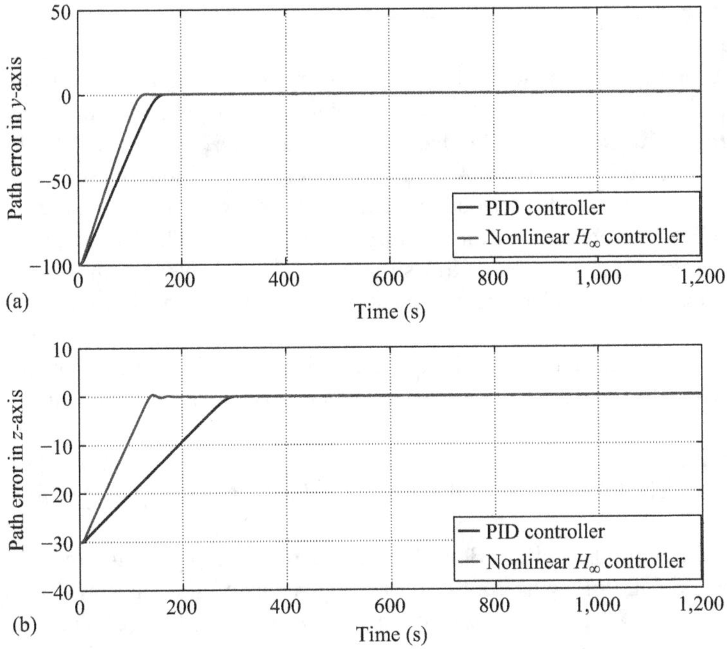

Figure 6.9 (a) path error in y-axis, (b) path error in z-axis during path following

Figure 6.8(b) that the nonlinear H_∞ controller exhibits an effective tracking of circular path by generating suitable control law for rudder plane. This path control can also be achieved by considering the coupling dynamics of AUV in which the path errors along x, y and z axes can be represented in S–F frame.

6.7 Concluding remarks

This chapter describes the design of a path-following control algorithm using nonlinear H_∞ technique. The nonlinear control algorithm is developed using L_2-gain analysis. A second-order polynomial-based nonlinear H_∞ state feedback control schemes are designed for both diving and steering planes in order to achieve the 3D path following task. The state feedback control law is obtained by solving an HJI inequality formulated using L_2-gain analysis. The control law provides excellent performance in tracking the desired circular path by attenuating the disturbances. This control law provides solutions regarding the issues associated with internal stability and robustness. The tracking of path is achieved by minimizing the path error coordinates obtained from the difference between the body and S–F frame to zero asymptotically.

References

[1] Aguiar, A.P., and Hespanha, J.P.: 'Trajectory tracking and path-following of underactuated autonomous vehicles with parametric modeling uncertainty', *IEEE Transactions on Automatic Control*, 2007, **52**, (8), pp. 1362–1379.

[2] Refsnes, J.E., Sorensen, A.J., and Pettersen, K.Y.: 'Model-based output feedback control of slender-body underactuated AUVs: Theory and experiments', *IEEE Transactions on Control Systems Technology*, 2008, **16**, (5), pp. 930–946.

[3] Fossen, T.I.: 'Marine control systems: Guidance, navigation and control of ships, rigs and underwater vehicles' (Trondheim, Norway: Marine Cybernetics, 2002).

[4] Silvestre, C., and Pascoal, A.: 'Control of the INFANTE AUV using gain scheduled static output feedback', *Control Engineering Practice*, 2004, **12**, (12), pp. 1501–1509.

[5] Silvestre, C., and Pascoal, A.: 'Depth control of the INFANTE AUV using gain-scheduled reduced order output feedback', *Control Engineering Practice*, 2007, **15**, (7), pp. 883–895.

[6] Ghommam, J., Mnif, F., Benali, A., and Derbel, N.: 'Nonsingular Serret–Frenet based path following control for an underactuated surface vessel', *Journal of Dynamic Systems, Measurement and Control*, 2009, **131**, (2), pp. 1–8.

[7] Lapierre, L., and Jouvencel, B.: 'Robust nonlinear path-following control of an AUV', *IEEE Journal of Oceanic Engineering*, 2008, **33**, (2), pp. 89–102.

[8] Cheng, J., Yi, J., and Zhao, D.: 'Design of a sliding mode controller for trajectory tracking problem of marine vessels', *IET Control Theory Applications*, 2007, **1**, (1), pp. 233–237.

[9] Aliyu, M.D.S.: 'Nonlinear H_∞ control, Hamiltonian systems and Hamilton–Jacobi equations' (Boca Raton, FL: CRC Press, 2011).

[10] Doyle, J.C., Glover, K., Khargonekar, P.P., and Francis, B.A.: 'State-space solutions to standard H_2 and H_∞ control problems', *IEEE Transactions on Automatic Control*, 1989, **34**, (8), pp. 831–847.

[11] Ball, J.A., and Helton, J.W. 'H_∞ control for nonlinear plants: Connections with differential games'. In: Proceedings of the 28th IEEE Conference on Decision and Control. vol. 2, 1989. pp. 956–962.

[12] Ball, J.A., Helton, J.W., and Walker, M.L.: 'H_∞ control for nonlinear systems with output feedback', *IEEE Transactions on Automatic Control*, 1993, **38**, (4), pp. 546–559.

[13] Isidori, A., and Astolfi, A.: 'Disturbance attenuation and H_∞-control via measurement feedback in nonlinear systems', *IEEE Transactions on Automatic Control*, 1992, **37**, (9), pp. 1283–1293.

[14] Isidori, A., and Kang, W.: 'H_∞ control via measurement feedback for general nonlinear systems', *IEEE Transactions on Automatic Control*, 1995, **40**, (3), pp. 466–472.

[15] van der Schaft, A.J.: 'On a state space approach to nonlinear H_∞ control', *Systems & Control Letters*, 1991, **16**, (1), pp. 1–8.

[16] van der Schaft, A.J.: 'L_2-gain analysis of nonlinear systems and nonlinear state-feedback H_∞ control', *IEEE Transactions on Automatic Control*, 1992, **37**, (6), pp. 770–784.

[17] Willems, J.C.: 'Dissipative dynamical systems part I: General theory', *Archive for Rational Mechanics and Analysis*, 1972, **45**, (5), pp. 321–351.

[18] Tsiotras, P., Corless, M., and Rotea, M.A.: 'An L_2 disturbance attenuation solution to the nonlinear benchmark problem', *International Journal of Robust and Nonlinear Control*, 1998, **8**, pp. 311–330.

[19] Yazdanpanah, M.J., Khorasani, K., and Patel, R.V.: 'Uncertainty compensation for a flexible-link manipulator using nonlinear H_∞ control', *International Journal of Control*, 1998, **69**, (6), pp. 753–771.

[20] Kugi, A., and Schlacher, K.: 'Nonlinear H_∞ controller design for a DC-to-DC power converter', *IEEE Transactions on Control Systems Technology*, 1999, **7**, (2), pp. 230–237.

[21] Hioe, D., Hudon, N., and Bao, J.: 'Decentralized nonlinear control of process networks based on dissipativity: A Hamilton–Jacobi equation approach', *Journal of Process Control*, 2014, **24**, (3), pp. 172–187.

[22] Hu, S.S., and Chang, B.C. 'Design of a nonlinear H_∞ controller for the inverted pendulum system'. In: Proceedings of IEEE International Conference on Control Applications. vol. 2, 1998. pp. 699–703.

[23] Ferreira, H.C., Rocha, P.H., and Sales, R.M. 'Nonlinear H_∞ control and the Hamilton–Jacobi–Isaacs equation'. In: Proceedings of the 17th IFAC World Congress, Seoul, South Korea. vol. 17, 2008. pp. 188–193.

[24] Lapierre, L., and Soetanto, D.: 'Nonlinear path-following control of an AUV', *Ocean Engineering*, 2007, **34**, (11–12), pp. 1734–1744.

[25] Prestero, T.J. 'Verification of a six-degree of freedom simulation model for the REMUS autonomous underwater vehicle' (Department of Ocean and Mechanical Engineering, Massachusetts Institute of Technology, Cambridge, MA, 2001).

[26] Silvestre, C. 'Multi-objective optimization theory with applications to the integrated design of controllers/plants for autonomous vehicles' (Instituto Superior Técnico (IST), Robot. Dept. Lisbon, Portugal, 2000).

[27] Børhaug, E. 'Nonlinear control and synchronization of mechanical systems' (Norwegian University of Science and Technology, Norway, 2008).

Chapter 7

Energy optimal real-time trajectory re-planning of an uninhabited surface vehicle in a dynamically changing environment

Haibin Huang[1], Yufei Zhuang[1], and Sanjay Sharma[2]

Abstract

This chapter proposes a real-time energy optimal trajectory planning and tracking method for an uninhabited surface vehicle (USV) in dynamically changing environment. The problem is first discretized by Legendre pseudospectral method to transform the trajectory planning problem into a nonlinear programming problem with less discrete points and higher accuracy compared with traditional discretization methods. Then a novel strategy is developed to re-plan the trajectory in real-time during the entire maneuvering. The re-planning strategy uses the values obtained in the last re-planning as its initial guess values and upper and lower bound, which can reduce the computational complexity dramatically. Simulation results show that this novel approach could save a considerable amount of equivalent energy compared with an existing traditional method.

7.1 Introduction

The application of uninhabited surface vehicles (USVs) has gained a growing interest in recent years. Trajectory planning and tracking is one of the most important mission objectives for USV and involves movement of a USV from its current states to desired final states satisfying some performance index in a given time interval. This problem can be combined with other applications such as collision avoidance, dynamic positioning, station keeping and docking.

Many trajectory planning algorithms have been proposed in the relevant literature. Line-of-sight (LOS) and its variations are the most common method for the trajectory planning of both USVs and autonomous underwater vehicles (AUVs) [1]. Ataei and Yousefi-Koma [2] used a multi-objective optimization method to find a set of

[1]School of Information Science and Engineering, Harbin Institute of Technology, Weihai, China
[2]School of Engineering, University of Plymouth, UK

Pareto-optimal solutions, and then proposed a three-dimensional guidance strategy inspired from the LOS guidance strategy to track the optimal paths. Whilst Naeem *et al.* [3] proposed a strategy combined with a manual biasing scheme and waypoint guidance by LOS for USV collision avoidance based on the International Regulations for Preventing Collisions at Sea (COLREGs) rules. The A* algorithm is a popular algorithm used for trajectory planning. Eichhorn [4] combined A* algorithm and an optimal navigation formula and presented a time optimal path planning method considering the ocean current for an AUV. Hernandez *et al.* [5] utilized a topological information based method to generate paths in a workspace, and then developed a homotopic A* algorithm to search the path using the generated paths.

Intelligent optimization algorithms are applied increasingly to trajectory planning. Zeng *et al.* [6] presented a time optimal method based on a shell space decomposition scheme integrated with a quantum-behaved particle swarm optimization (QPSO) algorithm for trajectory planning of an AUV in a variable environment. Aghababa [7] used five intelligent optimization algorithms to solve the energy optimal trajectory planning for AUV collision avoidance.

A trajectory tracking method can be used to complete the maneuvering for generated trajectory. Ashrafiuon *et al.* [8] presented a sliding-mode method for a USV. A first-order sliding surface was introduced in terms of surge tracking errors, and a second-order sliding surface was used in terms of sway tracking errors. Kim *et al.* [9] proposed an integral sliding-mode controller for an AUV, which is exponentially stable with unknown disturbances. Whereas Serrano *et al.* [10] designed a trajectory tracking controller based on linear equations for under-actuated surface vessels. The tracking errors can converge to zero with this method. Lefeber *et al.* [11] developed a global state-feedback control law for under-actuated ship. This cascade approach divided the problem into two separated subsystems for the yaw and surge control. Skjetne and Fossen [12] presented a three-step backstepping method for ships maneuvering onto curves in the plane.

Most of the trajectory planning methods mentioned above can only be used off-line, and real-time re-planning cannot be achieved due to its computational complexity during the maneuvering. Another problem is that most trajectory tracking methods aim to suppress interference caused by environmental forces causing loss of substantial energy to deal with the interference suppression. A USV normally has limited power supply and the energy consumption determines its usage duration. It is thus important to design an energy optimal method for USV trajectory planning and tracking. This chapter aims to design a real-time energy optimal re-planning strategy to control a USV to move from its current position and orientation to a designed position and orientation. This method involves the known environmental forces into the proposed strategy to optimize the energy in maneuvering. These known environmental forces are considered as thrust forces rather than not as disturbances and do not require suppression.

The rest of this chapter is organized as follows: Section 7.2 introduces the mathematical equations representing the motion of a USV, ocean current and wind load. Section 7.3 describes the pseudospectral method, which translates the trajectory planning problem into a nonlinear programming problem. Section 7.4 proposes a re-planning strategy to make sure the actual trajectory is energy optimal with environmental forces. Section 7.5 shows the simulation results of the proposed algorithm, and Section 7.6 provides the conclusions.

7.2 Mathematics representation

7.2.1 Motion equations

The reference frames used in this chapter are illustrated in Figure 7.1.

In this study, it is assumed that the center of buoyancy and the center of mass are the same, and the motion of heave, pitch and roll are negligible. The nonlinear equations of motion for a USV thus can be written as follows:

$$\dot{\boldsymbol{\eta}} = \boldsymbol{R}(\psi)\boldsymbol{v}, \tag{7.1}$$

$$\boldsymbol{M}\dot{\boldsymbol{v}} + \boldsymbol{C}(\boldsymbol{v})\boldsymbol{v} + \boldsymbol{D}(\boldsymbol{v})\boldsymbol{v} = \boldsymbol{f} + \boldsymbol{f}_{\mathrm{Env}}. \tag{7.2}$$

Here, $\boldsymbol{\eta} = [x, y, \psi]^{\mathrm{T}}$ denotes the position and orientation vector with coordinates in the earth-fixed frame, $\boldsymbol{v} = [u, v, r]^{\mathrm{T}}$ denotes the linear and angular velocity vector with coordinates in the body-fixed frame. $\boldsymbol{R}(\psi)$ is the translation matrix defined as

$$\boldsymbol{R}(\psi) = \begin{bmatrix} \cos\psi & -\sin\psi & 0 \\ \sin\psi & \cos\psi & 0 \\ 0 & 0 & 1 \end{bmatrix}. \tag{7.3}$$

$\boldsymbol{M}$ is the system inertia matrix including added mass, $\boldsymbol{C}(\boldsymbol{v})$ is the matrix of Coriolis and centripetal terms including added mass, $\boldsymbol{D}(\boldsymbol{v})$ is damping matrix, $\boldsymbol{f}$ is the vector of

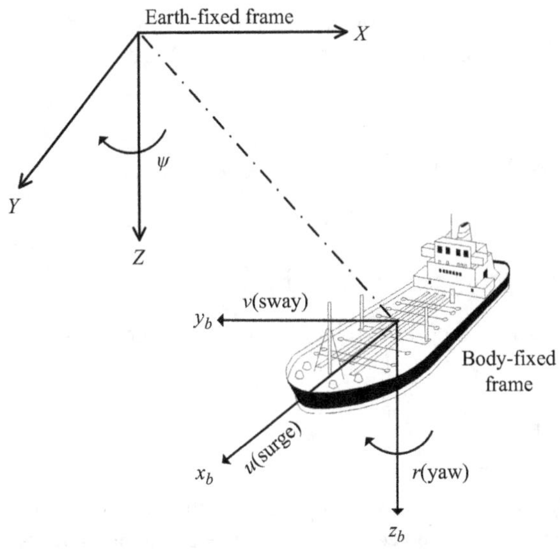

Figure 7.1 Definition of the earth-fixed and body-fixed frames

control inputs and $\boldsymbol{f}_{\text{Env}}$ is the environmental forces and moments acting on the vessel. These matrixes can also be decomposed as follows:

$$\boldsymbol{M} = \boldsymbol{M}_{RB} + \boldsymbol{M}_A = \begin{bmatrix} m - X_{\dot{u}} & 0 & 0 \\ 0 & m - Y_{\dot{v}} & mx_G - Y_{\dot{r}} \\ 0 & mx_G - Y_{\dot{r}} & I_z - N_{\dot{r}} \end{bmatrix}, \tag{7.4}$$

where m is the mass of USV, $X_{\dot{u}}$, $Y_{\dot{v}}$, $Y_{\dot{r}}$ and $N_{\dot{r}}$ are the parameters for added mass effects.

$$\boldsymbol{C}(\boldsymbol{v}) = \boldsymbol{C}_{RB}(\boldsymbol{v}) + \boldsymbol{C}_A(\boldsymbol{v}), \tag{7.5}$$

where $\boldsymbol{C}_{RB}$ is the Coriolis and centripetal matrix of the rigid-body, $\boldsymbol{C}_A$ is the hydrodynamic Coriolis and centripetal matrix.

The nonlinear damping can be neglected for a low-speed USV, hence only linear damping will be considered here. The matrix $\boldsymbol{D}$ is given by

$$\boldsymbol{D} = \begin{bmatrix} -X_u & 0 & 0 \\ 0 & -Y_v & -Y_r \\ 0 & -N_v & -N_r \end{bmatrix}, \tag{7.6}$$

$$\boldsymbol{f}_{Env} = \boldsymbol{f}_{\text{wind}} + \boldsymbol{f}_{\text{wave}} + \boldsymbol{f}_{\text{current}} + \boldsymbol{f}_{\text{other}}, \tag{7.7}$$

where $\boldsymbol{f}_{\text{wind}}$, $\boldsymbol{f}_{\text{wave}}$ and $\boldsymbol{f}_{\text{current}}$ are the disturbance forces and moments caused by wind, wave and current, $\boldsymbol{f}_{\text{other}}$ is the force and moment caused by other random disturbances.

7.2.2 Ocean currents

According to Fossen [13], the forces and moment generated by current could be included into the dynamic model, thus (7.1) and (7.2) can be rewritten as

$$\dot{\eta} = \boldsymbol{R}(\psi)\boldsymbol{v}_r + \boldsymbol{v}_c^E, \tag{7.8}$$

$$\boldsymbol{M}\dot{\boldsymbol{v}}_r + \boldsymbol{C}(\boldsymbol{v}_r)\boldsymbol{v}_r + \boldsymbol{D}(\boldsymbol{v}_r)\boldsymbol{v}_r = \boldsymbol{f} + \boldsymbol{f}_{\text{wind}} + \boldsymbol{f}_{\text{wave}} + \boldsymbol{f}_{\text{other}}, \tag{7.9}$$

where $\boldsymbol{v}_r$ is a relative velocity vector

$$\boldsymbol{v}_r = \boldsymbol{v} - \boldsymbol{v}_c = [\, u - u_c \quad v - v_c \quad r \,]^{\text{T}}, \tag{7.10}$$

where $\boldsymbol{v}_c$ is the horizontal current velocity vector in surge and sway, defined as

$$\boldsymbol{v}_c = \begin{bmatrix} u_c \\ v_c \\ 0 \end{bmatrix} = \begin{bmatrix} V_c \cos(\beta_c - \psi) \\ V_c \sin(\beta_c - \psi) \\ 0 \end{bmatrix}, \tag{7.11}$$

where V_c and β_c are the current velocity and direction respectively.

Note that the current load vector has already been included by the relative velocity vector in the nonlinear equation of motion, so the $\boldsymbol{f}_{current}$ is implicit in (7.9).

7.2.3 Wind load model

For this study, only mean and slowly varying wind loads are considered as in most papers. The resultant wind forces and moment acting on a USV can be defined in terms of relative wind speed V_R and angle γ_R

$$V_R = \sqrt{u_R^2 + v_R^2}, \tag{7.12}$$

where

$$u_R = V_w \cos(\gamma_R) - u, \tag{7.13}$$

$$v_R = V_w \sin(\gamma_R) - v, \tag{7.14}$$

$$\gamma_R = \psi_w - \psi, \tag{7.15}$$

where V_w and ψ_w are the wind speed and direction. The force and moment caused by wind then can be defined as

$$\boldsymbol{\tau}_{\text{wind}} = \begin{bmatrix} X_{\text{wind}} \\ Y_{\text{wind}} \\ N_{\text{wind}} \end{bmatrix} = \begin{bmatrix} \frac{1}{2} C_X(\gamma_R)\rho_\omega V_R^2 A_T \\ \frac{1}{2} C_Y(\gamma_R)\rho_\omega V_R^2 A_L \\ \frac{1}{2} C_N(\gamma_R)\rho_\omega V_R^2 A_L L \end{bmatrix}, \tag{7.16}$$

where ρ_ω is the density of air in kg/m^3, A_T and A_L are the transverse and lateral projected areas in m^2, and L is the overall length of the ship in m. C_X and C_Y are the force of coefficients and C_N is the moment coefficient, which can be fitted as follows:

$$C_X = A_0 + A_1\frac{2A_L}{L^2} + A_2\frac{2A_T}{B^2} + A_3\frac{L}{B} + A_4\frac{S}{L} + A_5\frac{C}{L} + A_6 M, \tag{7.17}$$

$$C_Y = B_0 + B_1\frac{2A_L}{L^2} + B_2\frac{2A_T}{B^2} + B_3\frac{L}{B} + B_4\frac{S}{L} + B_5\frac{C}{L} + B_6\frac{A_{ss}}{A_L}, \tag{7.18}$$

$$C_N = C_0 + C_1\frac{2A_L}{L^2} + C_2\frac{2A_T}{B^2} + C_3\frac{L}{B} + C_4\frac{S}{L} + C_5\frac{C}{L}, \tag{7.19}$$

where B is the breadth of the vessel, C is the distance from bow of centroid of lateral projected area, M is the number of distinct group of masts or kingposts seen in lateral. A_i and B_i $(i = 0, 1, \ldots, 6)$, and C_i $(i = 0, 1, \ldots, 5)$ can be calculated by Isherwood's table [13].

7.3 Trajectory planning using pseudospectral method

7.3.1 Problem statement

This chapter focuses on energy optimal trajectory planning for a fully actuated USV. The objective function is to find $\boldsymbol{f}(t)$ over a time interval $t \in [T_0, T_f]$, so that the equivalent energy consumption $\boldsymbol{J}$

$$J = \frac{1}{2}\int_{T_0}^{T_f} \boldsymbol{f}^{\mathrm{T}}\boldsymbol{f}\,\mathrm{d}t \tag{7.20}$$

is minimized. The control input vectors are assumed to use continuous low-thrust force and moment confined to lie within specified limits

$$\boldsymbol{f}(t) \leq |\boldsymbol{f}_{\max}|. \tag{7.21}$$

The states at the initial point and final point are constrained to be the specified conditions

$$\begin{aligned} \boldsymbol{\eta}_{T_0} &= \boldsymbol{\eta}_S, \quad \boldsymbol{\eta}_{T_f} = \boldsymbol{\eta}_F, \\ \boldsymbol{\nu}_{T_0} &= \boldsymbol{\nu}_S, \quad \boldsymbol{\nu}_{T_f} = \boldsymbol{\nu}_F, \end{aligned} \tag{7.22}$$

where $\boldsymbol{\eta}_S$ and $\boldsymbol{\eta}_F$ are the initial and final position and orientation vector, and $\boldsymbol{\nu}_S$ and $\boldsymbol{\nu}_F$ are the initial and final linear and angular velocity vector respectively.

7.3.2 Legendre pseudospectral method

The pseudospectral method is a newly developed class of methods for solving optimal control problems. In pseudospectral method, the state and control vectors are discretized at specified points using a structure of global orthogonal polynomials, which makes the optimal control problem easily to be solved with high accuracy [14].

Let $L_N(t)$ denote the Legendre polynomial of order N, and $\dot{L}_N(t)$ is the first-order derivative of it. Let $t_i, i = 0, 1, 2, \ldots, N$ be the zeros of $(t^2 - 1)\dot{L}_N(t)$, with $t_0 = -1$, $t_N = 1$. These points are called Legendre–Gauss–Lobatto (LGL) points, which serve as the system collocation points to select Nth order Lagrange interpolating polynomials [15]

$$\phi_i(t) = \frac{1}{N(N+1)L_N(t_i)} \cdot \frac{(t^2 - 1)\dot{L}_N(t)}{t - t_i}, \quad i = 0, 1, \ldots, N \tag{7.23}$$

and $\phi_i(t)$ satisfies the relationship $\phi_i(t_j) = \delta_{ij}$.

For a given function $F(t)$ defined in $[-1, 1]$, the Nth degree interpolation polynomial is

$$F^N(t) := \sum_{i=0}^{N} F(t_i)\phi_i(t). \tag{7.24}$$

The integral of $F^N(t)$ is

$$\int_{-1}^{1} F^N(t)dt := \sum_{i=0}^{N} F(t_i)w_i, \tag{7.25}$$

where

$$w_i = \frac{2}{N(N+1)} \cdot \frac{1}{[L_N(t_i)]^2}. \tag{7.26}$$

The derivative of $F^N(t)$ is

$$\dot{F}(t_i) = \sum_{j=0}^{N} D_{ij}F(t_j), \tag{7.27}$$

where $D := (D_{ij})$ is an $(N+1) \times (N+1)$ matrix, given by

$$D = (D_{ij}) := \begin{cases} \dfrac{L_N(t_i)}{L_N(t_j)} \cdot \dfrac{1}{t_i - t_j}, & i \neq j \\ -\dfrac{N(N+1)}{4}, & i = j = 0 \\ \dfrac{N(N+1)}{4}, & i = j = N \\ 0, & \text{otherwise} \end{cases}. \tag{7.28}$$

7.3.3 Discretization of the optimization problem

As mentioned above, the LGL points should be in the interval $[-1, 1]$, so that the original optimal problem should be restated from $[T_0, T_f]$ to $[-1, 1]$ with the transformation of the independent variable τ,

$$\tau = \frac{2t}{T_f - T_0} - \frac{T_f + T_0}{T_f - T_0}. \tag{7.29}$$

The position, velocity and control input vectors can be approximated with ϕ as

$$\boldsymbol{\eta}(t) \approx \boldsymbol{\eta}^N(t) = \sum_{j=0}^{N} \boldsymbol{\eta}_j \phi_j(t), \tag{7.30}$$

$$\boldsymbol{v}_r(t) \approx \boldsymbol{v}_r^N(t) = \sum_{j=0}^{N} \boldsymbol{v}_{rj} \phi_j(t), \tag{7.31}$$

$$\boldsymbol{f}(t) \approx \boldsymbol{f}^N(t) = \sum_{j=0}^{N} \boldsymbol{f}_j \phi_j(t) \tag{7.32}$$

and

$$\dot{\boldsymbol{\eta}}_i(t) \approx \sum_{j=0}^{N} D_{ij} \boldsymbol{\eta}_j, \quad i = 0, 1, \ldots, N, \tag{7.33}$$

$$\dot{\boldsymbol{v}}_{ri}(t) \approx \sum_{j=0}^{N} D_{ij} \boldsymbol{v}_{rj}, \quad i = 0, 1, \ldots, N. \tag{7.34}$$

The integral formulation of J can also be approximated as

$$\int_{T_0}^{T_f} \boldsymbol{f}^{\mathrm{T}} \boldsymbol{f} \, dt \approx \frac{T_f - T_0}{2} \sum_{j=0}^{N} \boldsymbol{f}_j^{\mathrm{T}} \boldsymbol{f}_j w_j. \tag{7.35}$$

According to (7.30)–(7.35), the trajectory planning problem can be transformed into a nonlinear programming problem to determine the position vector $\boldsymbol{\eta}_j$, velocity vector $\boldsymbol{v}_j$ and the control input vector $\boldsymbol{f}_j$ which can minimize the cost function J given by

$$J = \frac{T_f - T_0}{4} \sum_{j=0}^{N} \boldsymbol{\tau}_j^{\mathrm{T}} \boldsymbol{\tau}_j w_j \tag{7.36}$$

subject to the constraints

$$\sum_{j=0}^{N} D_{ij}\boldsymbol{\eta}_j - \frac{T_f - T_0}{2}\boldsymbol{R}(\psi_i)\boldsymbol{v}_{ri} = 0, \quad i = 0, 1, \ldots, N, \tag{7.37}$$

$$\sum_{j=0}^{N} D_{ij}\boldsymbol{v}_{rj} + \frac{T_f - T_0}{2}\boldsymbol{M}^{-1}(\boldsymbol{C}(\boldsymbol{v}_{ri})\boldsymbol{v}_{ri} + \boldsymbol{D}(\boldsymbol{v}_{ri})\boldsymbol{v}_{ri} - \boldsymbol{\tau}_i - \tau_{\text{wind}i}) = 0,$$
$$i = 0, 1, \ldots, N, \tag{7.38}$$

$$\boldsymbol{v}_i = \boldsymbol{v}_{ri} + \boldsymbol{v}_{ci} = \begin{bmatrix} u_{ri} + V_c\cos(\beta_c - \psi_{ri}) \\ v_{ri} + V_c\sin(\beta_c - \psi_{ri}) \\ r_{ri} \end{bmatrix}, \quad i = 0, 1, \ldots, N, \tag{7.39}$$

$$\begin{aligned} \boldsymbol{\eta}_{T_0} &= \boldsymbol{\eta}_S, \quad \boldsymbol{\eta}_{T_f} = \boldsymbol{\eta}_F, \\ \boldsymbol{v}_{T_0} &= \boldsymbol{v}_S, \quad \boldsymbol{v}_{T_f} = \boldsymbol{v}_F, \end{aligned} \tag{7.40}$$

$$\begin{aligned} \boldsymbol{\eta}_{\min} \leq \boldsymbol{\eta}_i &\leq \boldsymbol{\eta}_{\max} \\ \boldsymbol{v}_i &\leq |\boldsymbol{v}_{\max}| \quad i = 0, 1, \ldots, N, \\ \boldsymbol{\tau}_i &\leq |\boldsymbol{\tau}_{\max}| \end{aligned} \tag{7.41}$$

where (7.36) is obtained from (7.20), and (7.37)–(7.39) are obtained from (7.8)–(7.10), (7.40) is obtained from (7.20), and (7.41) is set according to the mission and the ability of the USV. Note that $\boldsymbol{\eta}_{\min} \neq -\boldsymbol{\eta}_{\max}$, this is useful for different applications, such as collision avoidance, station keeping and monitoring in a given area.

Let

$$\boldsymbol{X} = [\boldsymbol{x}_i \quad \boldsymbol{y}_i \quad \boldsymbol{\psi}_i \quad \boldsymbol{u}_i \quad \boldsymbol{v}_i \quad \boldsymbol{r}_i \quad \boldsymbol{\tau}_{\text{surge}i} \quad \boldsymbol{\tau}_{\text{sway}i} \quad \boldsymbol{\tau}_{\text{raw}i}] \quad i = 0, 1, \ldots, N. \tag{7.42}$$

Then $\boldsymbol{X}$ contains $9(N+1)$ variables to be determined to optimize the objective J. The number of the constraints described by (7.37) and (7.38) is $6(N+1)$. The number of the constraints described by (7.39) is 12. In (7.38), only the forces and moments caused by current and wind are considered, and the wave and other random environmental forces and moments are considered as disturbances. This translates the original trajectory planning problem into a nonlinear programming problem and can be solved by a proper optimization method, such as sequential quadratic programming, interior point method or intelligent optimization methods.

7.4 Optimization using particle swarm optimization

Most optimization method is sensitive to initial values, a feasible initial value within the convergence domain of the NLP solver can result in a good solution and reduce the time consumption. In this chapter, particle swarm optimization (PSO) is used to get a feasible initial value first. PSO is a stochastic optimization method invented by Kennedy and Eberhart in 1995 [16]. It is an evolutionary algorithm inspired by the social behavior of birds flocking or people grouping. In PSO, each possible solution is called a particle that is analogous to a bird in a flock. The objective of the particle population (called a swarm) is to find the global minimum of the fitness function (cost function). In each iteration, every particle updates itself by improving its own velocity, which is derived from the personal best solution (known as "pbest") and the global best position (known as "gbest") identified from previous iterations by the whole swarm. The basic PSO algorithm can be described as

$$v_{i,h}^{k+1} = \omega v_{i,h}^{k} + c_1 r_1^k (p_{i,h}^k - x_{i,h}^k) + c_2 r_2^k (p_{g,h}^k - x_{i,h}^k), \tag{7.43}$$

where $v_{i,h}^k$ and $x_{i,h}^k$ are the hth dimension velocity and position of particle i in the kth iteration; $p_{i,h}^k$ and $p_{g,h}^k$ are the hth dimension pbest and gbest of particle i in the kth iteration; ω is a weighting factor, known as inertia; c_1 is the cognitive weight and c_2 is the social weight; r_1^k and r_2^k are the two random numbers in the range of [0, 1]. The inertia weight ω is used to control the impact of previous velocity of the ith particle on the current velocity. A large ω makes the particle search more new areas, which facilitates global exploration. While a small ω makes the particle search current area, which facilitates local exploration. It is very important to select the ω to make a trade-off between the global and local exploration. The $c_1 r_1^k (p_{i,h}^k - x_{i,h}^k)$ only considers the ith particle's own experiences and will drive the particle to the direction of its own best experience; $c_2 r_2^k (p_{g,h}^k - x_{i,h}^k)$ represents the social interaction among all the particles, which will drag the ith particle to the best position found by all of the particles. The new position of a particle is then calculated as

$$\boldsymbol{X}_i^{k+1} = \boldsymbol{X}_i^k + \boldsymbol{V}_i^{k+1} \tag{7.44}$$

Here, $\boldsymbol{X}_i^k$ and $\boldsymbol{V}_i^k$ are the position and velocity vector of the ith particle in the kth iteration.

While updating, a high velocity will drive the particles out of bounds or cause divergence; thus, the velocity of the particle must be constrained. By setting $V_{\max,h}$ as the maximum velocity of the hth dimension, the velocity can then be updated using the following formulation:

$$v_{i,h}^{k+1} = \begin{cases} v_{i,h}^{k+1}, & \textit{if } \left|v_{i,h}^{k+1}\right| \le V_{\max,h} \\ V_{\max,h}, & \textit{if } v_{i,h}^{k+1} > V_{\max,h} \\ -V_{\max,h}, & \textit{if } v_{i,h}^{k+1} < -V_{\max,h} \end{cases}. \tag{7.45}$$

7.5 Re-planning strategy

7.5.1 Problem statement of re-planning

Trajectory tracking algorithms in most current literature suppresses disturbances and thus, the actual energy consumption is much higher than usual. Moreover, the accuracy is also compromised as the total numbers of discrete points are selected in small numbers to decrease the computational complexity than actually required in a real environment.

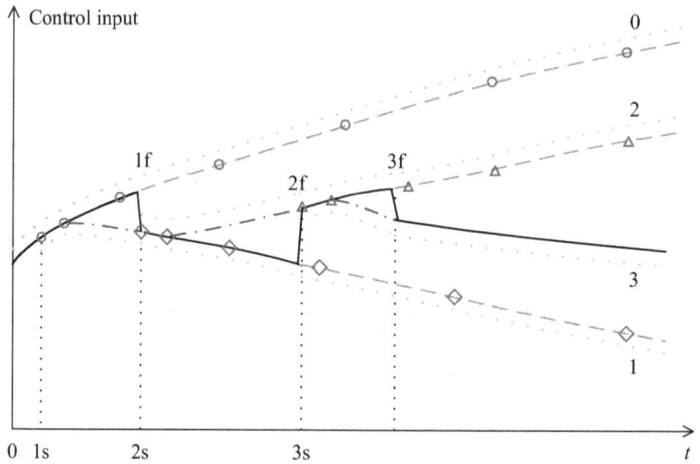

Figure 7.2 Schematic diagram for the changing of control input in each re-planning

According to the Bellman's principle, during the maneuvering, if the re-planned trajectory is optimal in every time interval, then the entire process is optimal. It can also be regarded as real-time closed-loop optimal feedback control [17]. The process for re-planning is similar with the method proposed above but with following differences:

(1) The re-planning strategy will be started at the first sample time, and repeated in the sequential sample times till the last re-planning is achieved.
(2) The new control inputs will be applied as soon as the new re-planning was finished. The current re-planning method begins with the control inputs, which are obtained during last re-planning. The schematic diagram for the changing of control input is shown in Figure 7.2. In this figure, it is assumed that only triple re-planning happened during maneuvering. Here, "0" means the data of original optimization, "1s" and "1f" mean the start time and the finish time of the first re-planning, and so on. The solid line is the actual control input, while the dotted lines are the control input optimized by each re-planning. The dashed lines are the control input in last re-planning and will be replaced by the new control input. The dash–dot lines are the control input obtained in current re-planning.
(3) Most optimization methods are sensitive with the initial guess values. During maneuvering, if the last re-planning is optimal, then the following re-planning trajectory could not depart from the last planning trajectory very much. Assuming that the new re-planning starts at $\tilde{T}_0$. Here, tilde is used to stand for the values of current re-planning. The position, velocity and control input values of last trajectory at the new LGL points of current re-planning time interval can be indicated as $\tilde{\boldsymbol{\eta}}_{ni}$, $\tilde{\boldsymbol{v}}_{ni}$ and $\tilde{\boldsymbol{\tau}}_{ni}$, $i = 0, 1, \ldots, N$. Using these values as the initial

guess values for the new re-planning, these initial guess values could make the optimization faster and could get a better solution compared with random initial guess values. Whilst, the solution should be constrained in a certain error range around $\tilde{\boldsymbol{\eta}}_{ni}$, $\tilde{\boldsymbol{\nu}}_{ni}$ and $\tilde{\boldsymbol{\tau}}_{ni}$. These constraints could narrow the feasible region so that the search will be speeded up and the computational time taken will be reduced.

(4) It is assumed that computation of the new re-planning finish at $\tilde{T}_f$, so the control inputs in $[\tilde{T}_0, \tilde{T}_f]$ are the solution of last re-planning. In order to make the control inputs continuous between different re-planning, the first two control inputs $\tilde{\boldsymbol{f}}_1$, $\tilde{\boldsymbol{f}}_2$ at new LGL points should be equal to the values obtained from the last re-planning. This is shown in 0, where the mark "o" denotes the location of LGL points from 1s to T_f, and the diamond mark denotes the location of LGL points from 2s to T_f, and so on. For example, in [1s, T_f], the control inputs in first two LGL points are the same as the original values, which makes the control inputs smoother.

The new re-planning problem at $\tilde{T}_0$ can thus be rewritten as

$$J = \frac{T_f - \tilde{T}_0}{4}\sum_{j=0}^{N} \tilde{\boldsymbol{f}}_j^{\mathrm{T}} \tilde{\boldsymbol{f}}_j w_j \tag{7.46}$$

subject to

$$\sum_{j=0}^{N} D_{ij}\tilde{\boldsymbol{\eta}}_j - \frac{T_f - \tilde{T}_0}{2}\boldsymbol{R}(\tilde{\psi}_i)\tilde{\boldsymbol{\nu}}_{ri} = 0, \quad i = 0, 1, \cdots, N, \tag{7.47}$$

$$\sum_{j=0}^{N} D_{ij}\tilde{\boldsymbol{\nu}}_{rj} + \frac{T_f - \tilde{T}_0}{2}\boldsymbol{M}^{-1}(\boldsymbol{C}(\tilde{\boldsymbol{\nu}}_{ri})\tilde{\boldsymbol{\nu}}_{ri} + \boldsymbol{D}(\tilde{\boldsymbol{\nu}}_{ri})\tilde{\boldsymbol{\nu}}_{ri} - \tilde{\boldsymbol{f}}_i - \boldsymbol{f}_{\mathrm{wind}i}) = 0, \qquad i = 0, 1, \ldots, N, \tag{7.48}$$

$$\tilde{\boldsymbol{\nu}}_i = \tilde{\boldsymbol{\nu}}_{ri} + \tilde{\boldsymbol{\nu}}_{ci} = \begin{bmatrix} \tilde{u}_{ri} + V_c\cos(\beta_c - \tilde{\psi}_{ri}) \\ \boldsymbol{\nu}_{ri} + V_c\sin(\beta_c - \tilde{\psi}_{ri}) \\ \tilde{r}_{ri} \end{bmatrix}, \quad i = 0, 1, \ldots, N, \tag{7.49}$$

$$\begin{aligned} \tilde{\boldsymbol{\eta}}_{\tilde{T}_0} &= \tilde{\boldsymbol{\eta}}_0, \quad \tilde{\boldsymbol{\eta}}_{T_f} = \boldsymbol{\eta}_F, \\ \tilde{\boldsymbol{\nu}}_{\tilde{T}_0} &= \tilde{\boldsymbol{\nu}}_0, \quad \tilde{\boldsymbol{\nu}}_{T_f} = \boldsymbol{\nu}_F, \end{aligned} \tag{7.50}$$

$$\tilde{\boldsymbol{f}}_1 = \tilde{\boldsymbol{f}}_{n1}, \quad \tilde{\boldsymbol{f}}_2 = \tilde{\boldsymbol{f}}_{n2}, \tag{7.51}$$

$$\begin{aligned} &\max(\boldsymbol{\eta}_{\min}, \tilde{\boldsymbol{\eta}}_i - \boldsymbol{\eta}_{\mathrm{err}}) \le \tilde{\boldsymbol{\eta}}_i \le \min(\boldsymbol{\eta}_{\max}, \tilde{\boldsymbol{\eta}}_i + \boldsymbol{\eta}_{\mathrm{err}}) \\ &\max(-\boldsymbol{\nu}_{\max}, \tilde{\boldsymbol{\nu}}_i - \boldsymbol{\nu}_{\mathrm{err}}) \le \tilde{\boldsymbol{\nu}}_i \le \min(\boldsymbol{\nu}_{\max}, \tilde{\boldsymbol{\nu}}_i + \boldsymbol{\nu}_{\mathrm{err}}) \quad i = 0, 1, \ldots, N. \\ &\max(-\boldsymbol{f}_{\max}, \tilde{\boldsymbol{f}}_i - \boldsymbol{f}_{\mathrm{err}}) \le \tilde{\boldsymbol{f}}_i \le \min(\boldsymbol{f}_{\max}, \tilde{\boldsymbol{f}}_i + \boldsymbol{f}_{\mathrm{err}}) \end{aligned} \tag{7.52}$$

The initial guess values and the feasible region introduced here are better than the first planning, so the re-planning will be much faster, and will be suitable for real-time energy optimal re-planning. There are $6(k+1)$ more equation constraints introduced by (48).

7.5.2 Problem reformulation in differentially flat outputs space

In order to feedback the state of USV real-time, it is required to complete the current re-planning before the arrival of next sample time. Thus, the biggest difficulty that lies in the re-planning process is to deal with the computational complexity more efficiently. In this study, differential flatness property of USV is employed, such that the number of unknowns to be optimized and the number of equality constraints will be both reduced greatly.

A dynamic system

$$\dot{x} = \boldsymbol{f}(\boldsymbol{x}, \boldsymbol{u}) \qquad \boldsymbol{x} \in \boldsymbol{R}^n, \quad \boldsymbol{u} \in \boldsymbol{R}^m \tag{7.53}$$

is differentially flat or just flat, if there are smooth maps $\boldsymbol{Y}, \boldsymbol{A}, \boldsymbol{B}$ defining on neighborhoods of $R^n \times (R^m)^{\rho+1}, (R^m)^{r+1}$ and $(R^m)^{r+2}$, such that

$$\begin{aligned} \boldsymbol{y} &= \boldsymbol{Y}(\boldsymbol{x}, \boldsymbol{u}, \dot{\boldsymbol{u}}, \ddot{\boldsymbol{u}}, \ldots, \boldsymbol{u}^{(\rho)}) \\ \boldsymbol{x} &= \boldsymbol{A}(\boldsymbol{y}, \dot{\boldsymbol{y}}, \ddot{\boldsymbol{y}}, \ldots, \boldsymbol{y}^{(r)}) \\ \boldsymbol{u} &= \boldsymbol{B}(\boldsymbol{y}, \dot{\boldsymbol{y}}, \ddot{\boldsymbol{y}}, \ldots, \boldsymbol{y}^{(r+1)}) \end{aligned} \tag{7.54}$$

wherein ρ and r are positive integers, $\boldsymbol{y}$ is called a set of flat outputs, and the components of $\boldsymbol{y}$ are not related by a differential relation [17–19]. The definition shows that if there exists a set of flat outputs with the same number of control inputs, then the state and control variables can both be parameterized in terms of flat outputs and a finite number of their derivatives.

By observing the mathematical model of USV in (7.8) and (7.9), a set of flat outputs can be easily found as $Y = [Y_1, Y_2, Y_3]^T = [x, y, \psi]^{\mathrm{T}}$, and the corresponding flat transforms can be obtained as

$$\begin{cases} u_i = \dot{x}_i \cos\psi_i + \dot{y}_i \sin\psi_i = \dot{Y}_{i,1} \cos Y_{i,3} + \dot{Y}_{i,2} \sin Y_{i,3} \\ v_i = \dot{y}_i \cos\psi_i - \dot{x}_i \sin\psi_i = \dot{Y}_{i,2} \cos Y_{i,3} - \dot{Y}_{i,1} \sin Y_{i,3}, \\ r_i = \dot{\psi}_i = \dot{Y}_{i,3}, \end{cases} \tag{7.55}$$

$$\begin{cases} \tau_{ui} = m\dot{u}_i - mv_i r_i + X_u u_i + X_{|u|u}|u_i|u_i \\ \quad = m(\ddot{Y}_{i,1} \cos Y_{i,3} + \ddot{Y}_{i,2} \sin Y_{i,3}) + (X_u + X_{|u|u}|\dot{Y}_{i,1} \cos Y_{i,3} + \dot{Y}_{i,2} \sin Y_{i,3}|) \\ \quad\quad \cdot (\dot{Y}_{i,1} \cos Y_{i,3} + \dot{Y}_{i,2} \sin Y_{i,3}) \\ \tau_{vi} = m\dot{v}_i + mu_i r_i + Y_v v_i + Y_{|v|v}|v_i|v_i \\ \quad = m(\ddot{Y}_{i,2} \cos Y_{i,3} - \ddot{Y}_{i,1} \sin Y_{i,3}) + (Y_v + Y_{|v|v}|\dot{Y}_{i,2} \cos Y_{i,3} - \dot{Y}_{i,1} \sin Y_{i,3}|) \\ \quad\quad \cdot (\dot{Y}_{i,2} \cos Y_{i,3} - \dot{Y}_{i,1} \sin Y_{i,3}) \\ \tau_{ri} = I_z \dot{r}_i + N_r r_i + N_{|r|r}|r_i|r_i \\ \quad = I_z \ddot{Y}_{i,3} + (N_r + N_{|r|r}|\dot{Y}_{i,3}|)\dot{Y}_{i,3} \end{cases} . \tag{7.56}$$

Further, the above original optimal problem can be converted and reformulated in flat outputs space by using flat transforms. It can be found the number of unknowns to be optimized has been reduced by two-thirds. And it is well known that most of time consumption for NLP solvers will be taken to deal with the equality constraints; however, since the flat transforms automatically satisfy the mathematical model of USV, the equality constraints corresponding to the model of USV in the re-formulated problem (7.8) and (7.9) can be totally eliminated. Hence, the total time consumption can be efficiently reduced, and the re-planning process can be completed real-time.

7.6 Simulation results

The simulations were conducted in a laptop with 1.8 GHz i7-4500U CPU and used a commercial software called KNITRO for optimization. This work used a USV at Plymouth University called *Springer* shown in Figure 7.3, which is approximately 4 m long and 0.6 tones weight. The start and end time can be set according to practical application, here, $T_0 = 0$ and $T_f = 200$ s. Other relative parameters were

$$M = \begin{bmatrix} 677.6 & 0 & 0 \\ 0 & 1161.1 & 0 \\ 0 & 0 & 836.5 \end{bmatrix}, \tag{7.57}$$

$$C = \begin{bmatrix} 0 & 0 & -600v \\ 0 & 0 & 600u \\ 600v & -600u & 0 \end{bmatrix}, \tag{7.58}$$

$$D = \begin{bmatrix} 17.0 & 0 & 0 \\ 0 & 29.2 & 1.0 \\ 0 & 1.0 & 19.8 \end{bmatrix}, \tag{7.59}$$

$$\begin{aligned} &\eta_S = \begin{bmatrix} 5 & 5 & \frac{\pi}{4} \end{bmatrix}^{\mathrm{T}}, \quad \eta_F = \begin{bmatrix} 45 & 45 & \frac{\pi}{4} \end{bmatrix}^{\mathrm{T}}, \\ &\nu_S = \begin{bmatrix} 0 & 0 & 0 \end{bmatrix}^{\mathrm{T}}, \quad \nu_F = \begin{bmatrix} 0 & 0 & 0 \end{bmatrix}^{\mathrm{T}} \end{aligned} \tag{7.60}$$

Figure 7.3 The Springer USV

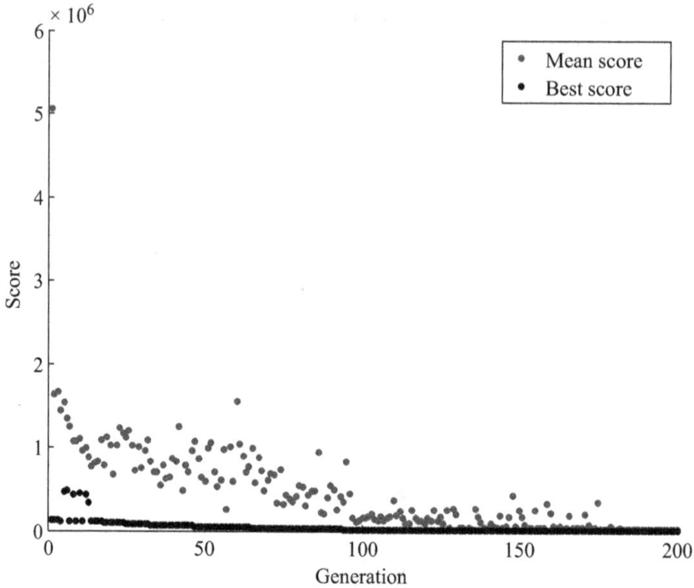

Figure 7.4 The history of the global best during iteration

$$\begin{aligned} \boldsymbol{\eta}_{\min} &= \begin{bmatrix} 0 \\ 0 \\ -\pi \end{bmatrix}, \quad \boldsymbol{\eta}_{\max} = \begin{bmatrix} 50 \\ 50 \\ \pi \end{bmatrix} \\ \boldsymbol{\nu}_{\max} &= \begin{bmatrix} 1.5 \\ 1.5 \\ 0.1\pi \end{bmatrix}, \quad \boldsymbol{\tau}_{\max} = \begin{bmatrix} 50 \\ 50 \\ 50 \end{bmatrix} \end{aligned} \tag{7.61}$$

7.6.1 Simulation results without disturbance

PSO was first used to find the initial value, where $c_1 = 0.8, c_2 = 1.2$ and $V_{\max} = 2, 40$ particles were employed to solve this problem. If the values of the fitness functions of all the swarms do not improve in the last 100 generations or till the generation maximum 200 is reached, stop the optimization, and the last global best position of the swarm will be the initial value of the next step. The history of the global best during iteration is shown in Figure 7.4. Figures 7.5–7.8 show the comparison results between the first optimization results and the final results using re-planning strategy without considering the time variant disturbance, where the mark "o" means the locations of the LGL points. Figures 7.9 and 7.10 show the objective function and computation time for every re-planning.

From Figure 7.6, it can be seen that the ψ of original trajectory is not acceptable in practical application. The orientation angle was almost reverse to the terminal

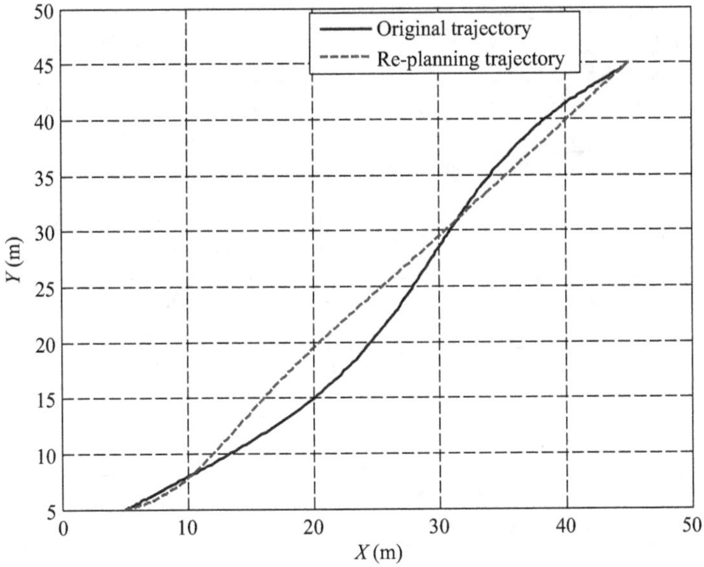

Figure 7.5 The trajectory comparison without disturbances

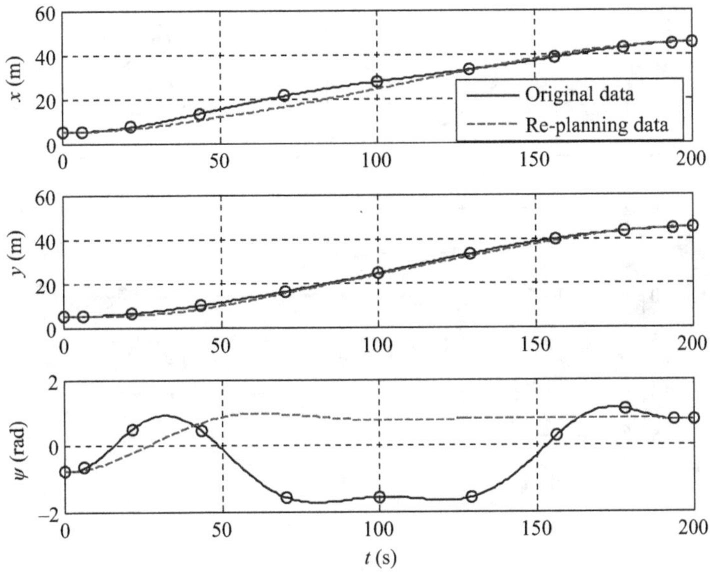

Figure 7.6 Position and heading comparison without disturbances

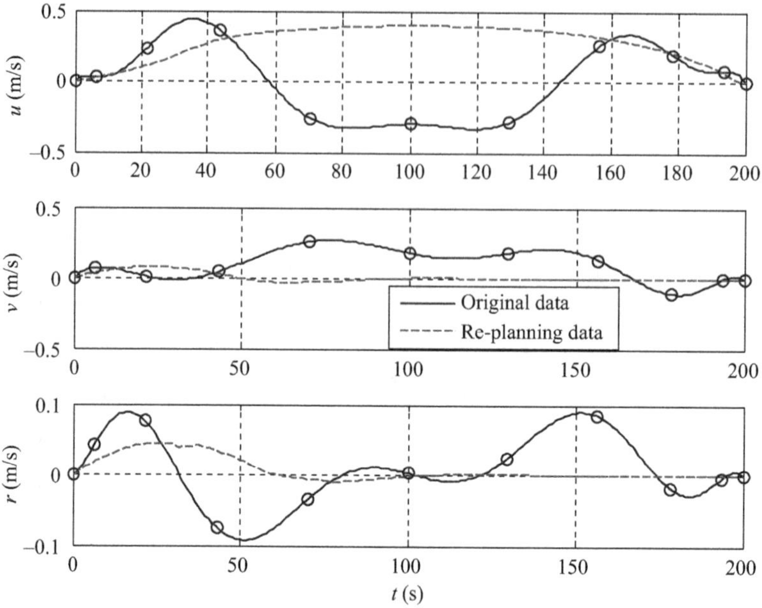

Figure 7.7 Velocity and angle velocity comparison without disturbances

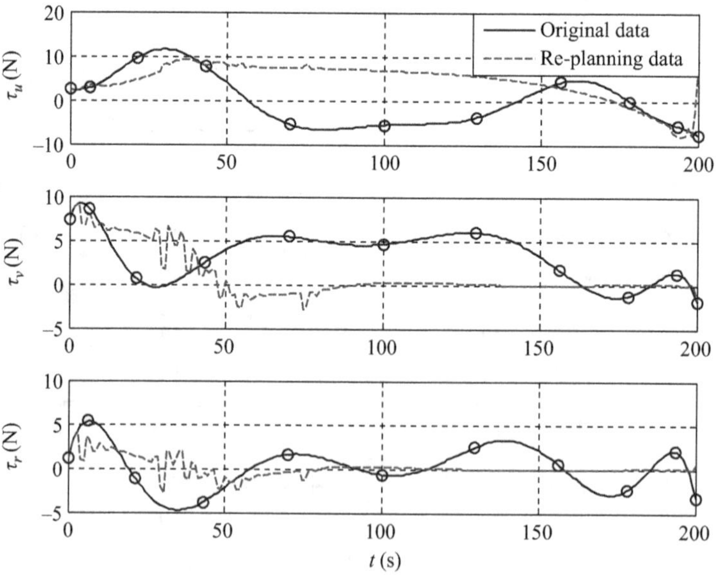

Figure 7.8 Control inputs comparison without disturbances

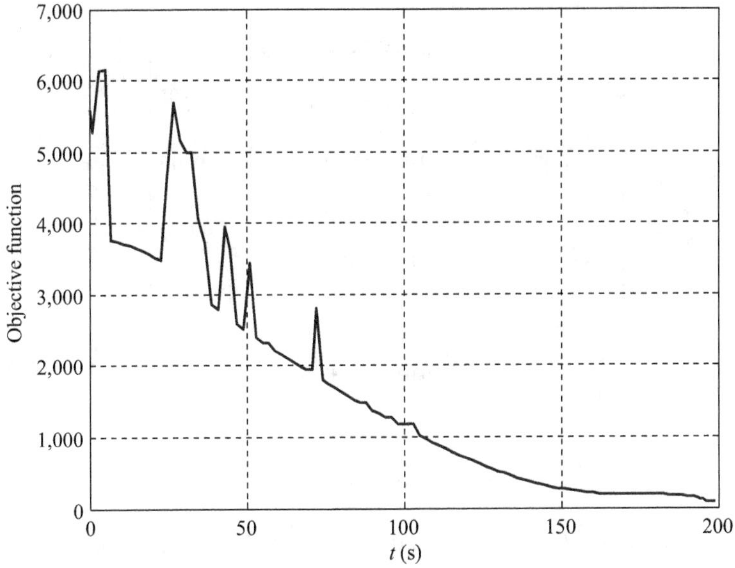

Figure 7.9 Objective function for every re-planning

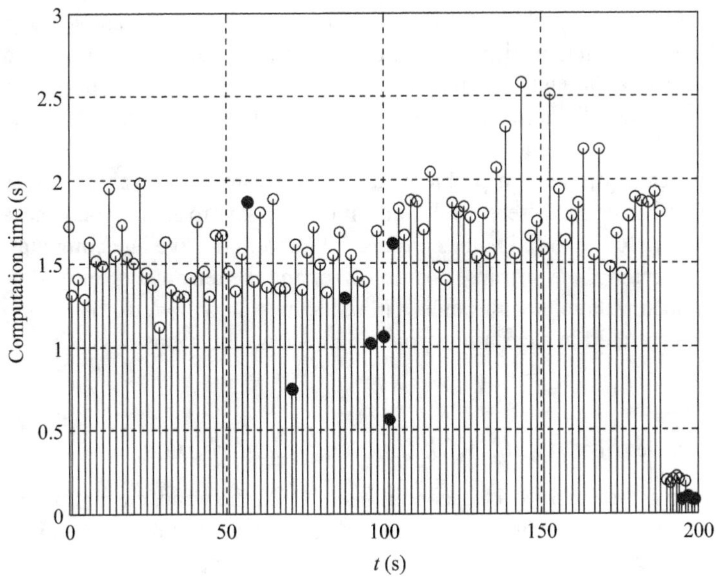

Figure 7.10 Computation time for every re-planning

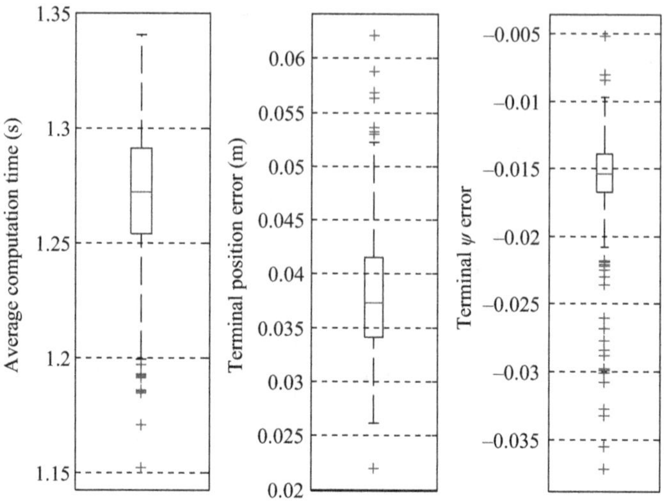

Figure 7.11 Results of 200-run Monte Carlo simulations

position during 70–130 s, and $u < 0$ during this time from Figure 7.7 which means that the USV was going backward. This is not acceptable even for full actuated USV. This result was caused by the small number of LGL points, but more LGL points will increase the computation time dramatically. It can be seen that the re-planning strategy can solve the problem during the entire maneuver, the ψ and u are all moved forward to the terminal position. It can be seen from Figure 7.10 that computational time of each re-planning is finished within 3 s, which is acceptable for practical USV applications. The solid dotted marks in Figure 7.10 indicate that the re-planning could not find the optimal solution.

Numerical optimization problems are sensitive to initial values, and different initial values can cause different solutions. It is necessary to analyse how the solutions vary according to the initial values. Figure 7.11 shows the average computation time of each re-planning, the terminal position error and the terminal error of ψ from 200 Monte Carlo simulations with random initial values. It is obvious that these data are mainly concentrated in a certain region, which shows the robustness of the proposed method.

7.6.2 Simulation results with time vary disturbance

Considering the normal working conditions of USV, the wind, current, wave, and all kinds of unknown external environmental disturbances was denoted as a time vary disturbance $d(t)$:

$$d(t) = \begin{bmatrix} 0.4\sin(0.3t) + 0.3\cos t - 1 \\ -0.15\sin(0.2)t + 0.3\cos(0.5t) + 1.5 \\ 0.3\sin t + 0.8\sin(0.4t) + 1 \end{bmatrix}. \tag{7.62}$$

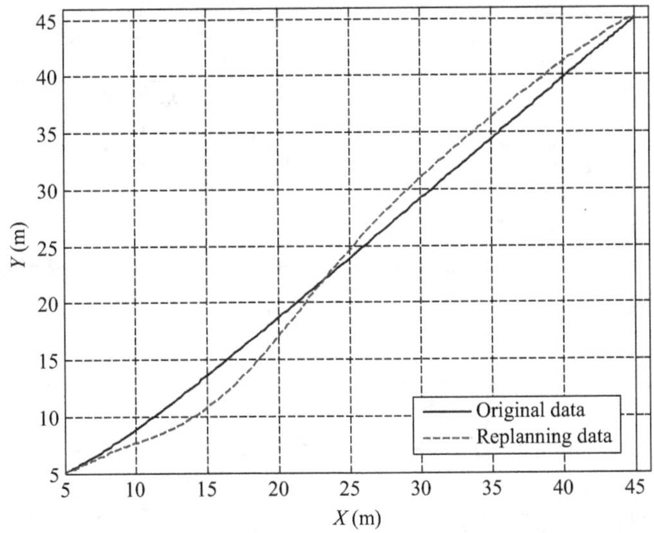

Figure 7.12 Trajectory comparison with time vary disturbance

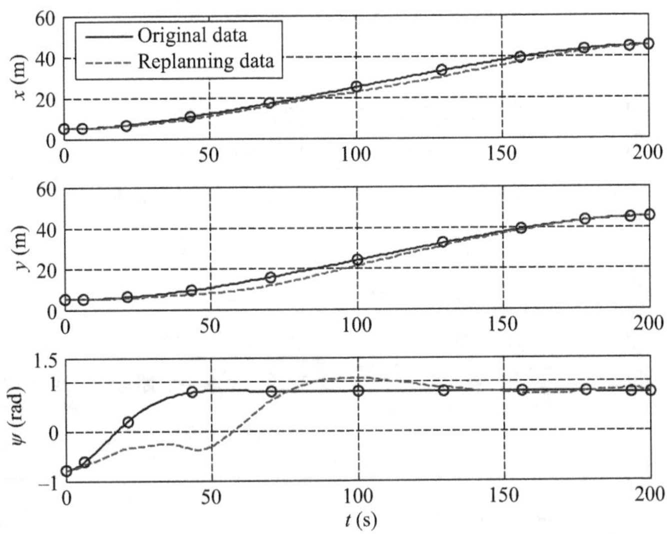

Figure 7.13 Position and heading comparison with time vary disturbance

The comparisons between the first optimization results and the final results with the time variant disturbance using re-planning strategy are shown in Figures 7.12–7.17. It can be seen that the USV can reach the target position even with time variant disturbance.

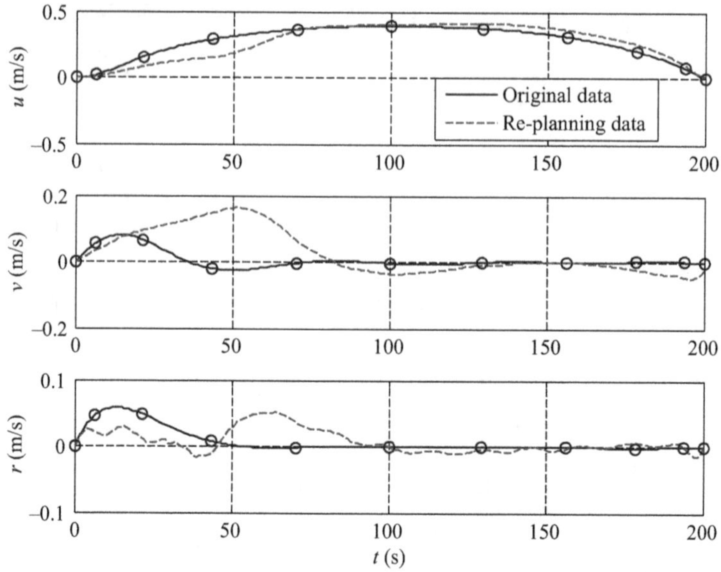

Figure 7.14 Velocity and angle velocity comparison with time vary disturbance

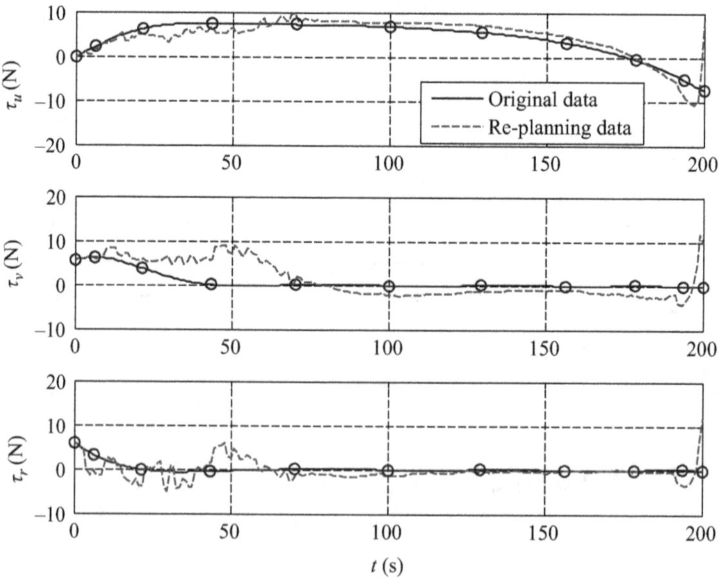

Figure 7.15 Control inputs comparison with time vary disturbance

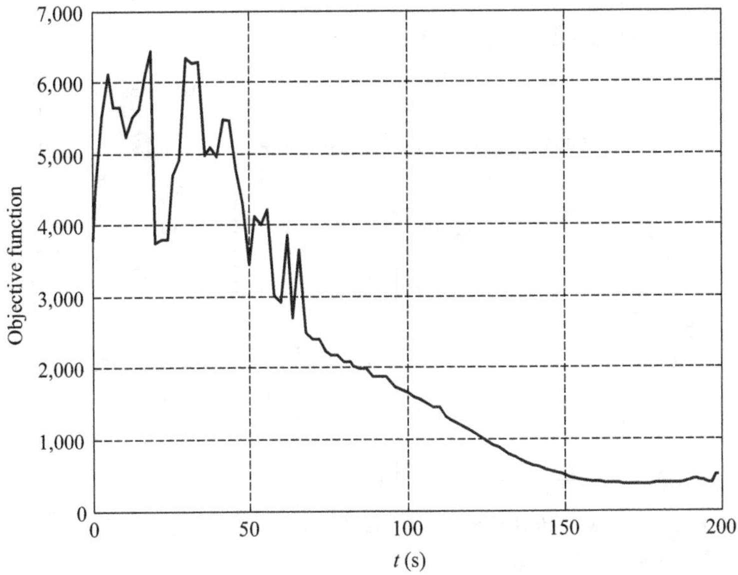

Figure 7.16 Objective function for every re-planning with time vary disturbance

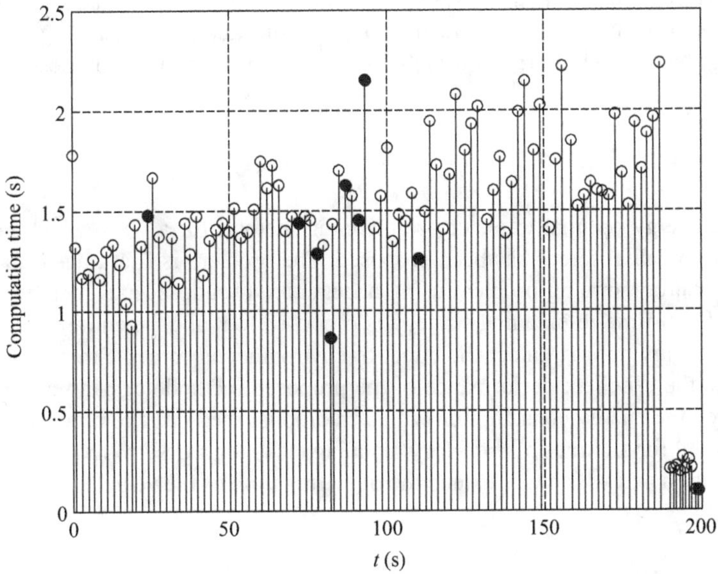

Figure 7.17 Computation time for every re-planning with time vary disturbance

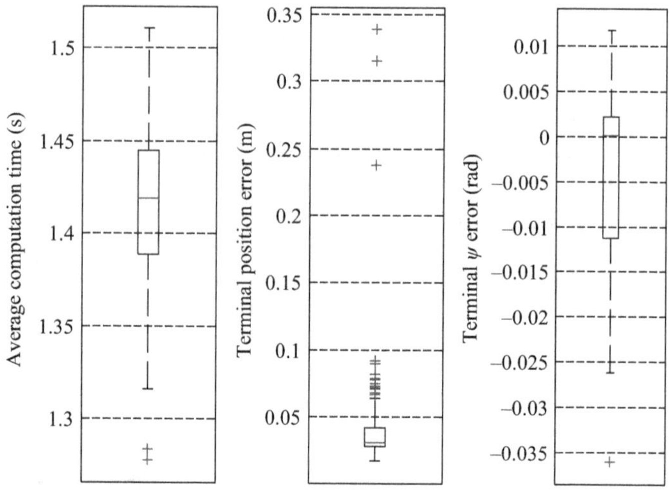

Figure 7.18 Results of 200-run Monte Carlo simulations with time vary disturbance

Figure 7.18 shows the average computation time of each re-planning, the terminal position error and the terminal error of ψ from 200 Monte Carlo simulations with random initial values when considering the time vary disturbance. The terminal position error is bigger than the result that time vary disturbance was not considered, but the error is acceptable for practical applications and these data are also mainly concentrated in a certain region which shows the robustness of the proposed method.

7.7 Conclusion

A novel energy optimal trajectory re-planning method for a USV, which accounts for environmental forces has been proposed. The problem is first transferred into a nonlinear programming problem using a Legendre pseudospectral method. Then a series of real-time re-planning strategy is proposed to make the entire maneuvering process energy optimal under the effect of environmental forces. Simulation results illustrate the excellent performance of the proposed methodology in real-time, and compared the energy consumption with a traditional method. The proposed method can be used for all kinds of trajectory planning tasks for a USV to reduce the energy consumption and which will extend the operating time of the vehicle.

Acknowledgments

This work was supported in part by the Fundamental Research Funds for the Central Universities under Grant no. HIT.KISTP.2014029, the Science and Technology

Foundation for the Universities in Shandong Province under Grant no. J14LN93, the Discipline Construction Foundation in Harbin Institute of Technology, Weihai under Grant no. WH20160103 and Education Research Project in Harbin Institute of Technology, Weihai under Grant no. BKQN201619.

References

[1] Campbell S., Naeem W., and Irwin G.W. 'A review on improving the autonomy of unmanned surface vehicles through intelligent collision avoidance maneuvers.' *Annual Reviews in Control* 2012;36:267–283.

[2] Ataei M., and Yousefi-Koma A. 'Three-dimensional optimal path planning for waypoint guidance of an autonomous underwater vehicle.' *Robotics and Autonomous Systems* 2015;67:23–32.

[3] Naeem W., Irwin G. W., and Yang A. 'COLREGs-based collision avoidance strategies for unmanned surface vehicles'. *Mechatronics* 2012;22:669–678.

[4] Eichhorn M. 'Optimal routing strategies for autonomous underwater vehicles in time-varying environment.' *Robotics and Autonomous Systems* 2015;67: 33–43.

[5] Hernandez E., Carreras M., Ridao P., Antich J., and Ortiz A. 'A Search-based Path Planning Algorithm with topological constraints. Application to an AUV.' *Proceeding of the 18th IFAC World Congress*; Milano, Italy, Aug–Sept 2011. Oxford: Elsevier; 2011. pp. 13654–13659.

[6] Zeng Z., Lammas A., Sammut K., He F., and Tang Y. 'Shell space decomposition based path planning for AUVs operating in a variable environment.' *Ocean Engineering* 2014;91:181–195.

[7] Aghababa M. P. '3D path planning for underwater vehicles using five evolutionary optimization algorithms avoiding static and energetic obstacles.' *Applied Ocean Research* 2012;38:48–62.

[8] Ashrafiuon H., Muske K. R., McNinch L. C., and Soltan R. A. 'Sliding-mode tracking control of surface vessels.' *IEEE Transactions on Industrial Electronic* 2008;55(11):4004–4012.

[9] Kim M., Joe H., Kim J., and Yu S., 'Integral sliding mode controller for precise maneuvering of autonomous underwater vehicle in the presence of unknown environmental disturbances.' *International Journal of Control.* 2015;88(10): 2055–2065.

[10] Serrano M. E., Scaglia G. J. E., Godoy S. A., Mut V., and Ortiz O. A. 'Trajectory tracking of underactuated surface vessels: A linear algebra approach.' *IEEE Transactions on Control System Technology* 2014;22(3):1103–1111.

[11] Lefeber E., Pettersen K. Y., and Nijmeijer H. 'Tracking control of an underactuated ship.' *IEEE Transactions on Control System Technology* 2003;11(1): 52–61.

[12] Skjetne R., and Fossen T. I. 'Nonlinear maneuvering and control of ships.' *Proceedings of MTS/IEEE Conference and Exhibition on OCEANS*; Honolulu, USA, Nov 2001. USA:IEEE;2001. pp. 1808–1815.

[13] Fossen T. I. *Guidance and Control of Ocean Vehicles*. Chichester: John Wiley & Sons Ltd; 1994. pp. 60–90.

[14] Huang H, and Zhuang Y. 'Optimal satellite formation reconfiguration using co-evolutionary particle swarm optimization in deep space.' *Acta Astronautica* 2015;(113):149–163.

[15] Gong Q., Fahroo F., and Ross I. M. 'Spectral algorithm for pseudospectral methods in optimal control.' *Journal of Guidance, Control, and Dynamics* 2008;31(3):460–471.

[16] Kennedy J., and Eberhart R. C. 'Particle swarm optimization.' *Proceedings of IEEE International Conference on Neural Networks*; Piscataway, USA, Nov–Dec 1995. USA:IEEE; 1995. pp. 1942–1948.

[17] Macfarland D., Swenson E., Black J., Cobb R., and Fosbury A. 'Near real-time closed-loop optimal control feedback for spacecraft attitude maneuvers.' *Proceedings of AIAA Modeling and Simulation Technologies Conference*; Illinois, USA, Aug 2009. USA: AIAA; 2009. pp. 1–19.

[18] Fliess M., Lévine J., Martin P., and Rouchon P. 'Flatness and defect of non-linear systems: Introductory theory and examples.' *International Journal of Control* 1995;61(6):1327–1361.

[19] Lévine J., 'On necessary and sufficient conditions for differential flatness.' *Applicable Algebra in Engineering, Communication and Computing* 2011;22(1):47–90.

Chapter 8

Cooperative path-following control with logic-based communications: theory and practice

Francisco C. Rego[1], *Nguyen T. Hung*[1], *Colin N. Jones*[2], *António M. Pascoal*[1], *and António Pedro Aguiar*[3]

Abstract

This chapter introduces an event-driven, logic-based communication system for decentralized control of a network of nonlinear systems (agents) with the objective of driving their outputs along predefined paths at desired speeds, while holding a desired formation pattern compatible with the paths. An extended cooperative path following (CPF) framework is adopted where communications among agents take place at discrete time instants, instead of continuously. The communication system takes into account explicitly the topology of the communications network, the fact that communications are discrete, and the cost of exchanging information among agents. The theoretical framework adopted allows for the consideration of communication losses and bounded delays. Conditions are derived under which the resulting multi-agent closed-loop system is input-to-state stable, that is stable and with guaranteed levels of performance in the presence of bounded external disturbances and measurement noise. The set-up derived is used to solve the problem of CPF control of multiple underactuated autonomous marine vehicles. The results of experimental field tests with a group of marine vehicles are presented and discussed.

8.1 Introduction

The advent of miniaturized sensors and actuators and the availability of powerful embedded computer systems have paved the way for the development of a new breed

[1]Department of Electrical and Computer Engineering, Laboratory of Robotics and Engineering Systems (LARSyS), ISR/IST, University of Lisbon, Portugal
[2]Institute of Mechanical Engineering (IGM), School of Engineering (STI), Automatic Control Laboratory, École Polytechnique Fédérale de Lausanne, Switzerland
[3]Department of Electrical and Computer Engineering, Faculty of Engineering, University of Porto, Portugal

of autonomous vehicles capable of working in cooperation towards the execution of challenging scientific and commercial missions at sea. At the core of the systems for multiple vehicle operations are those in charge of cooperative navigation and control. Among the latter, cooperative path following (PF) (CPF) systems are pervasive in a number of applications that include automated geotechnical surveying, marine habitat mapping, and deep sea surveying, to name but a few, see for example [1] and the references therein.

In its essence, the CPF problem can be briefly described as follows: given N autonomous vehicles and different spatial paths assigned to them, derive control laws to drive and maintain the vehicles on their paths with desired speed profiles, while adopting a specified, possibly time-varying formation. In the literature, [1,2] offer a theoretical overview of the subject and introduce techniques to solve the CPF problem. Different solutions to the CPF and similar problems can be found in [3–9]. Our key objective in this chapter is to solve the CPF problem under severe communication constraints, when only limited amounts of data can be exchanged among the vehicles per unit of time. Bandwidth limitations of this kind are particularly stringent in underwater applications since communication takes place over low bandwidth, short range communication channels that exhibit intermittent failures, multi-path effects, and delays.

A common strategy to solve the CPF problem consists of decoupling it in two subproblems: (i) a PF problem, where the goal is to derive closed loop control laws to drive each vehicle to and follow its assigned path while tracking a path-dependent speed profile and (ii) a multiple vehicle coordination problem, where the objective is to adjust the speed of each vehicle based on information exchanged with its neighbours, so as to achieve the desired formation pattern. The PF problem has been extensively addressed in the literature, see for example [10,11]. The coordination problem, however, requires work to overcome the problems that are naturally imposed by the communications network, both from a theoretical and practical standpoint. Some of the issues that stand in the way of CPF system implementation due to communication constraints have been addressed in the literature using graph theory to model the communications network and Lyapunov-based techniques to cope with intermittent communication failures and switching topologies; see for example [1,2,7].

This chapter extends the CPF framework discussed in [1] to take into account the fact that communication among vehicles occurs at discrete instants of time, instead of continuously. In this respect, the results of this chapter go further than those in [1], where communication failures and switching topologies were considered, but communications take place continuously. The goal of this chapter is also to reduce the frequency of information exchange among vehicles. To this end, we borrow some key ideas introduced in [12,13], where the authors tackled the problem of distributed control by resorting to local controllers that use the state of the corresponding local system, together with state estimates of the neighbouring systems it communicates with. The communication strategy adopted in the above references assumes that each system has an internal estimator of its own state, synchronized with that of its neighbours. Therefore, since each agent has access to its own state and its estimates as perceived by its neighbours, this setup allows each system to determine the estimation error of its neighbours by taking the difference between the actual and the estimated state. The communication logic consists of only transmitting information when the estimation error exceeds a certain threshold. With this method, communication occurs asynchronously at discrete instants of time.

The above strategy avoids communicating at a fixed rate and does so by triggering communications only when certain conditions are met. The same basic principle is inherent to different, yet related, work on event-triggered control and communications. The general idea of event-triggering, where a task such as broadcasting a variable or applying a control input only when a certain condition is satisfied, has been the subject of intensive work, as reported in [14–16] for multiple systems. In the context of control of multiple agents, it is important that each agent be able to compute its own triggering condition, see e.g. [12,13,17]. The work reported in [18–20] uses the principle of self-triggered communications to reach consensus of single or double integrators. The application of event-triggered communications to the problem of consensus of linear or nonlinear multi-agent systems is analysed in [21–23].

We present the theoretical framework behind CPF control with logic-based communications. We apply this concept to marine vehicles and prove stability of the devised control laws even in the presence of packet losses and delays. The present chapter is strongly motivated by and extends previous work by some of the authors in [24]. In particular, we provide full proofs of stability of the CPF algorithms that consider the interconnection between the coordination controller and the PF controller. Furthermore, in the stability proofs, we consider explicitly packet losses and delays in the communications and provide an upper bound for the delays that guarantee closed-loop system stability.

To show the efficacy of the CPF algorithm with logic-based communications that we propose, we implemented and tested the algorithm with three Medusas, a class of autonomous marine vehicles (AMVs) that have been employed extensively in many European union projects such as MORPH [25], CADDY [26], and WiMUST [27].

8.1.1 Chapter structure

The chapter is organized as follows: In Section 8.2, we propose a general control architecture for CPF with logic-based communications. In Section 8.3, the control objectives are formulated rigorously in terms of a CPF problem and conditions are derived under which the CPF control system (CPFCS) proposed solves this problem, even when communications are asynchronous. In Section 8.4, we give a specific example of a CPFCS design which solves the CPF problem for a general class of AMVs. Finally, in Section 8.5, we describe the results of field tests with three Medusa AMVs aimed at assessing the efficacy of the CPFCS architecture for multiple vehicle operations.

8.1.2 Notation

This section summarizes the notation used throughout the chapter. The symbol $\|\cdot\|$ represents the 2-norm of a vector of real numbers. The notation $|\cdot|$ represents the cardinality of a set. I_M represents an $M \times M$ identity matrix, and $\mathbf{1}$ represents a $N \times 1$ vector with ones in every entry. When clear from the context, the superscript of a variable, e.g. X^i, refers to the node index of that variable, where $i \in \mathcal{N} := \{1, \ldots, N\}$. In this context, when $i \in \mathcal{N} := \{1, \ldots, N\}$, the operator col $(\cdot)$ is defined as col $(X^i) := \left[X^{1^T}, \ldots, X^{N^T}\right]^T$, and the operator diag (X^i) yields a block diagonal matrix whose diagonal elements are $X^1, \ldots, X^N$. Given $a, b \in \mathbb{R}$, we use the notation $a \oplus b := \max(a, b)$. Given the interval $I \subset [0, \infty)$, $\|u\|_I$ denotes the essential supremum norm of a signal $u : [0, \infty) \to \mathbb{R}^n$, that is $\|u\|_I := \operatorname{ess\,sup}_{t \in I} \|u(t)\|$. For a piecewise continuous signal $v : \mathbb{R} \to \mathbb{R}^m$, with $m \geq 1$, we will use the notation $v(t^+) := \lim_{s \to t^+} v(s)$ and $v(t^-) := \lim_{s \to t^-} v(s)$. We denote the class of continuous differentiable functions up to some order n, of dimension m, as $\mathscr{C}_m^n$. A continuous function $\alpha : [0, \infty) \to [0, \infty)$ is said to belong to class $\mathscr{K}$, or $\alpha \in \mathscr{K}$ if it is strictly increasing and $\alpha(0) = 0$. If in addition $\lim_{r \to \infty} \alpha(r) = \infty$, then α is said to be of class $\mathscr{K}_\infty$. A continuous function $\beta : [0, \infty) \times [0, \infty) \to [0, \infty)$ is said to belong to class $\mathscr{KL}$, or $\beta \in \mathscr{KL}$, if, for each fixed t, the mapping $\beta(r; t)$ belong to class $\mathscr{K}$ with respect to r and, for each fixed r, the mapping $\beta(r; t)$ is decreasing with respect to t and $\beta(r; t) \to 0$ as $t \to \infty$. For a function $f : \mathbb{R} \to \mathbb{R}^n$, we define the notation $f^{[l]}(x) := \left[f(x), (\partial f / \partial x)(x), \ldots, (\partial^l f / \partial x^l)(x)\right]$.

We consider a very general setup for networked control, consisting of (i) a set of nodes (agents) $\mathcal{N}$, with cardinality $N := |\mathcal{N}|$ and (ii) a directed communication network between nodes $(\mathcal{N}, \mathscr{A})$, where $\mathscr{A} \subseteq \mathcal{N} \times \mathcal{N}$ is the set of node pairs describing the connections among the nodes (each standing for a data link), i.e. node i can transmit data to node j if and only if $(i, j) \in \mathscr{A}$. Let the in-neighbour set, $\mathcal{N}^i$, be the index set of the agents that transmit data to agent i, i.e. $\mathcal{N}^i := \{j : (j, i) \in \mathscr{A}\}$. Conversely, the out-neighbour set of agent i, defined as $\{j : (i, j) \in \mathscr{A}\}$, is the set of agents which receive data from agent i. We also define the network maximum degree as $d^* := \max_{i \in \mathcal{N}} \left(|\mathcal{N}^i|\right)$. For the general setup proposed in Section 8.3, we consider that the network is directed. However, in the following sections that deal with the particular controller design for AMVs, for practical reasons we consider that the network is undirected, i.e. $(j, i) \in \mathscr{A}$ implies that $(i, j) \in \mathscr{A}$. In the case of an undirected network, the in-neighbours of an agent are the same as the out-neighbours of that agent, and therefore we will simply refer to them as neighbours.

We define a path as an ordered sequence of nodes such that any pair of consecutive nodes in the sequence is an edge of the graph, i.e. a directed path of length n from node i to node j is a sequence of edges $(i_1, i_2), (i_2, i_3), \ldots, (i_{n-1}, i_n) \in \mathscr{A}$, where $i_1 = i$, $i_n = j$. A node of a network is globally reachable if it can be reached from any other node by traversing a path. A network is connected if every node is globally reachable. The adjacency matrix of a graph, denoted $\boldsymbol{A}$, is a square matrix with rows and columns indexed by the nodes such that the i,j-entry of $\boldsymbol{A}$ is 1 if $j \in \mathcal{N}^i$ and zero otherwise. The degree matrix D of a graph $(\mathcal{N}, \mathscr{A})$ is a diagonal matrix where the i-entry equals $|\mathcal{N}^i|$, the cardinality of the set of neighbours of i, $\mathcal{N}^i$. The Laplacian of a graph is defined as $\boldsymbol{L} := \boldsymbol{D} - \boldsymbol{A}$. Thus, every row sum of $\boldsymbol{L}$ equals zero, that is, $\boldsymbol{L}\mathbf{1} = \mathbf{0}$. It is well known that if $(\mathcal{N}, \mathscr{A})$ is connected, $\boldsymbol{L}$ has a simple eigenvalue at zero with an associated eigenvector $\mathbf{1}$ and the remaining eigenvalues are all positive, see e.g. [28].

8.2 Cooperative path-following control system architecture

We propose a CPF control architecture for a group of N decoupled agents, Σ_i; $i \in \mathcal{N}$ modelled by general systems of the form

$$\Sigma_i: \quad \dot{x}_i = F_i(x_i, u_i, w_i), \tag{8.1a}$$

$$y_i = H_i(x_i, u_i, v_i), \tag{8.1b}$$

$$z_i = J_i(x_i), \tag{8.1c}$$

where $x_i \in \mathbb{R}^{n_i}$ denotes the state of agent i, $u_i \in \mathbb{R}^{m_i}$ its control input, $z_i \in \mathbb{R}^{q_i}$ is the output to be controlled, $y_i \in \mathbb{R}^{p_i}$ is the measured noisy output available for control, w_i is an input disturbance, and v_i is the measurement noise. Loosely speaking, in the absence of disturbances and measurement noise, the output z_i is a variable that we require to reach and follow a desired feasible spatial path $z_{d_i}: \mathbb{R} \to \mathbb{R}^{q_i}$ parametrized by $\gamma_i \in \mathbb{R}$. For example, in the case of a robot moving in 2D, z_i is the position of the centre of mass of the vehicle which we require to converge to and follow a desired geometric path z_{d_i} with a given speed assignment along the path. In the above, F_i, H_i, and J_i are appropriately defined functions.

We consider an architecture for CPF control that consists of a control system for each agent which communicates with its set of neighbours. The innovation consists of introducing a communication system which considers asynchronous communications among agents. The objective of the resulting CPFCS is to drive, in the presence of disturbances and measurement noise, the output of each agent z_i to converge to and remain inside a tube centred around the desired path $z_{d_i}(\gamma_i)$, while ensuring that its rate of progression $\dot{\gamma}_i$ also converges to and remains inside a tube centred around the desired speed profile $v_r(\gamma_i)$. Notice that $v_r(\gamma_i)$, together with the path parametrization function, define the desired speed profile along the path in Cartesian coordinates. Additionally, the CPFCS must also guarantee that the path variables γ_i; $i \in \mathcal{N}$ are synchronized, that is, all the coordination errors $\gamma_i - \gamma_j$; $i,j \in \mathcal{N}$ converge to and remain inside a ball around the origin, thus ensuring approximate consensus of the path variables. The path variables γ_i are often called parameterizing variables, PF variables or, given their role in the coordination of agents, coordination states. By enforcing coordination of the PF variables, a proper choice of the latter will enforce desired geometric formations along the paths, see [1] for a discussion of this and related issues. For proper formation control, upon achieving consensus on the coordination states, we wish the time derivatives of γ_i to be equal for all $i \in \mathcal{N}$. This is achieved by guaranteeing that $v_r(\cdot)$ is the same for all the agents. Notice that the speed $v_r(\gamma_i)$ is not the actual vehicle speed in Cartesian coordinates, for it expresses the desired rate at which parameter γ_i changes.

The architecture for a general CPFCS proposed in this chapter is shown in Figure 8.1. The architecture consists of three interconnected subsystems for each agent:

PF controller Σ_i^{pf} – This dynamical system admits as inputs the agent's measured output y_i and its output is the agent's input u_i, computed so as to make it follow

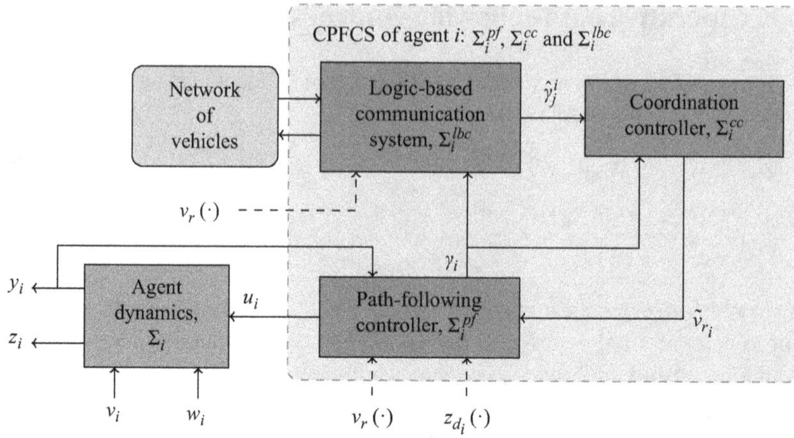

Figure 8.1 Cooperative path-following control system (CPFCS) architecture with logic-based communication system

the path at the assigned speed, asymptotically. In preparation for its connection with the coordination controller, this system produces also a path variable γ_i. Furthermore, it accepts corrective speed action from the coordination controller via the signal $\tilde{v}_{r_i}$. This corrective action is aimed at synchronizing the vehicles along their paths. Notice that the dynamics of the parameterizing variable γ_i are defined internally at this stage and play the role of an extra design knob to tune the performance of the PF control law as will become clear later. This system has knowledge of the desired spatial path $z_{d_i}(\cdot)$ and speed profile $v_r(\cdot)$. The manner by which this system depends on the desired path and speed profile will become clear in the next section.

Coordination controller Σ_i^{cc} – This is a distributed control law whose inputs are the path variable γ_i, and estimates $\hat{\gamma}_j^i$ of the coordination states γ_j; $j \in \mathcal{N}^i$. Its output is the speed correction signal $\tilde{v}_{r_i}$, which is used to synchronize agent i with its neighbours.

Logic-based communication system Σ_i^{lbc} – This dynamical system serves as an interface between agent i and its neighbours. To this end, it receives and sends appropriate information to the latter asynchronously and computes estimates of the path variables $\hat{\gamma}_l$. The information communicated to the neighbours and the communication triggering condition (CTC) depend only on local information, namely on the path variable γ_i. Its output is an estimate $\hat{\gamma}_j$ of the coordination states of the neighbouring agents γ_j; $j \in \mathcal{N}^i$. This communication system computes its estimates based on the received information from the neighbours. This system has knowledge of the desired speed profile $v_r(\cdot)$. Again, the manner by which this system depends on the desired speed profile will become clear in the next section.

In order to assess the conditions required for CPFCS design, its objectives are rigorously defined in the form of a CPF problem with logic-based communications in the next section. Our work builds upon the results obtained in [1], where the CPF problem was defined and conditions were derived under which the underlying PF and coordination controllers solve the CPF problem. In the present work, in Section 8.3, we generalize the framework developed in [1] to solve the CPF problem, considering bounded estimation errors of the path variables. We also derive a method aimed at guaranteeing that the estimation error is bounded, even when asynchronous communications take place among the agents.

8.3 Problem statement

The purpose of the PF and coordination controllers is to solve the CPF problem, assuming that each agent estimates the coordination state of its in-neighbours with bounded estimation error. In this section, we review the concepts of PF and coordination controllers introduced in [1] and consider the case where the in-neighbours' path-variables are not transmitted continuously but are instead estimated with information given through a logic-based communication system. We apply a small gain theorem to derive conditions under which the interconnection of the PF controller and the coordination controller yields an input to state stable (ISS) system.

8.3.1 Path-following problem

A PF controller can be considered to have two assignments, the geometric task and the dynamic task, see for example [11]. The geometric task consists of driving the agent output z_i to a desired path z_{d_i} parametrized by a continuous scalar variable γ_i which, as mentioned before, we call the PF variable, path variable, parameterizing variable, or coordination state. The dynamic task consists of forcing the PF variables γ_i to a certain dynamic behaviour. More specifically, the dynamic task considered in this section is a speed assignment where we require the parameterizing variables to have a desired speed $v_r(\gamma_i)$. Stated mathematically, the geometric task consists of driving to a value close to zero the PF errors defined as

$$e_i := z_i - z_{d_i}(\gamma_i),$$

and the dynamic task consists of driving to close to zero the speed errors defined as

$$e_{\dot{\gamma}_i} := \dot{\gamma}_i - v_r(\gamma_i) - \tilde{v}_{r_i},$$

where, as explained before, $\tilde{v}_{r_i}$ is the speed correction signal. For the sake of clarity and rigor, in what follows, we give a formal definition of the output PF problem. This is instrumental in understanding the conditions that a PF controller must satisfy in order to perform successfully the geometric and dynamic tasks.

To formulate the problem of PF, we must take into consideration pairs of spatial paths and speed profiles along them. Allowable pairs (z_{d_i}, v_r) of a path $z_{d_i}(\cdot)$ and a speed profile $v_r(\cdot)$ must be compatible with the vehicle dynamics, in the sense that a pair is allowable if there is an input such that the vehicle will follow its assigned path at a desired speed. Moreover, the controller design methods may require certain assumptions on the paths and speed profiles. To formalize these constraints, we consider that there exists a class of admissible mission profiles $\mathcal{M}_i \subseteq \mathcal{C}^0_{n_i} \times \mathcal{C}^0_1$ for each agent $i \in \mathcal{N}$. We say that a mission is admissible if $(z_{d_i}, v_r) \in \mathcal{M}_i$. The class $\mathcal{M}_i$ depends on the type of systems Σ_i to be controlled and the control design methods and should contain any mission specification that might be required of the system. For the sake of generality, we purposefully do not define the set $\mathcal{M}_i$ at this point, since different controller design methods and different types of agents have in general different classes of admissible missions. As an example, in the controller design considered in the next section, the class of admissible missions is defined in Section 8.4.5.

Definition 8.1 (PF controller). *Consider a speed profile v_r and a path z_{d_i} such that $(z_{d_i}, v_r) \in \mathcal{M}_i$. We define a PF controller as a system Σ_i^{pf} specified by appropriately defined functions F_i^{pf}, H_i^{pf} and H_i^{γ}, where H_i^{γ} is continuous, and integers $n_v, n_z \in \mathbb{N}$ as follows:*

$$\Sigma_i^{pf} : \dot{x}_i^{pf} = F_i^{pf}\left(x_i^{pf}, y_i, \tilde{v}_{r_i}, \mathbf{v}_r^{[n_v]}(\gamma_i), \mathbf{z}_{d_i}^{[n_z]}(\gamma_i)\right), \tag{8.2a}$$

$$u_i = H_i^{pf}\left(x_i^{pf}, y_i, \tilde{v}_{r_i}, \mathbf{v}_r^{[n_v]}(\gamma_i), \mathbf{z}_{d_i}^{[n_z]}(\gamma_i)\right), \tag{8.2b}$$

$$\gamma_i = H_i^{\gamma}\left(x_i^{pf}\right), \tag{8.2c}$$

where the notation for $\mathbf{v}_r^{[n_v]}(\cdot)$ and $\mathbf{z}_{d_i}^{[n_z]}(\cdot)$ is defined in the notation section, Section 8.1.2.

Remark 8.1. *In the PF controller of Section 8.4.2, we have $n_v = 1$ and $n_z = 2$. However, one might have different values of n_v and n_z for other types of PF controllers, as seen in [11].*

Definition 8.2 (error signal). *Consider an agent Σ_i with dynamics* (8.1a)–(8.1c), *a controller Σ_i^{pf} with dynamics* (8.2a)–(8.2c), *and a prescribed speed profile v_r and a path z_{d_i} such that $(z_{d_i}, v_r) \in \mathcal{M}_i$. We define an error signal*

$$\mathbf{e}_i := F_i^{\mathbf{e}}\left(x_i, x_i^{pf}, \tilde{v}_{r_i}, \mathbf{v}_r^{[n_v]}(\gamma_i), \mathbf{z}_{d_i}^{[n_z]}(\gamma_i)\right),$$

where $F_i^{\mathbf{e}}$ is an appropriately defined function such that the PF and speed errors satisfy the condition

$$\|e_i\| \oplus \|e_{\dot{\gamma}_i}\| \le \sigma^e\left(\|\mathbf{e}_i\|\right), \tag{8.3}$$

where $\sigma^e \in \mathcal{K}_\infty$.

Remark 8.2. *A possible construction of the above error signal is given by $\mathbf{e}_i := \left[e_i^T,\ e_{\dot{\gamma}_i}\right]^T$. However, the definition of the PF problem may require other definitions of the error signal such as the one in Lemma 8.1, Section 8.4.2.*

Definition 8.3 (output PF problem). *Consider an agent Σ_i with dynamics* (8.1a)–(8.1c) *and a controller Σ_i^{pf} with dynamics* (8.2a)–(8.2c). *The controller Σ_i^{pf} is said* *to solve the output PF problem if, for every prescribed speed profile v_r and path z_{d_i} such that $(z_{d_i}, v_r) \in \mathcal{M}_i$, the error signal $\boldsymbol{e}_i$ of Definition 8.5 is ISS with respect to w_i, v_i, and $\tilde{v}_{r_i}$, that is there exist functions $\sigma_w^e, \sigma_v^e, \sigma_{\tilde{v}_r}^e \in \mathcal{K}_\infty$, and $\beta^e \in \mathcal{KL}$ such that*

$$\begin{aligned} \|\boldsymbol{e}_i(t)\| \leq \; & \beta^e\left(\|\boldsymbol{e}_i(0)\|, t\right) \\ & \oplus \sigma_w^e\left(\|w_i\|_{[0,t]}\right) \oplus \sigma_v^e\left(\|v_i\|_{[0,t]}\right) \oplus \sigma_{\tilde{v}_r}^e\left(\|\tilde{v}_{r_i}\|_{[0,t]}\right). \end{aligned} \tag{8.4}$$

Recalling that the geometric task amounts to imposing a (small) bound on the PF errors e_i and the dynamic task consists of forcing a bound on the speed errors $e_{\dot{\gamma}_i}$, it follows from condition (8.3) that both tasks are satisfied if the signal $\boldsymbol{e}_i$ can be bounded. The ISS condition (8.4) implies that, regardless of the initial condition, asymptotically (as t goes to infinity) $\boldsymbol{e}_i$ goes to a neighbourhood of $\mathbf{0}$ whose size depends only on the magnitudes of the process noise w_i, measurement noise v_i, and speed correction $\tilde{v}_{r_i}$.

8.3.2 Coordination control problem

Besides meeting the requirements of PF, that is making each agent follow a desired path z_{d_i} (albeit with a small asymptotic error) at some required speed v_r, we also require coordination of the entire group of agents so as to achieve a desired formation pattern compatible with the paths adopted. Ideally, any two agents $i, j \in \mathcal{N}$ are coordinated, are synchronized, or have reached agreement, if and only if $\gamma_i - \gamma_j = 0$. As explained before, the choice of the path parameters γ_i is crucial to define the type of formation that the vehicles will adopt. Since nullifying the coordination errors $\gamma_i - \gamma_j$ is a very strict requirement, we require instead that the coordination errors became bounded by a small number after some settling time. In preparation for a rigorous analysis of the conditions required for the coordination controller to coordinate the agents, we now define the coordination control problem formally.

We start by recalling that the coordination controller acts on the PF controller Σ_i^{pf}, through the speed correction signal $\tilde{v}_{r_i}$, with the objective of synchronizing the agents along their paths and admits as inputs the PF variable γ_i. Finally, in what concerns the interaction with the communication system, notice that the latter is responsible for providing the coordination controller with the values of the PF variables of the in-neighbours γ_j; $j \in \mathcal{N}^i$. Since we aim to save the number of communications among agents, the actual values of the PF variables of the in-neighbours are not available continuously, and the output of the communication system consists of estimates of the PF variables $\hat{\gamma}_j^i$; $j \in \mathcal{N}^i$. We define the estimation error as $\tilde{\gamma}_j^i := \hat{\gamma}_j^i - \gamma_j$.

Definition 8.4 (coordination controller). *The class of coordination controllers that we consider is described by control laws Σ_i^{cc}; $i \in \mathcal{N}$ of the form*

$$\Sigma_i^{cc} : \tilde{v}_{r_i} = H_i^{cc}\left(\gamma_i, \hat{\gamma}_j^i; \, j \in \mathcal{N}^i\right), \tag{8.5}$$

where H_i^{cc} is an appropriately defined function.

Definition 8.5 (coordination error signal). *Consider a set of N coordination controllers Σ_i^{cc}; $i \in \mathcal{N}$ given by Definition 8.4. We define a coordination error signal*

$\xi := F^{\xi}(\gamma_i;\ i \in \mathscr{N})$, where F^{ξ} is an appropriately defined function that verifies the conditions

$$\max_{i\in\mathscr{N};j\in\mathscr{N}^i}\left(\|\gamma_i-\gamma_j\|\right)\leq\sigma^{\xi}\left(\|\xi\|\right),\tag{8.6a}$$

$$\max_{i\in\mathscr{N}}\left(\|\tilde{v}_{r_i}\|\right)\leq\sigma_{\xi}^{\tilde{v}_r}\left(\|\xi\|\right)\oplus\max_{i\in\mathscr{N};j\in\mathscr{N}^i}\left(\sigma_{\gamma}^{\tilde{v}_r}\left(\|\tilde{\gamma}_j^i\|\right)\right),\tag{8.6b}$$

where $\sigma^{\xi}, \sigma_{\xi}^{\tilde{v}_r}, \sigma_{\gamma}^{\tilde{v}_r} \in \mathscr{K}_{\infty}$.

Remark 8.3. *Note that a possible construction of the coordination error signal ξ is*

$$\xi := \mathrm{col}\left(\gamma_i-\gamma_j;\ i\in\mathscr{N}, j\in\mathscr{N}^i\right).$$

However, the definition of the coordination problem may require other definitions of the coordination error signal such as the one in Lemma 8.2, Section 8.4.3.

Definition 8.6 (coordination problem). *Consider a set of N agents $\Sigma_i;\ i \in \mathscr{N}$ with dynamics* (8.1a)–(8.1c) *equipped with PF controllers $\Sigma_i^{pf};\ i \in \mathscr{N}$ of Definition 8.1 and coordination controllers $\Sigma_i^{cc};\ i \in \mathscr{N}$ of Definition 8.4, together with the error signals $\boldsymbol{e}_i;\ i \in \mathscr{N}$ and the coordination error signal ξ of Definitions 8.2 and 8.5, respectively. We say that the coordination controllers $\Sigma_i^{cc};\ i \in \mathscr{N}$ solve the coordination problem if, for every prescribed speed profile v_r and paths $z_{d_i};\ i \in \mathscr{N}$ satisfying $(z_{d_i}, v_r) \in \mathscr{M}_i;\ i \in \mathscr{N}$, there exist functions $\beta^{\xi} \in \mathscr{KL}$ and $\sigma_{\gamma}^{\xi}, \sigma_{e}^{\xi} \in \mathscr{K}_{\infty}$ such that the evolution of the coordination error signal ξ satisfies the inequality*

$$\|\xi(t)\|\leq\beta^{\xi}\left(\|\xi(0)\|, t\right)\oplus\max_{i\in\mathscr{N};j\in\mathscr{N}^i}\left(\sigma_{\gamma}^{\xi}\left(\|\tilde{\gamma}_j^i\|_{[0,t]}\right)\oplus\sigma_{e}^{\xi}\left(\|\boldsymbol{e}_i\|_{[0,t]}\right)\right).\tag{8.7}$$

In plain terms, our approach to coordination amounts to imposing an upper bound on the coordination errors $\|\gamma_i - \gamma_j\|;\ \forall i \in \mathscr{N}; j \in \mathscr{N}^i$. From the properties (8.6a) and (8.6b), the coordination errors can be bounded if a bound can be found for the error signal ξ. The ISS condition (8.7) implies that if the error signal $\boldsymbol{e}_i$ and the estimation errors $\tilde{\gamma}_j^i$ are bounded, then, after a sufficiently large time, we can compute a bound on ξ, which depends on the bounds on the error signals and the estimation errors. The impact of these bounds on the controller performance, given by $\sigma_{e}^{\xi}\left(\|\boldsymbol{e}_i\|_{[0,t]}\right)$ and $\sigma_{\gamma}^{\xi}\left(\|\tilde{\gamma}_j^i\|_{[0,t]}\right)$, is expected to be small when compared to the precision required for coordination control. This was found to be the case in the experiments of Section 8.5.

8.3.3 Cooperative path-following

At this point, it is important to note that the coordination control problem and the PF problem are not independent because the dynamics of ξ and $\boldsymbol{e}$ are interconnected. This interconnection can be clearly seen in (8.4) and (8.7) of Definitions 8.3 and 8.6, respectively. Therefore, to ensure that the interconnection of the PF and coordination controllers solves the geometric and dynamic tasks and coordination is achieved, it is required to solve the two problems simultaneously, taking explicitly into account the connection among them. The CPF problem defined next, if solved, guarantees that the objectives specified in terms of the dynamic and geometric tasks and the coordination problem are achieved simultaneously in the presence of bounded process and measurement noise.

Definition 8.7 (CPF problem). *Consider a set of N agents Σ_i; $i \in \mathcal{N}$ with dynamics* (8.1a)–(8.1c)*, equipped with PF controllers Σ_i^{pf}; $i \in \mathcal{N}$ and coordination controllers Σ_i^{cc}; $i \in \mathcal{N}$ defined by* (8.2a)–(8.2c) *and* (8.5)*, respectively, and the error signals $\boldsymbol{e}_i$; $i \in \mathcal{N}$ and coordination error signal ξ of Definitions 8.2 and 8.5, respectively. We say that the PF and coordination controllers, Σ_i^{pf}; $i \in \mathcal{N}$ and Σ_i^{cc}; $i \in \mathcal{N}$, solve the CPF problem if, for every prescribed speed profile v_r and paths z_{d_i}; $i \in \mathcal{N}$ satisfying $(z_{d_i}, v_r) \in \mathcal{M}_i$; $i \in \mathcal{N}$, $\bar{\boldsymbol{e}} := \left[\mathrm{col}\,(\boldsymbol{e}_i)^T, \xi^T\right]^T$ is ISS with respect to w_i, v_i, and $\tilde{\gamma}_j^i$, that is if there exist functions $\sigma_\gamma^{\bar{e}}, \sigma_w^{\bar{e}}, \sigma_v^{\bar{e}} \in \mathcal{K}_\infty$, and $\beta^{\bar{e}} \in \mathcal{KL}$ such that*

$$\begin{aligned} \|\bar{\boldsymbol{e}}(t)\| &\leq \beta^{\bar{e}}\left(\|\bar{\boldsymbol{e}}(0)\|, t\right) \\ &\oplus \max_{i \in \mathcal{N}; j \in \mathcal{N}^i} \left(\sigma_w^{\bar{e}}\left(\|w_i\|_{[0,t]}\right) \oplus \sigma_v^{\bar{e}}\left(\|v_i\|_{[0,t]}\right) \oplus \sigma_\gamma^{\bar{e}}\left(\|\tilde{\gamma}_j^i\|_{[0,t]}\right)\right). \end{aligned} \tag{8.8}$$

The geometric and dynamic tasks are satisfied and coordination is achieved if e_i, $e_{\dot{\gamma}_i}$ and $\gamma_i - \gamma_j$; $j \in \mathcal{N}^i$ are bounded. The conditions (8.3), (8.6a) and (8.6b) guarantee that if the error signal $\bar{\boldsymbol{e}}$ is bounded then e_i, $e_{\dot{\gamma}_i}$ and $\gamma_i - \gamma_j$; $j \in \mathcal{N}_i$ are bounded. Finally, the ISS condition (8.8) guarantees that if v_i and w_i are bounded, then after a sufficiently large time, we can compute a bound on $\bar{\boldsymbol{e}}$ which depends only on v_i, w_i, and the estimation error $\tilde{\gamma}_i^j$; $j \in \mathcal{N}^i$. Therefore, if the CPF problem is solved, for bounded measurement and process noise (v_i and w_i, respectively), the PF and coordination controllers satisfy the geometric and dynamic tasks and achieve coordination.

The following theorem gives conditions under which a set of PF and coordination controllers solves the CPF problem, taking into account the interconnection of the PF and coordination controllers which solve the PF and coordination problems, respectively.

Theorem 8.1 (CPF). *Consider a set of N agents Σ_i; $i \in \mathcal{N}$ with dynamics* (8.1a)–(8.1c)*, equipped with PF controllers Σ_i^{pf}; $i \in \mathcal{N}$ and coordination controllers Σ_i^{cc}; $i \in \mathcal{N}$ defined by* (8.2a)–(8.2c) *and* (8.5)*, respectively. Assume that the PF and coordination controllers solve the PF and coordination problems, i.e. inequalities* (8.3), (8.4), (8.6a) *and* (8.6b)*. Suppose further that $\tilde{\gamma}_j^i := \hat{\gamma}_j^i - \gamma_j$ are bounded for all $i \in \mathcal{N}$ and $j \in \mathcal{N}^i$ and*

$$\sigma_{\tilde{v}_r}^{e} \circ \sigma_{\xi}^{\tilde{v}_r} \circ \sigma_e^{\xi}(r) < r, \ \forall r > 0. \tag{8.9}$$

Then, the PF and coordination controllers, Σ_i^{pf} and Σ_i^{cc}, solve the CPF problem.

Proof. From (8.4) and (8.7), we conclude that the PF and coordination controllers can be viewed as two interconnected ISS systems with outputs $\boldsymbol{e}$ and $\tilde{v}_r := \mathrm{col}\left(\tilde{v}_{r_i}\right)$. A straightforward application of the small-gain theorem in [29,30] implies that if (8.9) holds, then the whole interconnected system is ISS. We can then conclude that (8.8) holds because $\tilde{\gamma}_j^i$ for all $i \in \mathcal{N}$ and $j \in \mathcal{N}^i$ is bounded. □

This theorem states that the set of PF and coordination controllers solves the CPF problem, that is the interconnection of the PF controller and the coordination controller is ISS if the estimation errors are bounded, all the PF controllers and coordination controllers solve the PF and coordination problems, and inequality (8.9) holds. In the particular example of Section 8.4, inequality (8.9) is trivially satisfied since $\sigma^e_{\tilde{v}_r} \equiv 0$. For an example of CPF control design where the interconnection must be explicitly considered, see [1].

8.3.4 Logic-based communication system

In many applications that require the use of multiple marine vehicles working in cooperation, underwater communications have tight bandwidth limitations. For this reason. the latter must be addressed explicitly in the design of cooperative controls systems, such as those in charge of CPF, by reducing the rate at which data are transmitted among agents. To this end, in what follows, we assume that instead of having each agent send continuously, the PF variables $\gamma_i;\ i \in \mathcal{N}$ to its out-neighbours, messages are exchanged through a communication system at discrete instants of time asynchronously, using some logic that defines a CTC. Since the triggering condition that we consider is based on the perceived estimates in the out-neighbours of the local generalized path-variable γ_i, computed used only transmitted information, we consider the estimators to be an integral part of the communication system, henceforth called a logic-based communication system.

To guarantee that the coordination control error is smaller than some upper bound, we need to ensure that the communication system produces estimates of the in-neighbours' path variables $\hat{\gamma}^i_j;\ j \in \mathcal{N}^i$ with a bounded estimation error $\tilde{\gamma}^i_j := \hat{\gamma}_j - \gamma_j$. To meet this goal, one of the key contributions of this chapter is the introduction, in the general set-up for CPF, of a logic-based communication system capable of keeping the estimation errors $\tilde{\gamma}^j_i$ bounded.

Inspired by the setup proposed in [13], the communication system of each agent i is composed by a bank of estimators of the in-neighbours' PF variables $\hat{\gamma}^i_j;\ j \in \mathcal{N}^i$ and a communication logic block. The estimators run in open-loop most of the time but are sometimes reset (not necessarily periodically) to correct their states when measurements are received through the network.

The communication logic is responsible for assessing for each agent $i \in \mathcal{N}$, how well each of the out-neighbour agents can predict the coordination state γ_i through $\hat{\gamma}^j_i;\ i \in \mathcal{N}^j$, formally defined as the estimate of γ_i obtained by agent j, which agent i can compute using the data transmitted to agent j. Using that information, the communication logic is responsible to decide when it should communicate the actual measured value to its out-neighbours so as to guarantee that $\left\| \tilde{\gamma}^j_i(t) \right\| \leq \varepsilon$, for all j such that $i \in \mathcal{N}^j$, and all $t \geq 0$, for some positive scalar $\varepsilon > 0$.

In this section, we describe a general logic-based communications system using the principle of synchronized estimators found in [20].

Definition 8.8 (logic-based communication system). *Consider a reference speed* $v_r(\cdot)$. *Let* $t^{ij}_k, k \geq 0$ *indicate the instants of data transmission between agent i and* *a generic out-neighbour j, which occur when a CTC is satisfied. The logic-based communication system at agent i, Σ_i^{lbc}, is formally defined as the composition of the subsystems Σ_{ij}^{lbc}; $i \in \mathcal{N}^j$ and Σ_{li}^{lbc}; $l \in \mathcal{N}^i$, where Σ_{ij}^{lbc} is defined by the functions F_{ij}^{lbc}, H_{ij}^{lbc}, and J_{ij}^{lbc} as*

$$\Sigma_{ij}^{lbc}: \begin{cases} x_{ij}^{lbc}\left(t^{+}\right) = J_{ij}^{lbc}\left(x_{ij}^{lbc}, x_{li}^{lbc}, \gamma_i; l \in \mathcal{N}^i\right), & \text{for } t = t_k^{ij}, k \geq 0, \\ \dot{x}_i^{lbc} = F_i^{lbc}\left(x_i^{lbc}, v_r\left(\hat{\gamma}_i^j\right)\right), & \text{otherwise}, \end{cases} \tag{8.10a}$$

$$\hat{\gamma}_i^j = H_{ij}^{lbc}\left(x_{ij}^{lbc}\right), \tag{8.10b}$$

with H_{ij}^{lbc} continuous and J_{ij}^{lbc}, satisfying, for any admissible values of x_{ij}^{lbc}, x_{li}^{lbc}, and γ_i, the following relation:

$$H_{ij}^{lbc}\left(J_{ij}^{lbc}\left(x_{ij}^{lbc}, x_{li}^{lbc}, \gamma_i; l \in \mathcal{N}^i\right)\right) = \gamma_i.$$

Moreover, the communication system of agent i sends a message containing x_{li}^{lbc}; $l \in \mathcal{N}^i$, and γ_i, or a function of these data to its out-neighbours ($j \in \mathcal{N}$ such that $i \in \mathcal{N}^j$) at times t_k^{ij} when the following CTC holds

$$\|\tilde{\gamma}_j^i\| = \varepsilon. \tag{8.11}$$

In the definition above, information is sent from agent i at instants t_k^{ij} and x_{ij}^{lbc} is updated instantaneously. Accordingly, at instants t^{ji}; $j \in \mathcal{N}^i$ information is received by agent i from its in-neighbour j. While no data are sent or received, the communication system only uses internal information to compute the path variables estimates.

Notice that, Σ_i^{lbc} is composed by the subsystems Σ_{ij}^{lbc}; $j \in \mathcal{N}^i$ and by synchronized copies of the subsystems Σ_{ji}^{lbc}; $i \in \mathcal{N}^j$ contained in the out-neighbour agents j. By ensuring that at transmission times t_k^{ij}, on agent i and j the state x_{ij}^{lbc} is updated at the same time with the same rule, we guarantee that the estimates are kept synchronized. The purpose of this setup is that it allows each agent to compute the estimation error of its own path variable $\tilde{\gamma}_i^j$, which allows for the computation of the triggering condition (8.11).

We can now show that the logic-based communication systems of Definition 8.8. yields bounded estimation errors.

Theorem 8.2. *Given the set of logic-based communication systems Σ_i^{lbc}; $i \in \mathcal{N}$ of Definition 8.8, then for every $i \in \mathcal{N}$*

$$\left\|\tilde{\gamma}_j^i(t)\right\| \leq \varepsilon, \qquad \forall j \in \mathcal{N}^i,\ t \geq 0 \tag{8.12}$$

Proof. From Definition 8.8, one has that for any transmission time t_k^{ij}, $\left\|\tilde{\gamma}_i^j\left(t_{k+1}^{ij}\right)\right\| = \varepsilon$ and $\left\|\tilde{\gamma}_i^j\left(t_k^{ij+}\right)\right\| = 0$. Moreover, since during the interval $\left(t_k^{ij}, t_{k+1}^{ij}\right)$ x_{ij}^{lbc} and x_i^{pf} are differentiable and because H_{ij}^{lbc} and H_i^{γ} are continuous, $\tilde{\gamma}_j^i$ is continuous in that interval. From the continuity of $\tilde{\gamma}_j^i$ and the fact that $\left\|\tilde{\gamma}_i^j(t)\right\| \neq \varepsilon$ for $t \in \left(t_k^{ij}, t_{k+1}^{ij}\right)$, we observe that (8.12) is satisfied for $t \in \left(t_k^{ij}, t_{k+1}^{ij}\right)$. Applying the same reasoning for any transmission time t_k^{ij}, we show that (8.12) is satisfied for $t \geq 0$. □

The communication system introduced in Definition 8.8 represents a general architecture for non-delayed communications without packet losses. To introduce robustness with respect to packet losses, we will require that the communication systems send a reply whenever they receive a message to the sender of that same message. Therefore, in the remainder of this chapter, we assume that the communication network is undirected, that is if $i \in \mathcal{N}^j, j \in \mathcal{N}^i$.

It should also be noted that in Definition 8.8, we allow for the estimates of the PF variable of agent i, γ_i, on its out-neighbours, $\hat{\gamma}_i^j$; $i \in \mathcal{N}^j$, to be different from each other, that is it is admissible that for some $j \neq l$ such that $i \in \mathcal{N}^j \cup \mathcal{N}^l$, $\hat{\gamma}_i^j \neq \hat{\gamma}_i^j$. However, in the communication system of Section 8.4, all estimates of the PF variable of an agent are equal, that is for every two agents j and l such that $i \in \mathcal{N}^j \cup \mathcal{N}^l$, we have $\hat{\gamma}_i^j = \hat{\gamma}_i^l$ (except at certain moments in the case of communication losses). An example of communication system where the values of $\hat{\gamma}_i^j$; $i \in \mathcal{N}^j$ might be different from each other can be seen in [24].

8.4 Controller design: CPF for multiple AMVs

In this section, we describe how the general framework described in the previous section can be applied to the design of a CPF system for multiple vehicles. The design exercise is done for a group of AMVs manoeuvring in 2D.

8.4.1 Vehicle model

We start by describing the mathematical model of a class of AMVs (agents) used for motion control design. We write the kinematic equations of motion of a vehicle moving in the horizontal plane by using a global inertial coordinate frame $\{\mathcal{U}\}$ and a body-fixed coordinate frame $\{\mathcal{B}\}$ with the origin at the vehicle's centre of mass, yielding

$$\dot{x} = u_w \cos(\psi) - v_w \sin(\psi) + v_{c_x}, \tag{8.13a}$$

$$\dot{y} = u_w \sin(\psi) + v_w \cos(\psi) + v_{c_y}, \tag{8.13b}$$

$$\dot{\psi} = r, \tag{8.13c}$$

where u_w and v_w are body-fixed frame components of the vehicle's velocity with respect to the water, v_{c_x} and v_{c_y} are inertial frame components of the water current velocity, assumed to be constant, x and y are the inertial Cartesian coordinates of the vehicle's centre of mass, and ψ defines its orientation (heading angle). The kinematic equations (8.13a)– (8.13c) can be written in compact form by defining $\boldsymbol{p} := [x, y]^T$, $\boldsymbol{v}_{c_{xy}} := [v_{c_x}, v_{c_y}]^T$, and $\boldsymbol{v} := [u_w, v_w]^T$, leading to

$$\begin{aligned} \dot{\boldsymbol{p}} &= R(\psi)\boldsymbol{v} + \boldsymbol{v}_{c_{xy}} \\ \dot{\psi} &= r, \end{aligned} \tag{8.14}$$

where $R(\psi)$ is the orthonormal transformation matrix from $\{\mathscr{B}\}$ to $\{\mathscr{U}\}$, given by

$$R(\psi) := \begin{bmatrix} \cos(\psi) & -\sin(\psi) \\ \sin(\psi) & \cos(\psi) \end{bmatrix}. \tag{8.15}$$

To simplify notation, we denote $R(\psi)$ simply as R. In what follows, since we will be working mostly with body-fixed coordinates, it will be useful to express the water current in body-fixed coordinates as $\boldsymbol{v}_c := R^T \boldsymbol{v}_{c_{xy}}$. However, it should be noted that in this case, $\boldsymbol{v}_c$ is time varying, which is not the case for $\boldsymbol{v}_{c_{xy}}$.

Before proceeding into the description of the PF controller, we first consider the inner-loop dynamic task, which consists of making the vehicle's surge velocity u_w and heading rate r track-desired references, given by u_d and r_d, respectively, to be defined later. In what follows, we assume that the vehicle is equipped with inner loop controllers that satisfy the following stability assumption:

Assumption 8.1. *Let $\tilde{u} := u_w - u_d$ and $\tilde{r} := r - r_d$ be the speed and heading rate tracking errors, respectively, and v_w the sway velocity. We assume there exist positive constants $\varepsilon_{\tilde{r}}$, $\varepsilon_{\tilde{u}}$, ε_v such that for all $t \geq 0$*

$$\|\tilde{r}(t)\| \leq \varepsilon_{\tilde{r}}, \quad \|\tilde{u}(t)\| \leq \varepsilon_{\tilde{u}}, \quad \|v_w(t)\| \leq \varepsilon_{v_w}.$$

The constants $\varepsilon_{\tilde{r}}$, $\varepsilon_{\tilde{u}}$, and ε_v correspond to bounds on the tracking errors due to the dynamics of the vehicle, disturbances, or unforeseen behaviour due to plant modelling uncertainty. We also assume that each vehicle contains an observer of the water current velocity $\boldsymbol{v}_c$ which satisfies the following assumption:

Assumption 8.2. *Let $\tilde{\boldsymbol{v}}_c := \boldsymbol{v}_c - \hat{\boldsymbol{v}}_c$ be the estimation error of the water current velocity estimator. There exists a positive constant ε_{v_c} such that for all $t \geq 0$*

$$\|\tilde{\boldsymbol{v}}_c(t)\| \leq \varepsilon_{v_c}.$$

For an example of a water current estimator and its technical details, see [24]. Given the above-mentioned assumptions, the kinematics of the vehicle can be described as

$$\begin{aligned} \dot{\boldsymbol{p}} &= R\left(\begin{bmatrix} u_d + \tilde{u} \\ v_w \end{bmatrix} + \hat{\boldsymbol{v}}_c + \tilde{\boldsymbol{v}}_c\right), \\ \dot{\psi} &= r_d + \tilde{r} \end{aligned}$$

This design corresponds to the general form of an agent Σ given in (8.1a)–(8.1c) where the state x, measured output y, control input u, process disturbance w, and measurement noise v are explicitly expressed as

$$x := \begin{bmatrix} v_c \\ p \\ \psi \end{bmatrix}, \quad y := \begin{bmatrix} \hat{v}_c \\ p \\ \psi \end{bmatrix}, \quad u := \begin{bmatrix} u_d \\ r_d \end{bmatrix}, \quad w := \begin{bmatrix} \tilde{u} \\ \tilde{r} \\ v_w \end{bmatrix} \quad v := \tilde{\boldsymbol{v}}_c.$$

From Assumption 8.1, we have that $\|w\|_{[0,t]} \leq \varepsilon_w := \sqrt{\varepsilon_{\tilde{u}}^2 + \varepsilon_{\tilde{r}}^2 + \varepsilon_v^2}$ for $t \geq 0$, and from Assumption 8.2, we obtain $\|v\|_{[0,t]} \leq \varepsilon_v := \varepsilon_{v_c}$.

8.4.2 Path-following controller

To solve the problem of driving a vehicle along a desired path, the key idea exploited is to make the vehicle approach a virtual target that moves along the path. Let $z_d(\gamma)$ be the position of the target, and $v_r(\gamma)$ the desired rate of the progression of the path parameter γ. We decompose the motion-control problem into the inner-loop dynamic task mentioned in Assumption 8.1, and an outer-loop kinematic task, which consists of assigning references for surge speed and heading rate references $u := [u_d,\ r_d]^T$ and for the second derivative of the path parameter $\ddot{\gamma}$, in order to solve the PF problem.

Define the position error $e_{\mathscr{B}}$ as the difference between the positions of the vehicle and the virtual target expressed in the body frame $\{\mathscr{B}\}$, that is $e_{\mathscr{B}} := R^T\,(\boldsymbol{p} - z_d)$. The dynamics of $e_{\mathscr{B}}$ are described by

$$\dot{e}_{\mathscr{B}} = -S(r)e_{\mathscr{B}} + \begin{bmatrix} u_d + \tilde{u} \\ v_w \end{bmatrix} + \boldsymbol{v}_c - R^T\frac{\partial z_d}{\partial \gamma}\dot{\gamma}, \qquad S(r) = \begin{bmatrix} 0 & -r \\ r & 0 \end{bmatrix}, \tag{8.16}$$

where we used the fact that $\dot{R} = RS(r)$.

Following the methods adopted in [31], to make the control input $u := [u_d,\ r_d]^T$ appear in the position error dynamics, we introduce a constant design vector $\boldsymbol{\delta} := [\delta,\ 0]^T$, $\delta < 0$. From (8.16), defining $e_\delta := e_{\mathscr{B}} - \boldsymbol{\delta}$, simple computations show that the position error dynamics are given by

$$\dot{e}_\delta = -S(r)e_\delta + \Delta u - R^T\frac{\partial z_d}{\partial \gamma}\dot{\gamma} - \begin{bmatrix} 0 \\ \delta\tilde{r} \end{bmatrix} + \begin{bmatrix} \tilde{u} \\ v_w \end{bmatrix} + \hat{\boldsymbol{v}}_c + \tilde{\boldsymbol{v}}_c, \tag{8.17}$$

where

$$\Delta := \begin{bmatrix} 1 & 0 \\ 0 & -\delta \end{bmatrix}$$

In what follows, we assume that we explicitly control the virtual target speed error defined as

$$e_{\dot{\gamma}} := \dot{\gamma} - (v_r + \tilde{v}_r), \tag{8.18}$$

thereby introducing an additional control variable. We can observe that this design corresponds to defining the objective output as $z := p + R\boldsymbol{\delta}$, the PF controller Σ^{pf} with state $x^{pf} = \left[\gamma,\ e_{\dot{\gamma}}\right]^T$, and the PF error as $e := z - z_d(\gamma) = Re_\delta$.

Lemma 8.1 (PF controller). *Let $z_d \in \mathscr{C}_2^1$ be a desired path and $v_r \in \mathscr{C}_1^0$ a desired speed assignment. Consider the vehicle model described by* (8.13a)–(8.13c)*, in closed-loop with the output feedback control law composed by inner-loop controllers that satisfy Assumption 8.1, a water current estimator which satisfies Assumption 8.2, and the outer-loop controller given by*

$$\dot{e}_{\dot{\gamma}} = -k_{\dot{\gamma}}e_{\dot{\gamma}} + \frac{1}{c_{\dot{\gamma}}}e_\delta^T R^T\frac{\partial z_d}{\partial \gamma}(\gamma), \tag{8.19}$$

$$u = \Delta^{-1}\left(-K_k e_\delta - \hat{\boldsymbol{v}}_c + R^T\frac{\partial z_d}{\partial \gamma}(\gamma)(v_r + \tilde{v}_r)\right), \tag{8.20}$$

where $K_k := \text{diag}\left(k_x, k_y\right)$, *and the positive scalar design parameters* k_x, k_y, $k_{\dot{\gamma}}$, *and* $c_{\dot{\gamma}}$. *Then, the error vector*

$$\boldsymbol{e} := \begin{bmatrix} e_\delta \\ \sqrt{c_{\dot{\gamma}}} e_{\dot{\gamma}} \end{bmatrix} = \begin{bmatrix} R^T e \\ \sqrt{c_{\dot{\gamma}}} e_{\dot{\gamma}} \end{bmatrix}$$

is ISS with respect to w and v, that is there exist functions $\sigma_w^e, \sigma_v^e \in \mathscr{K}_\infty, \beta^e \in \mathscr{KL}$ *such that*

$$\|\boldsymbol{e}\| \leq \beta^e\left(\|\boldsymbol{e}(0)\|, t\right) \oplus \sigma_w^e\left(\|w\|_{[0,t]}\right) \oplus \sigma_v^e\left(\|v\|_{[0,t]}\right). \tag{8.21}$$

Therefore, the control laws (8.19) *and* (8.20) *solve the PF problem of Definition 8.3.*

Proof. See the Appendix. □

With this strategy, the evolution of the position of the virtual target z_d depends on the position error e_δ in which if the vehicle is ahead/behind the desired position, the virtual target moves faster/slower towards the position of the vehicle. That is, unlike trajectory tracking, where the virtual target moves along the path at the speed reference v_r, in PF the target dynamics actively aid in the convergence of the vehicle towards the path.

8.4.3 Coordination controller

Consider now the coordination control problem with a communication topology defined by a graph $(\mathscr{N}, \mathscr{A})$. We assume that the graph is undirected, that is the communication links are bidirectional, in which if $i \in \mathscr{N}^j, j \in \mathscr{N}^i$.

Remark 8.4. *The assumption of an undirected graph is important in this chapter for the case where packet losses exist and we require that the vehicles send an acknowledgement message to their in-neighbours acknowledging that a data message was received. If we assume that there are no packet losses, this assumption can be lifted.*

Using a Lyapunov-based design and backstepping techniques, we propose a decentralized feedback law for $\tilde{v}_{r_i}$ as a function of the information obtained from the neighbouring agents. Following [32], let Q be an $(N-1) \times N$ matrix with orthonormal rows that are each orthogonal to $\mathbf{1}$, that is

$$Q\mathbf{1} = \mathbf{0}, \qquad QQ^T = I_{N-1}, \qquad Q^T Q = \Pi := I_N - \frac{1}{N}\mathbf{1}\mathbf{1}^T. \tag{8.22}$$

We introduce the coordination or synchronization error vector

$$\zeta := Q\gamma, \tag{8.23}$$

where $\gamma := \text{col}\left(\gamma_i\right)$. From (8.18), the dynamics of the coordination subsystem can be written in vector form as

$$\dot{\gamma} = \bar{v}_r + \tilde{v}_r + \bar{e}_{\dot{\gamma}} \tag{8.24}$$

where $\bar{v}_r := \text{col}\,(v_r(\gamma_i))$, $\bar{e}_{\dot{\gamma}} := \text{col}\left(e_{\dot{\gamma}_i}\right)$, and $\tilde{v}_r := \text{col}\left(\tilde{v}_{r_i}\right)$. Consider the control Lyapunov function $V := (1/2)\xi^T\xi$. Computing its time-derivative yields

$$\dot{V} = \xi^T Q(\bar{v}_r + \tilde{v}_r + \bar{e}_{\dot{\gamma}}). \tag{8.25}$$

To make ξ ISS with respect to input $\bar{e}_{\dot{\gamma}}$, a natural choice would be $\tilde{v}_r = -k\boldsymbol{L}Q^T \xi = -k\boldsymbol{L}\Pi\gamma = -k\boldsymbol{L}\gamma$, where $\boldsymbol{L}$ is the Laplacian of the graph $(\mathcal{N}, \mathcal{A})$ and k is a positive scalar, or equivalently, $\tilde{v}_{r_i} = -k\sum_{j\in\mathcal{N}^i}(\gamma_i - \gamma_j)$ (the so-called neighbouring rule). To reduce the communication rate using a logic-based dynamical system, we will lift the assumption that each agent receives information from its neighbourhood continuously. We assume instead that it relies on the estimates $\hat{\gamma}_j^i$; $j \in \mathcal{N}^i$. Therefore, defining the estimation error $\tilde{\gamma}_j^i := \hat{\gamma}_j^i - \gamma_j$, the coordination control law becomes

$$\tilde{v}_{r_i} = -k\sum_{j\in\mathcal{N}^i}\left(\gamma_i - \hat{\gamma}_j^i\right) = -k\sum_{j\in\mathcal{N}^i}(\gamma_i - \gamma_j) + k\sum_{j\in\mathcal{N}^i}\tilde{\gamma}_j^i \tag{8.26}$$

or, in vector form, $\tilde{v}_r = -k\boldsymbol{L}Q^T\xi + k\bar{\gamma}$, where $\bar{\gamma} := \text{col}\left(\sum_{j\in\mathcal{N}^i}\tilde{\gamma}_j^i\right)$. The time derivative of V becomes

$$\dot{V} = -k\xi^T Q\boldsymbol{L}Q^T\xi + \xi^T Q\left(v_r + \bar{e}_{\dot{\gamma}} + k\bar{\gamma}\right). \tag{8.27}$$

In this case, the term $-k\xi^T Q\boldsymbol{L}Q^T\xi$ is negative definite provided that the graph that models the constraints imposed by the network topology among the agents is connected, see e.g. [28]. The following result is obtained.

Lemma 8.2 (coordination). *Consider a set of N agents Σ_i; $i \in \mathcal{N}$ with dynamics* (8.1a)–(8.1c)*, a set of controllers Σ_i^{pf}; $i \in \mathcal{N}$ with dynamics* (8.2a)–(8.2c)*, and the error signal of Definition 8.2, and assume that v_r is globally Lipschitz continuous with a Lipschitz constant smaller than or equal to l. If $(\mathcal{N}, \mathcal{A})$ is connected, then the coordination control of problem of Definition 8.6 is solved with the coordination control law* (8.26)*, that is ξ is ISS with respect to the inputs $\boldsymbol{e}_i$ and $\tilde{\gamma}_j^i$ for $k > l/\sigma_2$, where σ_2 is the second lowest singular value of $\mathbf{L}$. Stated equivalently, there exist functions $\beta^\xi \in \mathcal{KL}$, $\sigma_\gamma^\xi, \sigma_e^\xi \in \mathcal{K}_\infty$ such that the evolution of the coordination error signal ξ satisfies the inequality*

$$\|\xi(t)\| \leq \beta^\xi\left(\|\xi(0)\|, t\right) \oplus \max_{i\in\mathcal{N}: j\in\mathcal{N}^i}\left(\sigma_\gamma^\xi\left(\|\tilde{\gamma}_j^i\|_{[0,t]}\right) \oplus \sigma_e^\xi\left(\|\boldsymbol{e}_i\|_{[0,t]}\right)\right). \tag{8.28}$$

Proof. See the Appendix. □

8.4.4 Logic-based communication system

In this subsection, we present the logic-based communication system. We will start with the case where the communication links are ideal, that is, there are no delays or packet losses. We then move on to the case where there are bounded communication delays. Finally, we will describe a communication system that is robust to limited packet losses.

8.4.4.1 Ideal communication links

The logic-based communication system structure for a node i, with its neighbours j_1 to j_{d_i}, with $d_i := |\mathcal{N}^i|$ belonging to $\mathcal{N}^i$, is illustrated in Figure 8.2. The communication system is composed by observers of the PF variables of the neighbours j_1 to j_{d_i}, $\hat{\gamma}^i_{j_1}$ to $\hat{\gamma}^i_{j_{d_i}}$, and an observer of the own PF variable, $\hat{\gamma}^i_i$. The observers of the PF variables of the neighbours $\hat{\gamma}^i_j$ are reset when a message is received by the respective neighbour. Equivalently, the observer of the local PF variable, $\hat{\gamma}^i_i$, only uses data that are sent to the neighbours. The observer of the local PF variable, $\hat{\gamma}^i_i$, is used for the computation of the CTC in the following manner. The emitter compares the observed PF variable $\hat{\gamma}^i_i$ to the real γ_i, and if the norm of the difference reaches the value ε, a data message is sent to the neighbours containing the present γ_i.

Let t^i_k, $k > 0$ denote the instants of time at which agent i transmits information to the neighbours. Following the procedure described in Section 8.3, we propose for each agent i the following logic-based communication system:

$$\begin{cases} \dot{\hat{\gamma}}^j_i(t) = v_r(\hat{\gamma}_i), & \text{for } t^i_k < t < t^i_{k+1}, \\ \hat{\gamma}^j_i(t^{i+}_k) = \gamma_i(t^i_k), & \text{for } t = t^i_k. \end{cases}$$

Since the communication links are ideal, $\hat{\gamma}^i_i = \hat{\gamma}^j_i$ for all $j \in \mathcal{N}^i$ and all $t \in \mathbb{R}^+$.

We note that if we define $T_r(\gamma) := \int_0^\gamma (1/v_r(\sigma))d\sigma$ and its inverse as $\Gamma_r(t)$, that is $T_r(\Gamma_r(t)) := t$, we have for $t^i_k < t < t^i_{k+1}$ that $\hat{\gamma}_i(t) = \Gamma_r\left(t - t^i_k + T_r\left(\gamma_i\left(t^i_k\right)\right)\right)$. Therefore, the estimator can also be expressed as

$$\hat{\gamma}^j_i(t) = \Gamma_r\left(t - t^i_r(t) + T_r\left(\gamma_i\left(t^i_r(t)\right)\right)\right), \tag{8.29}$$

where $t^i_r(t) := \max_{k\in\mathbb{N}, t^i_k < t} t^i_k$ is the last time agent i sent a data message. When v_r is constant, as in the field tests of Section 8.5, the above expression simplifies to

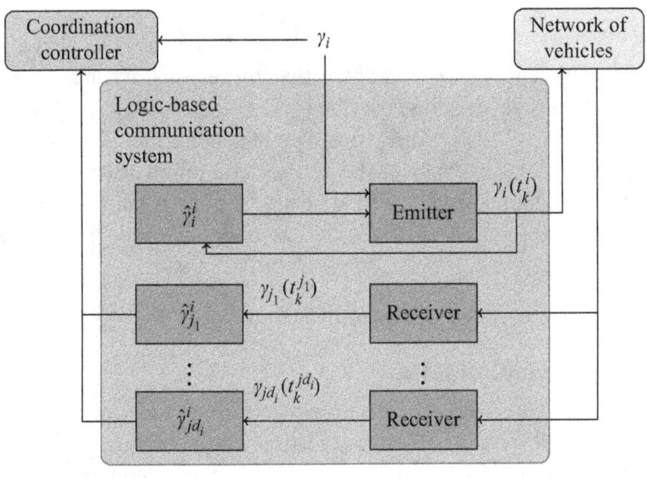

Figure 8.2 Logic-based communication system for ideal communication links

$\hat{\gamma}_i(t) = \gamma_i\left(t_r^i(t)\right) + \left(t - t_r^i(t)\right)v_r$. The choice of the estimator (8.29) is motivated by its simplicity and the fact that if, for some $t_0 > 0$, $\bar{e}_{\dot{\gamma}}(t) = \mathbf{0}$ and $\xi(t) = \mathbf{0}$ for $t > t_0$, if $\hat{\gamma}_i^j(t_{\dot{\gamma}}) = \gamma_i(t_{\dot{\gamma}})$, then $\hat{\gamma}_i^j(t) = \gamma_i(t) = 0$ for $t > t_0$. However, it is possible to choose more complex reproductions of the corresponding dynamic models, see [24,33].

To bound the estimation errors, we introduce the CTC $\|\tilde{\gamma}_i^j\| \geq \varepsilon$ where $\varepsilon > 0$, with $\tilde{\gamma}_i^j := \hat{\gamma}_i^j - \gamma_i$ (see Definition 8.8). In this setup, agent i transmits to j a data message with γ_i at time t_k^i when the CTC is satisfied. Note that the post reset value of $\tilde{\gamma}_i^j$ is $\tilde{\gamma}_i^j\left(t_k^i\right) = 0$. Consequently, at any time, $\|\tilde{\gamma}_i^j\| \leq \varepsilon$.

8.4.4.2 Delayed information

We now consider the case where the communication channels have bounded, time-varying, and non-homogeneous delays. Consider the following situation: agent i sends data at time t_k^i and agent j receives it at time $t_k^i + \tau_k^{ij}$. We assume that

$$\tau_k^{ij} \leq \bar{\tau}, \qquad \forall i, j, k$$

where the constant $\bar{\tau} > 0$ is known a priori. We also assume that all the vehicles run synchronized clocks. This is a realistic assumption in view of the recent availability of affordable chip atomic clocks [34]. The main idea is to keep the estimators always synchronized; therefore, both the emitter and the receiver only update the estimators at some time, with the same information, at $t_k^i + \bar{\tau}$.

Suppose that at time t_k^i, agent i transmits a message that contains the following data: $\{t_k^i, \gamma_l\}$. Then, the estimators $\hat{\gamma}_i$ in agent i and its neighbours $\mathcal{N}^i$ cannot be immediately updated. This is because we must guarantee that the value of the state estimate $\hat{\gamma}_i$ can be computed in all agents in $\mathcal{N}^i$. To this end, both estimates can only be updated at time $t = t_k^i + \bar{\tau}$. Upon receiving t_k^i, the coordination state estimate $\hat{\gamma}_i^j$ running in agent j should be updated at time $t = t_k^i + \bar{\tau}$ to

$$\hat{\gamma}_i^j\left(\left(t_k^i + \bar{\tau}\right)^+\right) = \Gamma_r\left(\bar{\tau} - t_k^i + T_r\left(\gamma_i\left(t_k^i\right)\right)\right)$$

In the case of a constant v_r, this simplifies to $\hat{\gamma}_i^j\left(\left(t_k^i + \bar{\tau}\right)^+\right) = \gamma_i\left(t_k^i\right) + \bar{\tau}v_r$. With the above procedure, we guarantee that the estimators are always synchronized. The estimator can thus be described as in (8.29) where $t_r^i(t)$ is redefined as $t_r^i(t) := \max_{k \in \mathbb{N}, t_k^i + \bar{\tau} < t} t_k^i$. Therefore, one can consider the ideal case mentioned previously as a particular instance of the delayed communication case with $\bar{\tau} = 0$.

Notice that in general, $\tilde{\gamma}_i^j\left(t_k^i + \bar{\tau}\right)$ will not be zero because $e_{\dot{\gamma}_i}$ and ξ may not be zero in the interval $\left[t_k^i, t_k^i + \bar{\tau}\right)$. Thus, we can only guarantee that this technique is valid if $\bar{\tau}$ so small that $\tilde{\gamma}_i^j$ satisfies $\tilde{\gamma}_i^j\left(\left(t_k^i + \bar{\tau}\right)^+\right) < \varepsilon$. The stability guarantees for the delayed case will be stated in Theorem 8.3.

8.4.4.3 Communication losses

We now address the case when the data messages are not always received, i.e. there are packet losses. To make the communication system robust to limited communication losses, we require each agent to send an acknowledgement message upon receiving a data message. The agent which sent the data message only updates his estimators $2\bar{\tau}$ time units after the data message has been sent in case the acknowledgement messages from all the agents was received, otherwise another data message with the same content is sent. This process is repeated until the agent receives the acknowledgement messages from all the neighbours. This guarantees that the receiving agents receive a data message at some point after the CTC is satisfied.

Consider the case where at time t_k^i agent i transmits to agent j a data message, which contains the following data: $\{t_k^i, \gamma_i\}$. Upon receiving the data message and sending an acknowledgement message, the coordination state estimate $\hat{\gamma}_i^j$ running in agent j should be updated at time $t = t_k^{ij} + 2\alpha_k^{ij}\bar{\tau}$, where α_k^{ij} is the number of messages containing $\{t_k^i, \gamma_i\}$ sent by agent i before agent j receives it, to

$$\hat{\gamma}_i^j\left(\left(t_k^i + 2\alpha_k^{ij}\bar{\tau}\right)^+\right) = \Gamma_r\left(2\alpha_k^{ij}\bar{\tau} - t_k^i + T_r\left(\gamma_i\left(t_k^i\right)\right)\right).$$

Equivalently, we define as α_k^{ii} the number of messages containing $\{t_k^i, \gamma_i\}$ that agent i sends after t_k^i, before at least one acknowledgement from all of its neighbours is received. The estimator can be represented formally as

$$\hat{\gamma}_i^j(t) = \Gamma_r\left(t - t_r^{ij}(t) + T_r\left(\gamma_i\left(t_r^{ij}(t)\right)\right)\right), \tag{8.30}$$

where $t_r^{ij}(t) := \max_{k\in\mathbb{N}, t_k^i + 2\alpha_k^{ij}\bar{\tau} < t} t_k^i$. In what follows, we assume that for a finite number $N_{\max}$ of consecutive data messages sent containing the same data, at least one acknowledgement message from each of the neighbours is received. That is, we assume that there exists an integer $N_{\max}$ such that for any $i \in \mathcal{N}$ and $j \in \mathcal{N}^i$ and $k \geq 0$, $\alpha_k^{ij} \leq \alpha_k^{ii} \leq N_{\max}$.

Note that if one agent receives a message while another misses it, then, unlike the case without communication losses, there is a brief period when the estimators are desynchronized, that is $\hat{\gamma}_i^j \neq \hat{\gamma}_i^l$ for $j \neq l$. However, the synchronization is recovered after the acknowledgement messages of all the neighbour agents are received. As in the case without communication losses, if $\tilde{\tau} := 2\bar{\tau}N_{\max}$ is sufficiently small so as to guarantee that $\tilde{\gamma}_i^j$ satisfies $\tilde{\gamma}_i^j\left(\left(t_k^i + 2\alpha_k^{ij}\bar{\tau}\right)^+\right) < \varepsilon$, it follows that the estimation error is bounded. We can now show that the proposed logic-based communication system, in the presence of communication losses and delays, yields bounded estimation error if certain conditions apply.

Theorem 8.3. *Consider a set of N agents Σ_i; $i \in \mathcal{N}$ with dynamics* (8.1a)–(8.1c), *equipped with CPFCS structures consisting of PF controllers Σ_i^{pf}; $i \in \mathcal{N}$ and coordination controllers Σ_i^{cc}; $i \in \mathcal{N}$ defined by,* (8.2a)–(8.2c) *and* (8.5), *respectively, that solve the CPF problem of Definition 8.7, together with the proposed logic-based communication system of this section in the presence of delayed information and communication losses. Further assume that $\sigma_\xi^{\tilde{v}_r}(s) = \alpha_\xi^{\tilde{v}_r}s$, $\sigma_\gamma^{\tilde{v}_r}(s) = \alpha_\gamma^{\tilde{v}_r}s$, $\sigma^e(s) = \alpha^e s$, and $\sigma_\gamma^{\bar{e}}(s) = \alpha_\gamma^{\bar{e}}s$, for appropriately defined positive constants $\alpha_\xi^{\tilde{v}_r}$, $\alpha_\gamma^{\tilde{v}_r}$, α^e, and $\alpha_\gamma^{\bar{e}}$. Then, for sufficiently small values of $\tilde{\tau} := 2\bar{\tau}N_{\max}$, $\|\bar{e}(0)\|$, $\max_{i\in\mathcal{N}}\|v_i\|_{[0,\infty)}$, and $\max_{i\in\mathcal{N}}\|w_i\|_{[0,\infty)}$, the CPFCS structures solve the CPF problem.*

In particular, we can compute continuous function $\bar{\alpha}^{\varepsilon} : [0, c) \to [1, \infty)$ *and* $\sigma^{\varepsilon} : [0, c) \to [0, \infty)$ *for some positive constant* c *with* $\bar{\alpha}^{\varepsilon}(0) = 1$ *and* $\sigma^{\varepsilon}(0) = 0$*, such that if*

$$\varepsilon > \sigma^{\varepsilon}(\tilde{\tau}) \left(\beta^{\bar{e}}(\|\bar{e}(0)\|, 0) \oplus \max_{i \in \mathcal{N}} \left(\sigma_w^{\bar{e}} \left(\|w_i\|_{[0,\infty)} \right) \oplus \sigma_v^{\bar{e}} \left(\|v_i\|_{[0,\infty)} \right) \right) \right) \tag{8.31}$$

holds then, for all $t \geq 0$*,*

$$\max_{i \in \mathcal{N}, j \in \mathcal{N}^i} \left\| \tilde{\gamma}_i^j(t) \right\| \leq \bar{\alpha}^{\varepsilon}(\tilde{\tau})\varepsilon. \tag{8.32}$$

Proof. See the Appendix. □

Remark 8.5. *In the ideal communications case where* $N_{\max} = \bar{\tau} = 0$*, we have that the proposed estimators of the communication system follow the structure in (8.10a) and (8.10b) with* $x_{ij}^{lbc} := \hat{\gamma}_j$*, and Theorem 8.2 guarantees that the estimation errors are bounded, and consequently from Theorem 8.1, if* (8.9) *is satisfied, the closed-loop system solves the CPF problem.*

Theorem 8.3 goes further than the ideal communications case considered in Section 8.3, since it also addresses the case where there are packet losses and delays.

Theorem 8.3 provides stability guarantees in the case of delays and packet losses. The following corollaries of Theorem 8.3 provide stability guarantees for the case of ideal communication links.

Corollary 8.1 (ideal communication). *The CPFCS structure considered in Theorem 8.3 solves the CPF problem for ideal communication links, i.e. when there are no delays or communication losses.*

Proof. Since the ideal communication case can be viewed as a particular case of that with delayed communication for which $\bar{\tau} = 0$, Theorem 8.3 provides a proof of stability of the overall system for the case with ideal communication links. Moreover, it follows from Theorem 8.3 that the boundedness assumptions on $\|\bar{e}(0)\|$, $\max_{i \in \mathcal{N}} \|v_i\|_{[0,\infty)}$, and $\max_{i \in \mathcal{N}} \|w_i\|_{[0,\infty)}$ are not required anymore since the condition (8.31) is always satisfied because $\sigma^{\varepsilon}(\tilde{\tau})(0) = 0$. □

8.4.5 Stability of the overall-closed loop system

Applying the previous results, we can now show that the CPFCS structure consisting of the PF controllers described in Section 8.4.2, the coordination controllers described in Section 8.4.3 and logic-based communication systems described in Section 8.4.4 solves, for the class of AMVs described in Section 8.4.1, the CPF problem globally in the case of ideal communication links and locally, i.e. for sufficiently small values of $\|\bar{e}(0)\|$, $\max_{i \in \mathcal{N}} \|v_i\|_{[0,\infty)}$ and $\max_{i \in \mathcal{N}} \|w_i\|_{[0,\infty)}$, the case of delayed communications. The class of admissible missions $\mathcal{M}_i$ is defined as follows: we say that $(z_{d_i}, v_r) \in \mathcal{M}_i$ if $z_d \in \mathcal{C}_2^1$ and $v_r \in \mathcal{C}_1^0$ is globally Lipschitz continuous with a Lipschitz constant lower or equal to a non-negative scalar l.

Theorem 8.4. *Consider the overall closed-loop system consisting of N AMVs with the dynamics given by (8.13a)–(8.13c). Assume the AMVs are equipped with inner loop controllers satisfying Assumption 8.1 and CPFCS structures consisting of PF controllers of the form* (8.19) *and* (8.20), *with* $k_x, k_y, k_\gamma, c_{\dot{\gamma}} > 0$, *and coordination controllers of the form* (8.26), *with* $k > l/\sigma_2$, *together with the proposed logic-based communication systems. For the case of ideal communication links or, in the presence of delayed information and/or communication losses, for sufficiently small values of* $\bar{\tau}$, $\|\bar{e}(0)\|$, $\max_{i\in\mathcal{N}} \|v_i\|_{[0,\infty)}$, *and* $\max_{i\in\mathcal{N}} \|w_i\|_{[0,\infty)}$, *the CPFCS structures solve the CPF problem, for every prescribed speed profile* v_r *and paths* z_{d_i}; $i \in \mathcal{N}$ *satisfying* $(z_{d_i}, v_r) \in \mathcal{M}_i$; $i \in \mathcal{N}$.

Proof. Lemma 8.1 implies that the set of Σ_i^{pf}; $i \in \mathcal{N}$ solves the PF problem with $\sigma^e_{\tilde{v}_r} \equiv 0$. Therefore, given Lemma 8.2 and since (8.9) holds, Theorem 8.1 implies that the CPFCSs solve the CPF problem. The conclusion of the theorem follows from a straightforward application of Theorem 8.3 and its Corollary 8.1. □

8.5 Field tests with AMVs

This section describes the results of field trials with AMVs that illustrate the efficacy of the proposed CPF algorithms with logic-based communications derived previously.

8.5.1 Test set-up

The field tests were performed with three Medusa-class AMVs built at IST. We, henceforth, refer to the vehicles as black, red, and yellow Medusas, corresponding to their colours. See Figure 8.3 for a close up and an aerial view of the vehicles while manoeuvring in coordination at sea. For simplicity of systems implementation, in the tests, the vehicles operated at the surface, acting as proxies for autonomous underwater vehicles. Each Medusa vehicle was equipped with a navigation system using global positioning system–real-time kinematic (GPS-RTK) that allowed for the computation of its position. In the set-up adopted, the vehicles communicated over Wi-Fi. The software modules were implemented using the Robot Operating System, in Python, and ran on an EPIC computer board (model NANO-PV-D5251). All Medusa-class vehicles are equipped with two thrusters on starboard and portside, to generate longitudinal forces and torque about the vertical axis. Motor drivers allow the onboard computer to assign a revolutions per minute (RPM) control signal to each thruster, which will in turn generate a desired thrust force. The Medusas are also equipped with low-level control systems that the CPF algorithms build upon. We now describe the surge speed and heading rate control loops.

Surge speed: The surge speed controller is responsible for making the vehicle track a desired surge speed command. Estimates of the state to be controlled can be obtained from a navigation filter which uses GPS data by projecting the velocity with respect to the ground on the longitudinal axis of the vehicle, and subtracting the projection of the water current vector estimate on the longitudinal axis of the vehicle.

(a) (b)

Figure 8.3 The Medusa AMVs and an aerial snapshot during the tests: (a) the Medusa-class of vehicles and (b) the three Medusas performing a cooperative manoeuver at sea

A proportional–integral (PI) control law computes the commanded common mode (CM) for the two longitudinal thrusters, where the CM is defined as the mean of the commands for the two thrusters, as a percentage of the maximum RPMs, defined as $CM(\%) := ((\text{Left RPMs}(\%) + \text{Right RPMs}(\%))/2)$, so as to make the vehicle move at the desired surge speed.

Heading rate: The heading rate controller is responsible for making the vehicle track a desired heading rate command. Estimates of the heading rate are performed using a filter with data from an attitude and heading reference system. A PI control law computes the commanded differential mode (DM) for the two longitudinal thrusters, where the DM is defined as the mean of the commands for the two thrusters, as a percentage of the maximum RPMs, defined as $DM(\%) := ((\text{Left RPMs}(\%) - \text{Right RPMs}(\%))/2)$, to make the vehicle rotate at the desired heading rate.

For further details on the technology used in the Medusa vehicles, we refer the reader to [35]. The communication topology between the vehicles used during the field tests is one where the red vehicle communicates with the black and the yellow vehicles, but the yellow and black vehicles do not communicate among themselves. Furthermore, in order to emulate the conditions encountered underwater, communications occur asynchronously, as defined by the logic-based communication system. The type of mission performed in all the reported trials was the following:

- Nominal path for the red vehicle: A lawn mowing path with 30 m length for straight line segments and 12 m radius for circumference segments
- Formation: Alongside alignment, with 5 m of separation and the red vehicle at the centre following the nominal path
- Normalized speed along the nominal path: $v_r = 0.5$ m/s

In each test, a saturation of the RPM control signal in one of the vehicles (the Medusa-black vehicle) was temporarily enforced in order to test the resilience of the proposed algorithms to transient events that may have a negative impact on coordination. In the following plots, when displaying the dimensionless PF variable, the latter corresponds to the arc-length in meters along the nominal path, i.e. the path taken by the red vehicle.

In the field tests, the vehicles performed coordinated PF with logic-based communications with values of $\varepsilon = 0.2$ m, $\varepsilon = 0.6$ m, and $\varepsilon = 1.4$ m. The current was negligible and thus the current estimator used, akin to that proposed in [24] but using a simple quasi-steady relationship between command mode thruster activity and surge speed and assuming negligible sideslip, had almost no effect on the system. The implementation of the general architecture proposed for CPF relied on the inner loop controllers installed on board the vehicles, together with the systems in charge of PF, coordination, and logic-based communications. Adopting the theoretical framework for integrated system design, tuning of all system parameters, was done in simulation using a dynamic model of the Medusa vehicles. The reader will find in [36] a description of a model of the Medusa vehicles, together with details of the MATLAB-based simulation used for mission simulation and parameter tuning. The following values for the most relevant parameters were used on all the vehicles during the trials:

- PF controller: $k_x = k_y = 0.35$, $k_{\dot{\gamma}} = 1$, $c_{\dot{\gamma}} = 0.5$, $\delta = 0.2$ m.
- Coordination controller: $k = 1$.

In what follows, due to space limitations, we focus our attention on the mechanism adopted for logic-based communications.

8.5.2 Results

This section contains the results of the field tests described before. For all the performed tests in this section, the saturation of the propeller rotation in the black vehicle, occurs between 400 and 550 s from the beginning of the mission, which corresponds to the period from the beginning of the third circular segment to the middle of the fourth straight line segment, see Figure 8.4.

8.5.2.1 Test with $\varepsilon = 0.2$

Figures 8.4 and 8.5 show the vehicle paths and the evolution of the PF variables for $\varepsilon = 0.2$. The PF variable is in meters.

One can observe from Figures 8.4 and 8.5 that coordination is maintained throughout the test, with the vehicles running at a lower speed during the enforced saturation. This is because the coordination scheme will temporarily force the red and yellow vehicles to slow down in an attempt to remain in formation with the black vehicle. Once the saturation is removed, all vehicles speed up to coordinate at the same normalized, desired speed. The communication events among the vehicles and the estimation error of the vehicles are plotted in Figure 8.6. From this figure, one can observe large periods without communications after the initial period when the vehicles are still converging to the paths from their initial positions and before the enforced saturation. However, the overall number of communications is still high. During the enforced saturation, the vehicles move at a constant speed which is lower than their reference. Therefore, between communications, the estimation error norm increases at a constant rate and the vehicles communicate at a fixed rate. In this

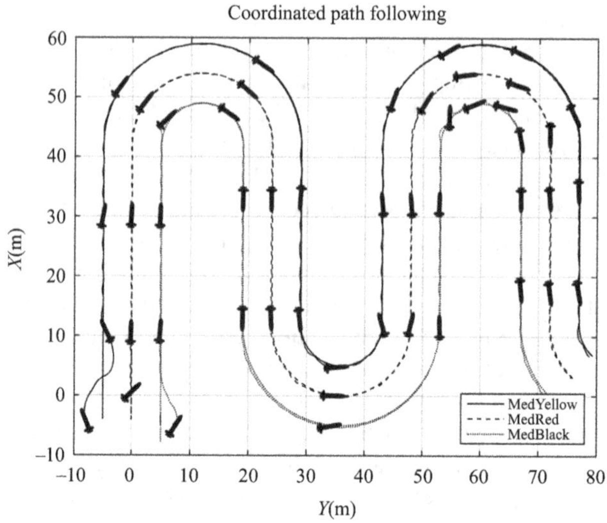

Figure 8.4 Vehicle paths for logic-based communications with ε = 0.2

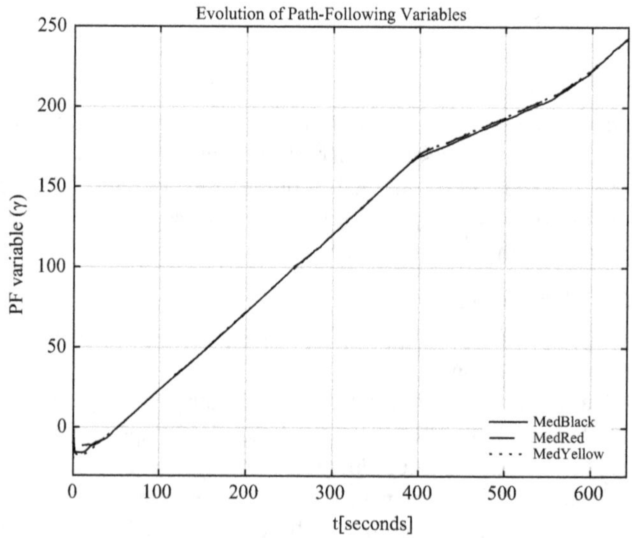

Figure 8.5 Path-following variables for logic-based communications with ε = 0.2

case, the period between communications is approximately 1 s. As expected, since we do not have packet losses in this test, the estimation error of all the vehicles is almost always below 0.2, and only passes slightly 0.2 due to the sampling rate of the computer, set at 0.2 s.

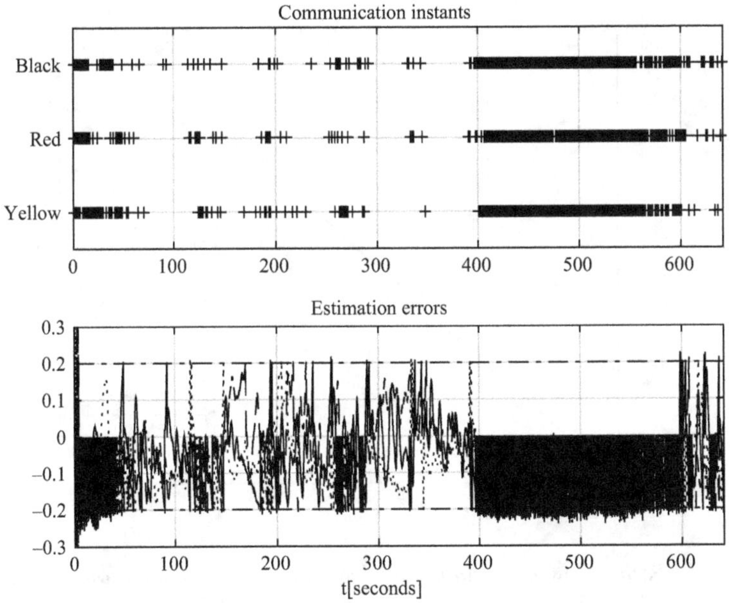

Figure 8.6 Communication events and estimation error on the red vehicle for $\varepsilon = 0.2$. Crosses mark the times at which communications take place

8.5.2.2 Test with $\varepsilon = 0.6$

For $\varepsilon = 0.6$, the PF variables of the three vehicles are show in Figure 8.7. Figure 8.8 contains the communication instants and the estimation errors of the PF variables.

From Figure 8.7, a slightly worse performance is visible during the saturation than for the case with $\varepsilon = 0.2$. From Figure 8.8, one can observe that after the initial convergence and before the enforced saturation, the communications are sporadic and occur only at the instants when the vehicles enter or leave the circular segments. This is because, in order to maintain coordination on the circular segments, the outer vehicle must move faster than the nominal speed and the inner vehicle more slowly, also with respect to the nominal normalized speed. As a consequence, the outer and inner vehicles are required to respectively accelerate and decelerate instantaneously to follow their desired paths.

8.5.2.3 Test with $\varepsilon = 1.4$

For $\varepsilon = 1.4$, the vehicle PF variables are shown in Figure 8.9. The communication events among the vehicles are shown in Figure 8.10.

Throughout most of the mission, the vehicles are well coordinated. However, it is apparent that during the enforced saturation, we have a much worse performance than in the previous cases, with the black vehicle further behind the other two vehicles. In this case, we obtain a very low number of exchanged messages after the vehicles

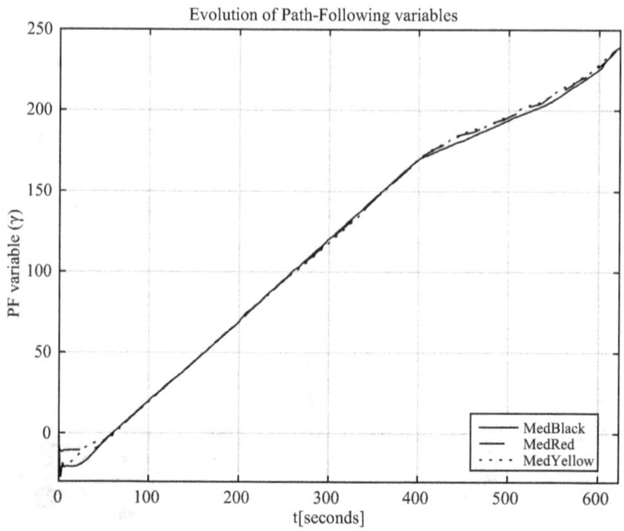

Figure 8.7 Path-following variables for logic-based communications with $\varepsilon = 0.6$

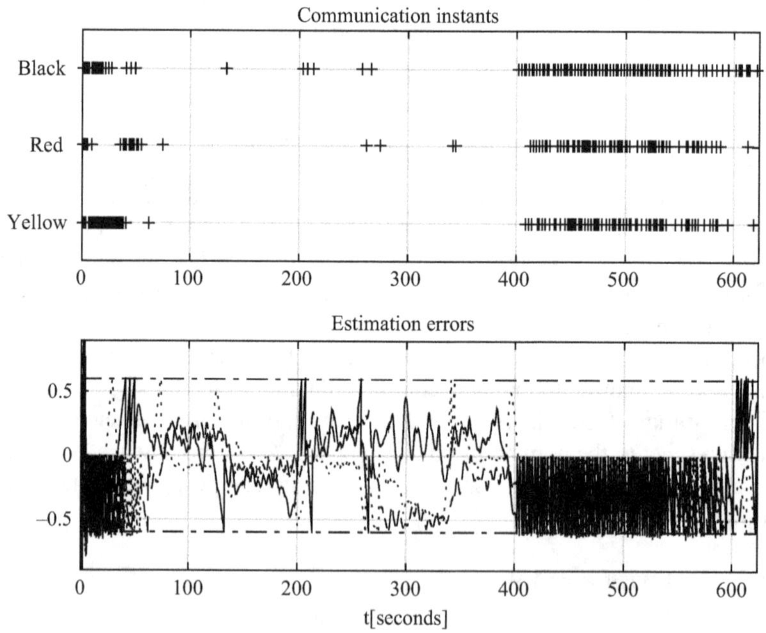

Figure 8.8 Communication events and estimation errors for $\varepsilon = 0.6$

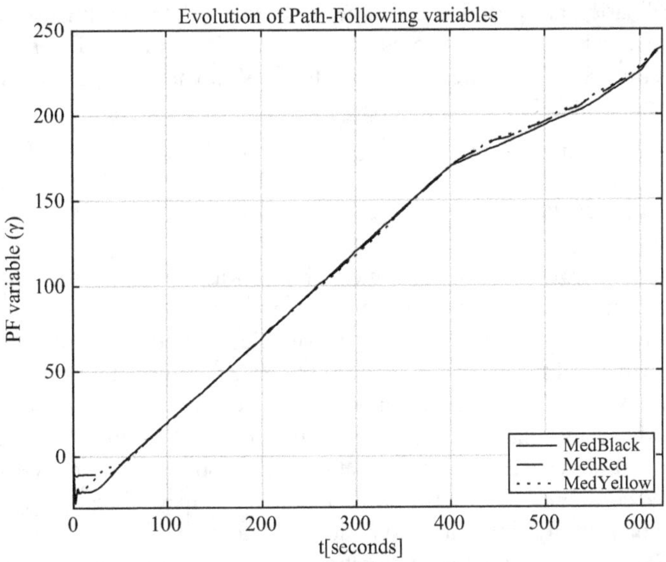

Figure 8.9 Path-following variables for logic-based communications with ε = 1.4

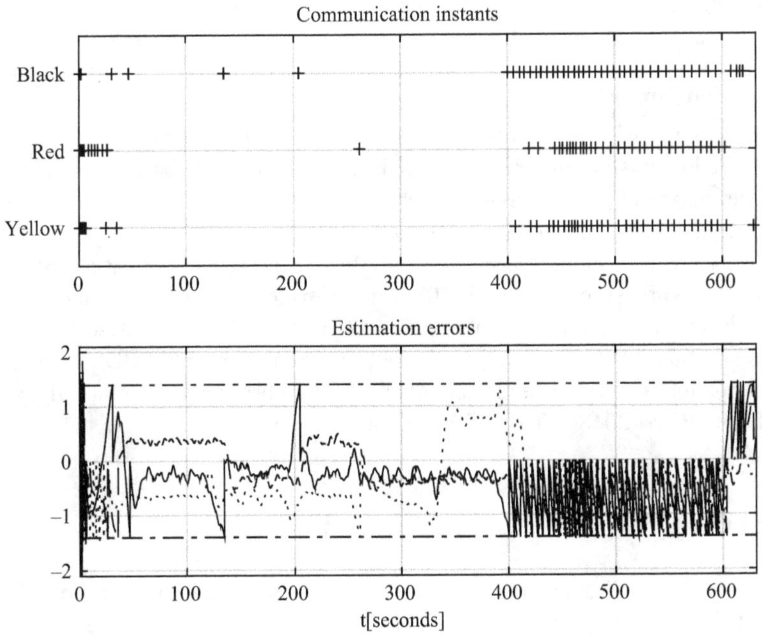

Figure 8.10 Communication events and estimation errors for ε = 1.4

reach coordination and before the saturation. In total, after coordination and before saturation, we can count only two communications from the black vehicle and one from the yellow. This illustrates clearly the trade-off involved in performance versus communication rates.

8.6 Conclusions

A distributed control system for multiple vehicles was introduced that makes use of logic-based communications with low bandwidth requirements to solve robustly the CPF problem, i.e. to steer and maintain the vehicles at predefined paths at a desired common normalized speed profile, while holding a given formation pattern. The control system takes into account the topology of the communication network between vehicles and the fact that communication takes place at discrete instants of time, instead of continuously. The communication logic also considers packet losses and delays in the communication network. Using the mathematical tools of graph theory and Lyapunov stability, conditions were derived under which the overall closed loop system is ISS. To show how the theoretical framework adopted can be applied in practice, tests were done with a group of three autonomous marine robots. The tests showed that the framework holds considerable promise for real system implementation. Future work will address the implementation of proposed CPF strategy for a larger number of vehicles exchanging messages over hybrid acoustic/optical modems.

Acknowledgements

We thank the DSOR team, composed by Jorge Ribeiro, Henrique Silva, Manuel Rufino, Pedro Góis, Miguel Ribeiro, João Botelho, and Luís Sebastião for their efforts in the development of the MEDUSA vehicles and their support and collaboration on the planning and execution of our sea trials.

This work was supported by the H2020 EU Marine Robotics Research Infrastructure Network (Project ID 731103), the MarineUAS project under the Marie Curie Sklodowska grant agreement No. 642153, and the Portuguese FCT Project UID/EEA/5009/2013. The first author benefited from grant SFRH/BD/51929/2012 of the Foundation for Science and Technology (FCT), Portugal, and grants BL23/2018-IST-ID and BL24/2018-IST-ID of the FCT-CMU + Atlântico project.

Appendix A

Proof of Lemma 8.1. The proof exploits the use of the Lyapunov function defined as $V := (1/2)\|\boldsymbol{e}\|^2 = (1/2)\left(\|e_\delta\|^2 + c_{\dot\gamma} e_{\dot\gamma}^2\right)$.

Equations (8.17) and (8.20) yield

$$\dot{\boldsymbol{e}} = -S(r)e_\delta - K_k e_\delta - R^T \frac{\partial z_d}{\partial \gamma} e_{\dot\gamma} - \begin{bmatrix} 0 \\ \delta\tilde{r} \end{bmatrix} + \begin{bmatrix} \tilde{u} \\ v_w \end{bmatrix} + \tilde{\boldsymbol{v}}_c.$$

Because $x^T S(r)x = 0;\ \forall x \in \mathbb{R}^2$, the time derivative of V is given by

$$\dot{V} = e_\delta^T \left(-K_k e_\delta - \begin{bmatrix} 0 \\ \delta \tilde{r} \end{bmatrix} + \begin{bmatrix} \tilde{u} \\ v_w \end{bmatrix} + \tilde{\boldsymbol{v}}_c \right) - k_{\dot{\gamma}} c_{\dot{\gamma}} e_{\dot{\gamma}}^2$$

$$= -e_\delta^T K_k e_\delta - c_{\dot{\gamma}} k_{\dot{\gamma}} e_{\dot{\gamma}}^2 + e_\delta^T \left(\begin{bmatrix} \tilde{u} \\ v_w \end{bmatrix} - \delta \begin{bmatrix} 0 \\ \tilde{r} \end{bmatrix} + \tilde{\boldsymbol{v}}_c \right)$$

Defining $K_e := \operatorname{diag}\{k_x, k_y, k_{\dot{\gamma}}\}$ we obtain

$$\dot{V} \leq -\boldsymbol{e}^T K_e \boldsymbol{e} + \|\boldsymbol{e}\| \left((1+\delta)\|w\| + \|v\|\right) \tag{A.1}$$

The rest of the proof involves dominating all positive terms of (A.1) by $-\boldsymbol{e}^T K_e \boldsymbol{e} \leq -\min(k_x, k_y, k_{\dot{\gamma}})\|\boldsymbol{e}\|^2$. To do this, notice from (A.1) that, defining a positive scalar θ in the interval $(0, 1)$,

$$\dot{V} \leq -(1-\theta)\min(k_x, k_y, k_{\dot{\gamma}})\|\boldsymbol{e}\|^2 + \|\boldsymbol{e}\| \left((1+\delta)\|w\| + \|v\| - \theta \min(k_x, k_y, k_{\dot{\gamma}})\|\boldsymbol{e}\|\right)$$

and, therefore, defining

$$\sigma_w^e(x) := \frac{2(1+\delta)x}{\theta \min(k_x, k_y, k_{\dot{\gamma}})}, \qquad \sigma_v^e(x) := \frac{2x}{\theta \min(k_x, k_y, k_{\dot{\gamma}})},$$

one obtains

$$\dot{V} \leq -(1-\theta)\min(k_x, k_y, k_{\dot{\gamma}})\|\boldsymbol{e}\|^2, \forall \|\boldsymbol{e}\| \geq \sigma_w^e(\|w\|) \oplus \sigma_v^e(\|v\|),$$

from which (8.21) follows with $\beta^e(x,t) := xe^{-(1-\theta)\min(k_x,k_y,k_{\dot{\gamma}})t}$. □

Proof of Lemma 8.2. Defining $\bar{\gamma} := (1/N)\mathbf{1}^T\gamma$, γ can be decomposed as $\gamma = \mathbf{1}\bar{\gamma} + Q^T\xi$. Using this fact and recalling that $v_r(\gamma)$ is assumed to be globally Lipschitz, we can conclude that

$$\xi^T Q \bar{v}_r = \xi^T Q \Pi\left(\mathbf{1}v_r(\bar{\gamma}) - (\mathbf{1}v_r(\bar{\gamma}) - \bar{v}_r)\right) = \xi^T Q \Pi\left(\bar{v}_r - \mathbf{1}v_r(\bar{\gamma})\right) \leq l\|\xi\|^2.$$

From the definition of $\bar{e}_{\dot{\gamma}}$ and Definition 8.2, one has that

$$\|\bar{e}_{\dot{\gamma}}\| \leq \sqrt{N} \max_{i \in \mathcal{N}} \sigma^e(\|\boldsymbol{e}_i\|) \tag{A.2}$$

We now use the fact that, since $(\mathcal{N}, \mathcal{A})$ is connected, $\boldsymbol{L}$ has a simple eigenvalue at zero with an associated eigenvector $\mathbf{1}$ and the remaining eigenvalues are all positive, see e.g. [28]. Applying the Courant–Fischer theorem in [37] and noting that $Q\mathbf{1} = \mathbf{0}$, it follows that

$$\xi^T Q \boldsymbol{L} Q^T \xi \geq \sigma_2 \|\xi\|^2, \forall \xi \in \mathbb{R}^{N-1} \tag{A.3}$$

From (8.25), (A.2), (A.3), and the fact that $\|\bar{\gamma}\| \leq d^*\sqrt{N}\max_{i\in\mathcal{N}, j\in\mathcal{N}^i}\|\tilde{\gamma}_j^i\|$, we can bound the time derivative of V to obtain

$$\dot{V} \leq -k\xi^T Q\boldsymbol{L}Q^T\xi + l\|\xi\|^2 + \xi^T Q\bar{e}_{\dot{\gamma}} + k\xi^T Q\bar{\gamma}$$

$$\leq -k\sigma_2\|\xi\|^2 + l\|\xi\|^2 + \sqrt{N}\|\xi\| \max_{i\in\mathcal{N}, j\in\mathcal{N}^i}\left(\sigma^e(\|\boldsymbol{e}_i\|) + kd^*\|\tilde{\gamma}_j^i\|\right).$$

As a consequence,

$$\dot{V} < -(1-\theta)(k\sigma_2 - l)\|\xi\|^2$$

for

$$\|\xi\| \geq \max_{i\in\mathcal{N};j\in\mathcal{N}^i} \left(\sigma_\gamma^\xi \left(\|\tilde{\gamma}_j^i\|_{[0,t]}\right) \oplus \sigma_e^\xi \left(\|\boldsymbol{e}_i\|_{[0,t]}\right)\right),$$

and $1 > \theta > 0$, where $\sigma_e^\xi(s) := \dfrac{2\sqrt{N}}{(k\sigma_2 - l)\,\theta}\sigma^e(s)$, $\sigma_\gamma^\xi(s) := \dfrac{2\sqrt{N}kd^*}{(k\sigma_2 - l)\,\theta}s$. The conclusions of the lemma follow with $\beta^\xi(x,t) := xe^{-(1-\theta)(k\sigma_2-l)t}$ □

The following sequence of lemmas is instrumental in the proof of Theorem 8.3.

Lemma 8.3. *Consider the CPFCS strictures of Theorem 8.3 in the presence of delayed information and packet losses. If, for some $\bar{t} > 0$, for any time $t_k^i \leq \bar{t}$ the post-reset estimation errors $\tilde{\gamma}_i^j$; $j \in \mathcal{N}^i$ satisfy*

$$\left\|\tilde{\gamma}_i^j\left(\left(t_k^i + 2\alpha_k^{ij}\bar{\tau}\right)^+\right)\right\| < \varepsilon, \forall j \in \mathcal{N}^i, \tag{A.4}$$

then, for sufficiently small $\tilde{\tau}$,

$$\max_{i\in\mathcal{N},j\in\mathcal{N}^i} \left\|\tilde{\gamma}_i^j\right\|_{[0,\bar{t}+\tilde{\tau})} \leq \alpha^\varepsilon(\tilde{\tau})\varepsilon + \alpha^{\bar{e}}(\tilde{\tau})\|\bar{\boldsymbol{e}}\|_{[0,\bar{t}+\tilde{\tau})} \tag{A.5}$$

for continuous functions $\alpha^\varepsilon : [0, b) \to [1, \infty)$ and $\alpha^{\bar{e}} : [0, b) \to [0, \infty)$ and for some positive constant b with $\alpha^\varepsilon(0) = 1$ and $\alpha^{\bar{e}}(0) = 0$.

Proof. If (A.4) is satisfied for any time $t_k^i < \bar{t}$, then using the same arguments as in the proof of Theorem 8.2, one obtains that for any time $t_k^i < \bar{t}$

$$\left\|\tilde{\gamma}_i^j\left(t_k^i\right)\right\|_{\left(t_k^i+2\alpha_k^{ij}\bar{\tau},\, t_{k+1}^i\right]} \leq \varepsilon, \forall j \in \mathcal{N}^i. \tag{A.6}$$

Recall also that we assumed that $2\alpha_k^{ij}\bar{\tau} \leq \tilde{\tau}$ for all $i \in \mathcal{N}, j \in \mathcal{N}^i$ and $k \geq 0$. From the additional assumptions that $\sigma_\xi^{\tilde{v}_r}(s) = \alpha_\xi^{\tilde{v}_r}s$, $\sigma_\gamma^{\tilde{v}_r}(s) = \alpha_\gamma^{\tilde{v}_r}s$, and $\sigma^e(s) = \alpha^e s$, for appropriately defined positive constants $\alpha_\xi^{\tilde{v}_r}$, $\alpha_\gamma^{\tilde{v}_r}$ and α^e we have that

$$\begin{aligned}
\max_{i\in\mathcal{N}} \|\tilde{v}_{r_i}\|_{[0,\bar{t}+\tilde{\tau})} &\leq \alpha_\xi^{\tilde{v}_r}\|\xi\|_{[0,\bar{t}+\tilde{\tau})} + \alpha_\gamma^{\tilde{v}_r} \max_{i\in\mathcal{N},j\in\mathcal{N}^i} \left\|\tilde{\gamma}_i^j\right\|_{[0,\bar{t}+\tilde{\tau})} \\
&\leq \alpha_\xi^{\tilde{v}_r}\|\bar{\boldsymbol{e}}\|_{[0,\bar{t}+\tilde{\tau})} + \alpha_\gamma^{\tilde{v}_r} \max_{i\in\mathcal{N},j\in\mathcal{N}^i} \left\|\tilde{\gamma}_i^j\right\|_{[0,\bar{t}+\tilde{\tau})}, \\
\max_{i\in\mathcal{N}} \|e_{\dot{\gamma}_i}\|_{[0,\bar{t}+\tilde{\tau})} &\leq \alpha^e \max_{i\in\mathcal{N}} \leq \|\boldsymbol{e}_i\|_{[0,\bar{t}+\tilde{\tau})} \leq \|\bar{\boldsymbol{e}}\|_{[0,\bar{t}+\tilde{\tau})}.
\end{aligned}$$

From the definition of $\tilde{\gamma}_i^j$ and the fact that $\dot{\gamma}_i = v_r(\gamma_i) + \tilde{v}_{r_i} + e_{\dot{\gamma}_i}$ and $\dot{\hat{\gamma}}_i^j = v_r\left(\hat{\gamma}_i^j\right)$, when the derivative is defined, we have that for $t \in \left[t_k^i, t_k^i + 2\alpha_k^{ij}\bar{\tau}\right)$

$$\dot{\tilde{\gamma}}_i^j = v_r\left(\hat{\gamma}_i^j\right) - v_r(\gamma_i) - \tilde{v}_{r_i} - e_{\dot{\gamma}_i},$$

and, therefore,

$$\left\|\dot{\tilde{\gamma}}_i^j\right\| \le l\left\|\tilde{\gamma}_i^j\right\| + \left(\alpha_\xi^{\tilde{v}_r} + \alpha^e\right)\|\bar{e}\|_{[0,\bar{t}+2\bar{\tau})} + \alpha_\gamma^{\tilde{v}_r} \max_{o\in\mathcal{N},p\in\mathcal{N}^o} \|\tilde{\gamma}_o^p\|_{[0,\bar{t}+\bar{\tau})}.$$

Using the comparison lemma in [38] (Lemma 3.4), it follows that for $t \in \left[t_k^i, t_k^i + 2\alpha_k^{ij}\bar{\tau}\right)$, $\left\|\tilde{\gamma}_i^j(t)\right\| < \Gamma(t)$, where $\Gamma : \left[t_k^i, \infty\right) \to [\varepsilon, \infty)$ is defined by $\Gamma\left(t_k^i\right) = \varepsilon$ and

$$\dot{\Gamma} = l\Gamma + \left(\alpha_\xi^{\tilde{v}_r} + \alpha^e\right)\|\bar{e}\|_{[0,\bar{t}+\bar{\tau})} + \alpha_\gamma^{\tilde{v}_r} \max_{i\in\mathcal{N},j\in\mathcal{N}^i} \left\|\tilde{\gamma}_i^j\right\|_{[0,\bar{t}+\bar{\tau})}.$$

As a consequence,

$$\begin{aligned}\left\|\tilde{\gamma}_i^j\right\|_{\left[t_k^i,t_k^i+2\alpha_k^{ij}\bar{\tau}\right)} &\le \Gamma\left(t_k^i + \bar{\tau}\right)\\ &\le \varepsilon e^{l\bar{\tau}} + z\left(l,\bar{\tau}\right)\left(\left(\alpha_\xi^{\tilde{v}_r} + \alpha^e\right)\|\bar{e}\|_{[0,\bar{t}+\bar{\tau})} + \alpha_\gamma^{\tilde{v}_r} \max_{o\in\mathcal{N},p\in\mathcal{N}^o} \left\|\tilde{\gamma}_o^p\right\|_{[0,\bar{t}+\bar{\tau})}\right),\end{aligned}$$

where

$$z(l,t) := \begin{cases} \frac{e^{lt}-1}{l}, & \text{if } \; l > 0 \\ t, & \text{if } \; l = 0 \end{cases}.$$

From (A.6),

$$\max_{i\in\mathcal{N},j\in\mathcal{N}^i} \|\tilde{\gamma}_i^j\|_{[0,\bar{t}+\bar{\tau})} \le \max_{i\in\mathcal{N},j\in\mathcal{N}^i,t_k^i\le\bar{t}} \|\tilde{\gamma}_i^j\|_{\left[t_k^i,t_k^i+2\alpha_k^{ij}\bar{\tau}\right)}.$$

Thus, if $\bar{\tau}$ is small enough so that $\alpha_\gamma^{\tilde{v}_r} z(l,\bar{\tau}) < 1$, we have

$$\begin{aligned}\max_{i\in\mathcal{N},j\in\mathcal{N}^i} \|\tilde{\gamma}_i^j\|_{[0,\bar{t}+\bar{\tau})} &\le \varepsilon e^{l\bar{\tau}}\\ &\quad + z(l,\bar{\tau})\left(\left(\alpha_\xi^{\tilde{v}_r} + \alpha^e\right)\|\bar{e}\|_{[0,\bar{t}+\bar{\tau})} + \alpha_\gamma^{\tilde{v}_r} \max_{i\in\mathcal{N},j\in\mathcal{N}^i} \|\tilde{\gamma}_i^j\|_{[0,\bar{t}+2\bar{\tau})}\right).\end{aligned}$$

Finally,

$$\max_{i\in\mathcal{N},j\in\mathcal{N}^i} \|\tilde{\gamma}_i^j\|_{[0,\bar{t}+\bar{\tau})} \le \varepsilon \frac{e^{l\bar{\tau}}}{1 - \alpha_\gamma^{\tilde{v}_r} z(l,\bar{\tau})} + \frac{z(l,\bar{\tau})\left(\alpha_\xi^{\tilde{v}_r} + \alpha^e\right)}{1 - \alpha_\gamma^{\tilde{v}_r} z(l,\bar{\tau})}\|\bar{e}\|_{[0,\bar{t}+\bar{\tau})},$$

and (A.5) holds with

$$\alpha^\varepsilon(x) := \frac{e^{lx}}{1 - \alpha_\gamma^{\tilde{v}_r} z(l,x)}, \qquad \alpha^{\bar{e}}(x) := \frac{z(l,x)}{1 - \alpha_\gamma^{\tilde{v}_r} z(l,x)}\left(\alpha_\xi^{\tilde{v}_r} + \alpha^e\right),$$

for all $x \ge 0$. □

Lemma 8.4. *Consider the CPFCS structures of Theorem 8.3 in the presence of delayed information and packet losses. If, for some* $\bar{t} > 0$*, for any time* $t_k^i \leq \bar{t}$*, the post-reset values of the estimation errors* $\tilde{\gamma}_i^j$; $j \in \mathcal{N}^i$ *satisfy*

$$\left\| \tilde{\gamma}_i^j \left(\left(t_k^i + 2\alpha_k^{ij} \bar{\tau} \right)^+ \right) \right\| < \varepsilon, \forall j \in \mathcal{N}^i,$$

then, for sufficiently small $\tilde{\tau}$*, we can compute continuous functions* $\tilde{\alpha}^{\varepsilon} : [0, c) \to [1, \infty)$ *and* $\tilde{\alpha}^{\bar{e}} : [0, c) \to [0, \infty)$*, for some positive constant* c *with* $\tilde{\alpha}^{\varepsilon}(0) = 1$ *and* $\tilde{\alpha}^{\bar{e}}(0) = 0$*, such that for all* $0 \leq t < \bar{t} + \tilde{\tau}$

$$\max_{i \in \mathcal{N}, j \in \mathcal{N}^i} \left\| \tilde{\gamma}_i^j(t) \right\| \leq \tilde{\alpha}^{\bar{e}}(\tilde{\tau}) \breve{e} + \tilde{\alpha}^{\varepsilon}(\tilde{\tau}) \varepsilon, \tag{A.7}$$

where

$$\breve{e} := \beta^{\bar{e}}(\|\bar{e}(0)\|, 0) \oplus \max_{i \in \mathcal{N}} \left(\sigma_w^{\bar{e}} \left(\|w_i\|_{[0,\infty)} \right) \oplus \sigma_v^{\bar{e}} \left(\|v_i\|_{[0,\infty)} \right) \right).$$

Proof. From Definition 8.7 and the assumption that $\sigma_{\gamma}^{\bar{e}}(s) = \alpha_{\gamma}^{\bar{e}} s$, for an appropriately defined positive constant $\alpha_{\gamma}^{\bar{e}}$, we obtain

$$\|\bar{e}\|_{[0, \bar{t}+\tilde{\tau})} \leq \breve{e} + \alpha_{\gamma}^{\bar{e}} \max_{i \in \mathcal{N}, j \in \mathcal{N}^i} \|\tilde{\gamma}_i^j\|_{[0, \bar{t}+\tilde{\tau})}.$$

If $\tilde{\tau}$ is sufficiently small to satisfy the conditions of Lemma 8.3 and $\alpha_{\gamma}^{\bar{e}} \alpha^{\bar{e}}(\tilde{\tau}) < 1$, then (A.7) is satisfied for $0 \leq t < \bar{t} + \tilde{\tau}$ with

$$\tilde{\alpha}^{\varepsilon}(x) := \frac{\alpha^{\varepsilon}(x)}{1 - \alpha_{\gamma}^{\bar{e}} \alpha^{\bar{e}}(x)}, \qquad \tilde{\alpha}^{\bar{e}}(\tilde{\tau}) := \frac{\alpha^{\bar{e}}(x)}{1 - \alpha_{\gamma}^{\bar{e}} \alpha^{\bar{e}}(x)},$$

for all $x \geq 0$. □

Lemma 8.5. *Consider the CPFCS structures of Theorem 8.3 in the presence of delayed information and packet losses. For sufficiently small values of* $\tilde{\tau}$, $\|\bar{e}(0)\|$, $\max_{i \in \mathcal{N}} \|v_i\|_{[0,\infty)}$, *and* $\max_{i \in \mathcal{N}} \|w_i\|_{[0,\infty)}$ *such that* (8.31) *holds, if* $\max_{i \in \mathcal{N}, j \in \mathcal{N}^i} \left\| \tilde{\gamma}_i^j(t) \right\|$ *satisfies* (A.7) *for* $0 \leq t < \bar{t} + \tilde{\tau}$*, for some* $\bar{t} > 0$ *then, for any time* $t_k^i \leq \bar{t}$*, the post-reset value of the estimation errors* $\tilde{\gamma}_i^j$; $j \in \mathcal{N}^i$ *satisfy*

$$\left\| \tilde{\gamma}_i^j \left(\left(t_k^i + 2\alpha_k^{ij} \bar{\tau} \right)^+ \right) \right\| < \varepsilon, \forall j \in \mathcal{N}^i, \tag{A.8}$$

and (8.32) *holds for all* $0 \leq t \leq \bar{t} + \tilde{\tau}$.

Proof. As in Lemma 8.4, for simplicity of notation we denote

$$\breve{e} := \beta^{\bar{e}}(\|\bar{e}(0)\|, 0) \oplus \max_{i \in \mathcal{N}} \left(\sigma_w^{\bar{e}} \left(\|w_i\|_{[0,\infty)} \right) \oplus \sigma_v^{\bar{e}} \left(\|v_i\|_{[0,\infty)} \right) \right).$$

The post reset estimation error can be expressed as

$$\tilde{\gamma}_i^j\left(\left(t_k^i+2\alpha_k^{ij}\bar{\tau}\right)^+\right) = \hat{\gamma}_i^j\left(\left(t_k^i+2\alpha_k^{ij}\bar{\tau}\right)^+\right) - \gamma_i\left(t_k^i+2\alpha_k^{ij}\bar{\tau}\right)$$
$$= \Gamma_r\left(2\alpha_k^{ij}\bar{\tau} - t_k^i + T_r\left(\gamma_i\left(t_k^i\right)\right)\right) - \gamma_i\left(t_k^i+2\bar{\tau}\right).$$

Defining, on the interval $t \in \left[t_k^i,\ t_k^i+2\alpha_k^{ij}\bar{\tau}\right]$, the function

$$\eta(t) := \Gamma_r\left(t - t_k^i + T_r\left(\gamma_i\left(t_k^i\right)\right)\right) - \gamma_i(t),$$

we have $\eta\left(t_k^i\right)=0$ and $\eta\left(t_k^i+2\alpha_k^{ij}\bar{\tau}\right) = \tilde{\gamma}_i^j\left(\left(t_k^i+2\alpha_k^{ij}\bar{\tau}\right)^+\right)$. From the derivative of the inverse function, we obtain

$$\dot{\eta} = v_r\left(\Gamma_r\left(t-t_k^i+T_r\left(\gamma_i\left(t_k^i\right)\right)\right)\right) - \dot{\gamma}_i = v_r(\gamma_i+\eta) - v_r(\gamma_i) - \tilde{v}_{r_i} - e_{\dot{\gamma}_i}.$$

Taking norms and using the results from Lemmas 8.3 and 8.4 and their proofs, we obtain

$$\|\dot{\eta}\| \le l\|\eta\| + \left(\alpha_\xi^{\tilde{v}_r}+\alpha^e\right)\|\bar{e}\|_{[0,\bar{t}+\tilde{\tau})} + \alpha_\gamma^{\tilde{v}_r}\max_{i\in\mathcal{N},j\in\mathcal{N}^i}\left\|\tilde{\gamma}_i^j\right\|_{[0,\bar{t}+\tilde{\tau})}$$
$$\le l\|\eta\| + \left(\alpha_\xi^{\tilde{v}_r}+\alpha^e\right)\breve{e} + \left(\alpha_\gamma^{\tilde{v}_r}+\alpha_\gamma^{\bar{e}}\left(\alpha_\xi^{\tilde{v}_r}+\alpha^e\right)\right)\max_{i\in\mathcal{N},j\in\mathcal{N}^i}\left\|\tilde{\gamma}_i^j\right\|_{[0,\bar{t}+\tilde{\tau})}$$
$$\le l\|\eta\| + \left(\alpha_\gamma^{\tilde{v}_r}\tilde{\alpha}^{\bar{e}}(\tilde{\tau}) + \left(\alpha_\xi^{\tilde{v}_r}+\alpha^e\right)\left(1+\alpha_\gamma^{\bar{e}}\tilde{\alpha}^{\bar{e}}(\tilde{\tau})\right)\right)\breve{e}$$
$$+ \left(\alpha_\gamma^{\tilde{v}_r}+\alpha_\gamma^{\bar{e}}\left(\alpha_\xi^{\tilde{v}_r}+\alpha^e\right)\right)\tilde{\alpha}^{\varepsilon}(\tilde{\tau})\varepsilon,$$

Following the proof of Lemma 8.3, it can be seen that

$$\left\|\tilde{\gamma}_i^j\left(\left(t_k^i+2\alpha_k^{ij}\bar{\tau}\right)^+\right)\right\| = \left\|\eta\left(t_k^i+2\alpha_k^{ij}\bar{\tau}\right)\right\|$$
$$\le z(l,\tilde{\tau})\left(\alpha_\gamma^{\tilde{v}_r}\tilde{\alpha}^{\bar{e}}(\tilde{\tau}) + \left(\alpha_\xi^{\tilde{v}_r}+\alpha^e\right)\left(1+\alpha_\gamma^{\bar{e}}\tilde{\alpha}^{\bar{e}}(\tilde{\tau})\right)\right)\breve{e}$$
$$+z(l,\tilde{\tau})\left(\alpha_\gamma^{\tilde{v}_r}+\alpha_\gamma^{\bar{e}}\left(\alpha_\xi^{\tilde{v}_r}+\alpha^e\right)\right)\tilde{\alpha}^{\varepsilon}(\tilde{\tau})\varepsilon$$

Finally, $\left\|\tilde{\gamma}_i^j\left(\left(t_k^i+2\alpha_k^{ij}\bar{\tau}\right)^+\right)\right\| < \varepsilon$ if $\tilde{\tau}$ is sufficiently small such that

$$\left(\alpha_\gamma^{\tilde{v}_r}+\alpha_\gamma^{\bar{e}}\left(\alpha_\xi^{\tilde{v}_r}+\alpha^e\right)\right)\tilde{\alpha}^{\varepsilon}(\tilde{\tau})z(l,\tilde{\tau}) < 1,$$

and for sufficiently small values of $\breve{e}$ such that $\varepsilon > \tilde{\sigma}^{\varepsilon}(\tilde{\tau})\breve{e}$, with

$$\tilde{\sigma}^{\varepsilon}(x) := \frac{z(l,x)\left(\alpha_\gamma^{\tilde{v}_r}\tilde{\alpha}^{\bar{e}}(x) + \left(\alpha_\xi^{\tilde{v}_r}+\alpha^e\right)\left(1+\alpha_\gamma^{\bar{e}}\tilde{\alpha}^{\bar{e}}(x)\right)\right)}{1-\left(\alpha_\gamma^{\tilde{v}_r}+\alpha_\gamma^{\bar{e}}\left(\alpha_\xi^{\tilde{v}_r}+\alpha^e\right)\right)\tilde{\alpha}^{\varepsilon}(x)\,z\,(l,x)}.$$

Because $\log(1+x) \le x$ for $x \ge 0$, if we define

$$\sigma^{\varepsilon}(x) := \frac{1}{\log\left(1 + (1/\tilde{\sigma}^{\varepsilon}(x))\right)},$$

it follows by continuity that $\sigma^{\varepsilon}(0) := \lim_{x\to 0^+} \sigma^{\varepsilon}(x) = 0$ and $\tilde{\sigma}^{\varepsilon}(x) \le \sigma^{\varepsilon}(x)$. By requiring that $\varepsilon > \sigma^{\varepsilon}(\tilde{\tau})\breve{e}$, (8.32) holds with

$$\bar{\alpha}^{\varepsilon}(x) := \left(\tilde{\alpha}^{\bar{e}}(\tilde{\tau}) \log\left(1 + \frac{1}{\sigma^{\varepsilon}(\tilde{\tau})}\right) + \tilde{\alpha}^{\varepsilon}(\tilde{\tau})\right)\varepsilon.$$

Since $\bar{\alpha}^{\varepsilon}(0)$ is not defined, we define it formally by continuity as $\bar{\alpha}^{\varepsilon}(0) = \lim_{x\to 0^+} \bar{\alpha}^{\varepsilon}(x) = 1$, where we used the fact that $\lim_{x\to 0^+} \tilde{\alpha}^{\bar{e}}(x) \log(1 + (1/\sigma^{\varepsilon}(x))) = 0$. □

Proof of Theorem 8.3. Suppose that for some time t_k^i, for all previous data messages sent from any agent $j \in \mathcal{N}$ at times $t_o^j \le t_k^i$, the post-reset values of the estimation errors satisfy $\left\| \tilde{\gamma}_j^p \left(\left(t_o^j + 2\alpha_o^{jp}\tilde{\tau} \right)^+ \right) \right\| < \varepsilon$ for all $p \in \mathcal{N}^j$. This holds true if t_k^i is the time when the first data message is sent. If $\tilde{\tau}$, $\|\bar{e}(0)\|$, $\max_{i\in\mathcal{N}} \|v_i\|_{[0,\infty)}$, and $\max_{i\in\mathcal{N}} \|w_i\|_{[0,\infty)}$, are sufficiently small so as to satisfy the conditions of Lemmas 8.4 and 8.5, we have

$$\left\| \tilde{\gamma}_i^j \left(\left(t_k^i + 2\alpha_k^{ij}\tilde{\tau} \right)^+ \right) \right\| \le \tilde{\varepsilon} < \varepsilon, \ \forall j \in \mathcal{N}^i \tag{A.9}$$

and we can apply Lemmas 8.4 and 8.5 for the next received data message.

From Lemma 8.4, when $\dot{\tilde{\gamma}}_i^j(t)$ is defined, $\max_{i\in\mathcal{N}, j\in\mathcal{N}^i} \left\| \dot{\tilde{\gamma}}_i^j(t) \right\| \le v_{\dot{\tilde{\gamma}}}$, where $v_{\dot{\tilde{\gamma}}}$ and $\tilde{\varepsilon}$ in (A.9) depend on $\|\bar{e}(0)\|$, $\max_{i\in\mathcal{N}} \|v_i\|_{[0,\infty)}$, $\max_{i\in\mathcal{N}} \|w_i\|_{[0,\infty)}$, and ε but not on time. Therefore, the triggering condition $\left\| \tilde{\gamma}_i^j(t) \right\| = \varepsilon$ can only occur N times within $\frac{\varepsilon - \tilde{\varepsilon}}{v_{\dot{\tilde{\gamma}}}}$ time units. Repeating the same reasoning every transmission time t_k^i; $i \in \mathcal{N}$ and $k \ge 0$, the theorem holds by recursion. □

References

[1] Ghabcheloo R, Aguiar AP, Pascoal A, *et al.* Coordinated path-following in the presence of communication losses and time delays. SIAM Journal on Control and Optimization. 2009;48(1):234–265.

[2] Ihle IAF, Arcak M, and Fossen TI. Passivity-based designs for synchronized path-following. Automatica. 2006;43:1508–1518.

[3] Giulietti F, Pollini L, and Innocenti M. Autonomous formation flight. IEEE Control Systems. 2000;20(6):34–44.

[4] Stilwell DJ, and Bishop BE. Platoons of underwater vehicles. IEEE Control Systems. 2000;20(6):45–52.

[5] Ogren P, Egerstedt M, and Hu X. A control Lyapunov function approach to multiagent coordination. IEEE Transactions on Robotics and Automation. 2002;18(5):847–851.
[6] Jadbabaie A, Lin J, and Morse AS. Coordination of groups of mobile autonomous agents using nearest neighbor rules. IEEE Transactions on Automatic Control. 2003;48(6):988–1001.
[7] Moreau L. Stability of multiagent systems with time-dependent communication links. IEEE Transactions on Automatic Control. 2005;50(2): 169–182.
[8] Ma CQ, and Zhang JF. Necessary and sufficient conditions for consensusability of linear multi-agent systems. IEEE Transactions on Automatic Control. 2010;55(5):1263–1268.
[9] Dong W. On consensus algorithms of multiple uncertain mechanical systems with a reference trajectory. Automatica. 2011;47(9):2023–2028.
[10] Soetanto D, Lapierre L, and Pascoal A. Adaptive, non-singular path-following control of dynamic wheeled robots. vol. 2 of 42nd IEEE Conference on Decision and Control; 2003. p. 1765–1770.
[11] Skjetne R, Fossen TI, and Kokotović PV. Robust output maneuvering for a class of nonlinear systems. Automatica. 2004;40(3):373–383.
[12] Yook JK, Tilbury DM, and Soparkar NR. Trading computation for bandwidth: Reducing communication in distributed control systems using state estimators. IEEE Transactions on Control Systems Technology. 2002;10(4):503–518.
[13] Xu Y, and Hespanha JP. Communication Logic Design and Analysis for Networked Control Systems. In: Menini L, Zaccarian L, Abdallah CT, editors. Current Trends in Nonlinear Systems and Control. Systems & Control: Foundations & Applications. Birkhäuser, Boston; 2006. p. 495–514.
[14] Tabuada P. Event-triggered real-time scheduling of stabilizing control tasks. IEEE Transactions on Automatic Control. 2007;52(9):1680–1685.
[15] Aström KJ. Event Based Control. Analysis and Design of Nonlinear Control systems. Berlin, Heidelberg: Springer; 2008, pp. 127–147.
[16] Wang X, Sun Y, and Hovakimyan N. Asynchronous task execution in networked control systems using decentralized event-triggering. Systems & Control Letters. 2012;61(9):936–944.
[17] Battistelli G, Benavoli A, and Chisci L. Data-driven communication for state estimation with sensor networks. Automatica. 2012;48(5):926–935.
[18] Dimarogonas DV, and Johansson KH. Event-triggered control for multi-agent systems. 48th IEEE Conference on Decision and Control, 2009 held jointly with the 2009 28th Chinese Control Conference. IEEE; 2009. p. 7131–7136.
[19] Dimarogonas DV, Frazzoli E, and Johansson KH. Distributed event-triggered control for multi-agent systems. IEEE Transactions on Automatic Control. 2012;57(5):1291–1297.
[20] Seyboth GS, Dimarogonas DV, and Johansson KH. Event-based broadcasting for multi-agent average consensus. Automatica. 2013;49(1):245–252.
[21] Garcia E, Cao Y, and Casbeer DW. Cooperative control with general linear dynamics and limited communication: Centralized and decentralized

event-triggered control strategies. American Control Conference (ACC). IEEE; 2014. p. 159–164.

[22] Viel C, Bertrand S, Piet-Lahanier H, *et al.* New state estimator for decentralized event-triggered consensus for multi-agent systems. IFAC-PapersOnLine. 2016;49(5):365–370.

[23] Almeida J, Silvestre C, and Pascoal A. Synchronization of multiagent systems using event-triggered and self-triggered broadcasts. IEEE Transactions on Automatic Control. 2017;62(9):4741–4746.

[24] Vanni F, Aguiar AP, and Pascoal AM. Cooperative path-following of underactuated autonomous marine vehicles with logic-based communication. NGCUV'08-IFAC Workshop on Navigation, Guidance and Control of Underwater Vehicles; 2008. p. 1–6.

[25] Kalwa J, Pascoal A, Ridao P, *et al.* The European R&D-Project MORPH: Marine robotic systems of self-organizing, logically linked physical nodes. IFAC Proceedings Volumes. 2012;45(5):349–354.

[26] Mišković N, Bibuli M, Birk A, *et al.* CADDY-cognitive autonomous diving buddy: Two years of underwater human–robot interaction. Marine Technology Society Journal. 2016;50(4):54–66.

[27] Abreu P, Antonelli G, Arrichiello F, *et al.* Widely scalable mobile underwater sonar technology: An overview of the H2020 WiMUST project. Marine Technology Society Journal. 2016;50(4):42–53.

[28] Godsil C, and Royle GF. Algebraic Graph Theory. vol. 207. New York: Springer-Verlag; 2001.

[29] Jiang ZP, Teel AR, and Praly L. Small-gain theorem for ISS systems and applications. Mathematics of Control, Signals, and Systems (MCSS). 1994;7:95–120.

[30] Isidori A. Nonlinear Control Systems II. London, UK: Springer; 1999.

[31] Aguiar AP, and Hespanha JP. Trajectory-tracking and path-following of underactuated autonomous vehicles with parametric modeling uncertainty. IEEE Transactions on Automatic Control. 2007;52(8):1362–1379.

[32] Kaminer I, Pascoal A, Xargay E, *et al.* Time-Critical Cooperative Control of Autonomous Air Vehicles. Oxford, UK: Butterworth-Heinemann; 2017.

[33] Vanni FV. Coordinated Motion Control of Multiple Autonomous Underwater Vehicles. M.Sc. Thesis, Lisbon, Portugal: Instituto Superior Técnico; 2007.

[34] Kebkal K, Kebkal O, Glushko E, *et al.* Underwater acoustic modems with integrated atomic clocks for one-way travel-time underwater vehicle positioning. Underwater Acoustics Conference and Exhibition (UACE); 2017.

[35] Abreu PC, Botelho J, Góis P, *et al.* The MEDUSA class of autonomous marine vehicles and their role in EU projects. OCEANS 2016-Shanghai. IEEE; 2016. p. 1–10.

[36] Ribeiro J. Motion Control of Single and Multiple Autonomous Marine Vehicles. M.Sc. Thesis, Lisbon, Portugal: Instituto Superior Técnico; 2011.

[37] Horn RA, and Johnson CR. Matrix Analysis. Cambridge, UK: Cambridge University Press; 1990.

[38] Khalil HK. Nonlinear Systems. Upper Saddle River, NJ: Prentice-Hall, 1996.

Chapter 9
Formation control of autonomous marine vehicles

Basant Kumar Sahu[1], Bidyadhar Subudhi[2], and Sanjay Kumar Sharma[3]

Abstract

This chapter presents a brief review on numerous formation control strategies of multiple autonomous marine vehicle (AMV) such as autonomous underwater vehicle (AUV), unmanned surface vehicle (USV), surface vessels, etc. Classification of available formation control schemes for AMVs is presented that considers different structures, control strategies employed for formation, and coordination strategy. Advantages and disadvantages of the control strategies for multiple agents cooperative motion together with the network and communication issues in formation are also discussed.

9.1 Introduction

Formation control is an popular and important research topic in the field of multi-robot systems which focuses on controlling the cooperative motion including relative positions as well as orientations of autonomous vehicles which moves in a group as a whole. A multi-vehicle system is more work efficient in comparison to single vehicle and solves multitask problems easily.

Inspiration for formation control originates from natural sources such as teaming activities of living beings, schools of fishes, flocks of birds, colonies of bacteria, herds of wild beasts, etc. It is well known that a cooperative behavior has certain advantages. From the observation of the flocking of birds, it may be concluded that when the flying agents are in a flock, they able to avoid collision among themselves while maintaining a common heading. These delightful features of nature provide sufficient idea for mathematical understanding of cooperative behaviors of agents

[1]Department of Electrical Engineering, National Institute of Science and Technology, Berhampur, India
[2]Department of Electrical Engineering, National Institute Technology, Rourkela, India
[3]Intelligent Autonomous Control Systems, School of Engineering, Plymouth University, UK

while moving in a group. Reynolds hypothesized the program '*boid*' which explains the fundamental aspects of flocking behavior observed are: multiple agent separation, cohesion, and alignment [1,2]. In *boid*, every agent has its local control strategy but together performed to achieve a common goal. The three components are possessed by the flying strategy are: (i) alignment: steering toward the average heading of neighbors, (ii) separation: navigate for avoiding crowd, and (iii) cohesion: direct toward the average position of neighbors.

In various applications, a team of autonomous vehicles are intended to follow a predefined desired trajectory in spatial pattern. There are many advantages of deploying a number of vehicles in formation group over appointing single vehicle. It enhances the efficiency and robustness of the system with structure flexibility, reconfiguration ability, and provides redundancy [3]. For example, NASA and the Air Force employs formation flying of spacecraft which was the 21st-century missions [4,5].

In a number of applications formation control technique is used. These includes: security patrols, rescue, and search in risky environments. Formation of multiple vehicles is used in military missions for keeping specified formation for reconnaissance and area coverage. Similarly, formation helps for expanding the sensing capabilities and reduces the fuel consumption in small satellite clustering. Formation control of a robotic system is used for manipulation and transport of large objects, for toxic residual cleaning. If vehicles move in platoons with a desired velocity, in automated highway system (AHS), the transportation network throughput can be greatly increased. This also maintains a specified distance among vehicles [6]. For covering more area in less time, a group of multiple AUVs are employed in deep floor of oceans in separate levels with desired geometric fashions.

Generally, the formation control problems include the following typical features serially: (i) feasible formation assignment, (ii) formation shape maintenance, and (iii) switching among formations structures. Formation of a group of autonomous vehicles may be in rigid formation [7] or in flexible formation [8]. In flexible formation, there is a chance of change in structure of formation which is not possible in rigid formation.

Due to vast uses of formation control technique in different important fields, it is necessary to know various control issues of cooperative motion. A successful formation control of multiple agents can be achieved through a number of important steps which are as follows: choosing of appropriate agents to stay in group formation, fixation of destination and trajectory, control strategies and coordination strategies. As a whole, these steps are chronologically organized and are named as *Taxonomy* of formation structure. This is explained in Section 9.2.

The followings are the contributions of this chapter. An extensive review on formation control of multiple AMVs is presented. The various steps of taxonomy of formation structure are explained. Performances, merits, limitations of various coordination, and control strategies are compared so that one can choose the appropriate method according to the objectives.

The rest of the chapter is organized as follows. The taxonomy of formation control is described in Section 9.2. Section 9.3 discusses coordination strategies for formation control. Section 9.4 presents the control strategies of group of vehicles. Section 9.5 presents the network and communication issues. Section 9.6 describes about the different sub-problems of formation control. Conclusions are made in Section 9.7.

9.2 Classification of formation control techniques

Formation control techniques can be classified based on four stages, namely: (A) selection of vehicles with targets, (B) control abstraction, (C) coordination strategies, and (D) control strategies (Figure 9.1).

9.2.1 Selection of vehicles with targets

A vehicle is known as under-actuated if system has fewer number of inputs than degrees of freedom (DOF), over-actuated if system has higher number of actuators/inputs than that DOF or fully actuated if system has same number of inputs and DOF. In practical cases, the vehicles are chosen always non-holonomic. Non-holonomic vehicle is a system on which the non-integrable constraints act upon the motion. For example, if a vehicle has velocity constraint then the space positions cannot be found out by integrating that constraint. The maneuvering target may be dynamic or static. To reach at the target point, the formation group has to travel through the desired path.

9.2.2 Control abstraction

The control abstraction of formation control is divided into three layers [9]. These are *formation shape, robotic control* and *reference type*. The first level of *control abstraction* concerned on formation representation. An agent needs control itself to the right formation position through robotic control. As there is a control strategy to be explained to control the agents, the layer of robotic control is neglected here. In the reference type, the position of the agents is defined.

9.2.2.1 Formation shape representation

A set of edges assembled together to form a formation graph (formation shape). Figure 9.2 shows some useful scalable formation graphs. The most popular scalable forms are: the line, column, wedge, and double-platoon shapes. Here the circle in these figures represents an agent. The upward dashed arrows are the leader's moving directions.

9.2.2.2 Reference types

Right formation position of an agent is needed to maintain in the group. The method used by an agent to calculate its formation position which is called *reference type*. The three reference types for the double-platoon formation are shown in Figure 9.3. In Figures 9.2 and 9.3, the bubbles represent robots, and the relation between a robot and its referring robots is presented by a dotted arrows.

a. Single reference type

There are three types of single reference types which are explained below.

1. Predecessor (p) reference type: The first one is predecessor reference type which is shown in Figure 9.3(a). Here each vehicle predicts its own desired position on the basis of the position of the predecessor.

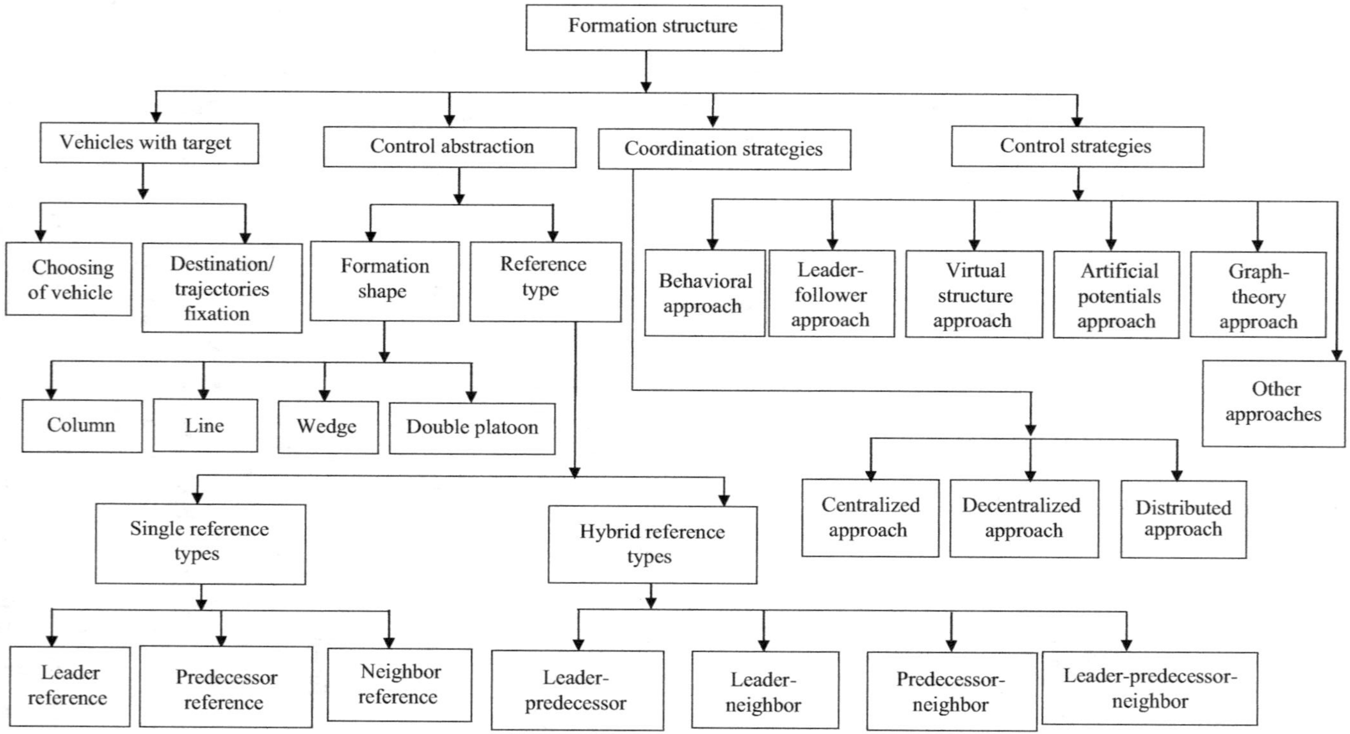

Figure 9.1 Classification of formation control of multiple AMVs

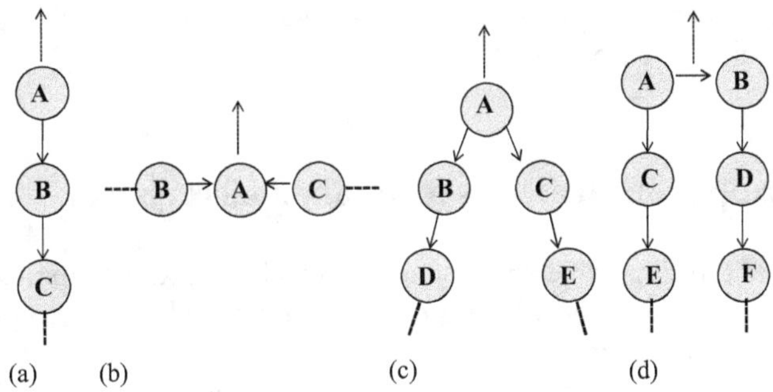

Figure 9.2 Various types of formation shapes: (a) column, (b) line, (c) wedge, and (d) double platoon

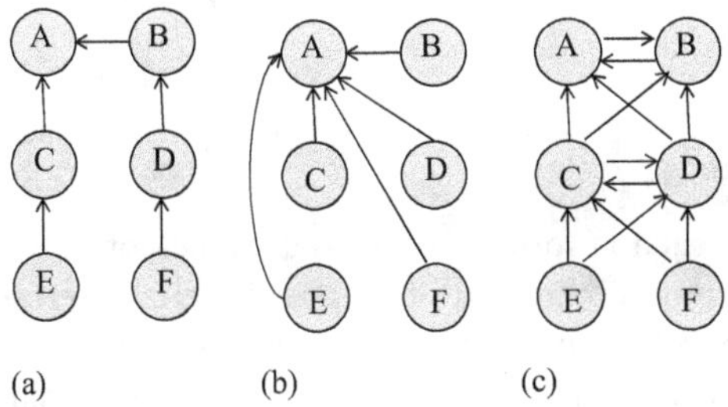

Figure 9.3 Different reference types: (a) predecessor, (b) leader, and (c) neighbor reference types

Observing this figure it is found that, the vehicle D is the predecessor of the vehicle F, and on the basis of location of the vehicle D, the vehicle F calculates its desired position.

2. Leader (l) reference type: In the leader reference type, on the basis of the position of the leader vehicle, each vehicle calculates its desired position which is shown in Figure 9.3(b). The mostly used method is the leader-follower approach. The followers which are just immediate after the leader/leaders play the role of local leaders for the other follower's. On the basis of locations of the leader as well as local leaders, a vehicle in the group computes its correct formation position.

3. Neighbor (n) reference type: Each vehicle calculates its desired position according to the position of its neighbor vehicles, which is shown in Figure 9.3(c). A vehicle detects its right, left, and front neighbor vehicles for keeping formation.

b. Hybrid reference types

The hybrid reference type is formulated from the combination of more than one type of single reference types and the root single reference types are known as *ancestor references*. There are four types of hybrid references (i.e. leader-neighbor (l-n), the leader-predecessor (l-p), predecessor-neighbor (p-n), and leader-predecessor-neighbor (l-p-n) reference types).

9.3 Coordination strategies of autonomous vehicles

The process of forming stable link among the vehicles is coordination. Due to development of coordination, the whole group will able to solve a complex task in an organized, better, and efficient way.

The coordinated structure for group of vehicles is formed by different types of formation feedback [4]. Formation error is calculated on the basis of formation function for a group of multiple vehicles to make the system stable [10]. There are three different coordination approaches for solving the problems of motion planning/control of a set of autonomous vehicles including mainly centralized, decentralized, and distributed approaches.

9.3.1 Centralized approach

The centralized approach is formulated with a single controller. All the agents are controlled with a central control processing unit which is presented in Figure 9.4. All the group members have idea about the desired shape to form during movement, desired path to be traveled, and location of formation. For achieving the formation control objective, the central controller used all the information available in the whole system for processing. Hence it possesses the best performance centrally.

For multiple vehicles, the centralized method of coordination is presented in Figure 9.4. All the decisions for the movement of the group are taken by the central controller.

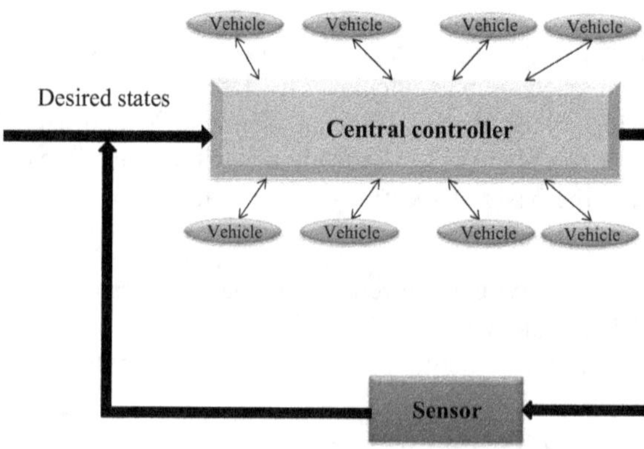

Figure 9.4 Centralized approach coordination of agents

9.3.1.1 Advantages

For a small group of multiple vehicles, the centralized formation approach is used more and is a good strategy. The whole team is monitored and controlled by a single computer, which act as the central processor and a single feedback sensor. The complex movement scheme can be performed easily.

9.3.1.2 Disadvantages

The centralized approaches have the main drawback in the computation complexity in executing the cooperating algorithm. With the number of robots, gradually the algorithm complexity raises. A large number of robots are communicated by a single robot which has decision-making power. It desires huge information and high power of computation to possess the super performance which is needed to achieve the formation goal. Due to its high dependency of the system on a single central controller, its performance is not robust.

9.3.2 Decentralized approach

The decentralized control approach structure is formulated by coordinated individual agent of the group. Here, every individual agent has controller and decision-making power, hence individual agent is autonomous. For a team having a large number of vehicles, the computational complexity increases and it needs large communication bandwidth for formation operation. In this condition, it is better to use decentralized formation control. Sometimes to keep decentralized coordination, the agents used the neighbor's information [11]. According to this, the decentralized control approach is the action taken in the distributed manner. The schematic control structure of decentralized controller is presented in Figure 9.5.

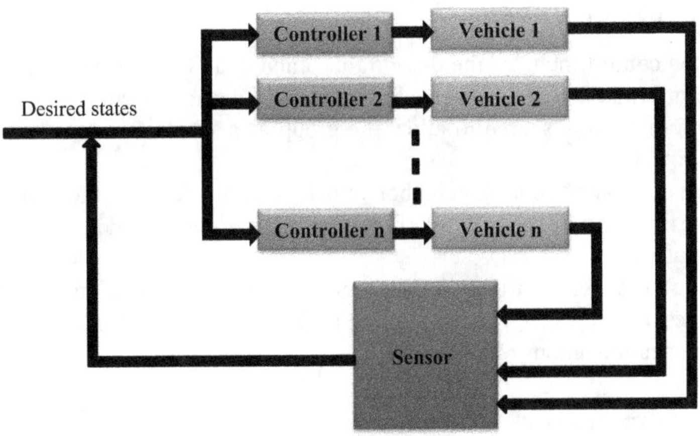

Figure 9.5 Decentralized approach coordination of AUVs

Figure 9.5 presents the decentralized approach of coordination of vehicles. In this method, each vehicle is controlled by a separate controller. Communications between neighbors are established to keep coordination between vehicles. In particular cases, communication among agents can be established for keeping coordination which is not limited to neighbors.

The controller may develop in centralized manner but the navigation functions are developed in decentralization manner [12].

9.3.2.1 Advantages

As a separate controller is used to operate individual agent, hence, this reduces both the information flow and the controller computational power significantly which leads the system to solve complex tasks efficiently. Due to this property, the flexibility scalability and robustness of the systems essentially rise. The dependency on any central controller decreases. Therefore, damage of any controller may not cause any failure of the system formation. This controller can be easily synthesized and implemented. If it is necessary, by this method, the human decision can supersede. It does not depend on the geometry of formation which leads to less faulty switching between formation structures. When the communication among the vehicles is limited, in that case the decentralized controller is mostly effective.

9.3.2.2 Disadvantages

The main disadvantage includes the prediction and control of the critical points. It has lack of good designing methods. Stability margins of the system decreases with increasing the number of vehicles. Less robust to external disturbances. Decentralized approaches are basically model-based controllers. There are multiple equilibrium points exist in a closed loop decentralized controller.

9.3.3 Distributed approach

Distributed control includes the decentralized and coupled controllers. It uses only neighbor information (communicated or measured) to generate control inputs to each agent. Information is shared to allow the group as a whole to solve the particular control problem.

In the distributed control, neighbor vehicles communication among the vehicles is possible through wireless networks [13]. The whole group is modeled as a networked control system.

Figure 9.6(a) demonstrates the distributed method of coordination strategy among the vehicles. Figure 9.6(b) is similar to Figure 9.6(a) with the presence of probability of communication among vehicles.

9.3.3.1 Advantages

With the knowledge of vehicle transmits only state information between neighbors, it is assumed that each state transmission is possible through only one packet of data. Assuming the delay in the feed-forward with feedback channels is negligible as each

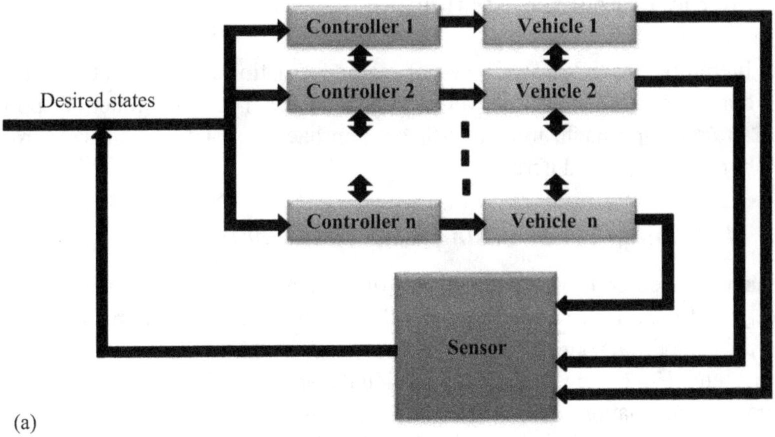

(a)

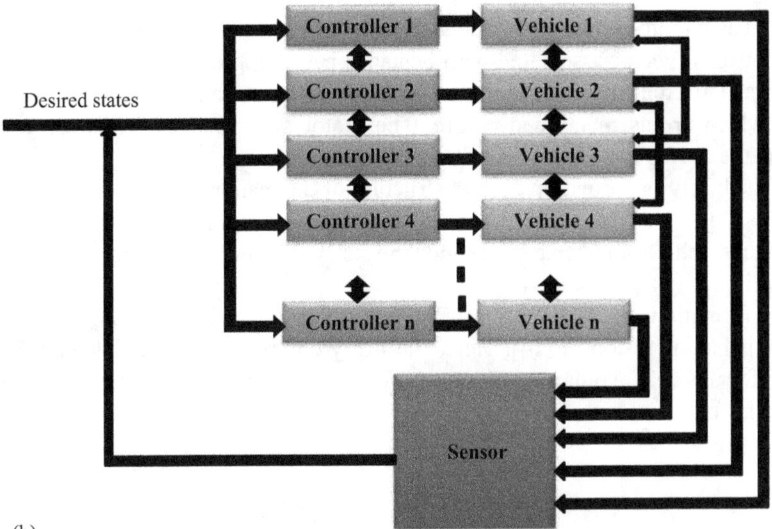

(b)

Figure 9.6 (a and b) Distributed approach coordination of vehicles

vehicle is directly connected to a sensor and an actuator. The interfacing of the vehicles with the sensor and an actuator are very fast comparing with the sample period. Easy for calculation and synthesize.

9.3.3.2 Disadvantages

It is difficult for solving any formation problem by individual vehicle in the formation group. As it uses the decentralized method, so all disadvantages of decentralized control are included in this method.

9.4 Formation control strategies

In the literature, there are five types of major formation control strategies such as behavioral approach, leader-follower approach, artificial potential function approach, virtual structure approach, and graph-theory approach are found. All these approaches are elaborately explained here.

9.4.1 Formation control using behavioral approach

In behavior-based control, each behavior of the final control is computed by using the weight of the relative importance of all individual agents. From behavior of each agent, the behavioral approach starts, e.g. obstacle avoidance, goal seeking, formation keeping, etc. Hence, by using the individual behavior of different vehicles more complex motion pattern can be generated. Behavioral-approach formation control structure is schematically presented in Figure 9.7.

A basic motor schema is associated with each agent in behavioral approach [8]. A significant vector associated with behavior-based approach is generated by each schema. Different motor schemas may be obstacle avoidance, collision avoidance, etc., which are as mentioned before. The motor schema behavior is assigned by taking the average vector weighted of control action.

The behavior formation control structure of the team of robots consists of three levels such as team behavior, robot role and behavior, and robot control. The role of individual vehicle and behavior defined the task performed by the individual robot in formation.

A formation control law of group of vehicles may be solved by using fuzzy logic-based behavior fusion in [14]. Here, the fuzzy control inputs are deviation in yaw angle for obstacle avoidance. This method is applicable for leader vehicle as well as for follower vehicles separately.

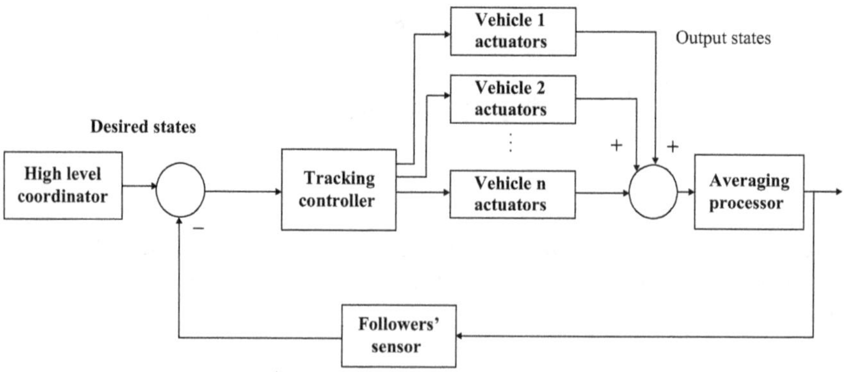

Figure 9.7 Schematic presentation of behavior approach and potential field approach

Null-space-based (NSB) formation controller is verified both by theoretically and experimentally by taking five Khepera II mobile robots in [15].

9.4.1.1 Advantages

The control strategies can be derived by itself naturally with multiple objectives. The vehicles are communicated with each other through explicit feedback network. With less significant communication behavior-based approach is decentralized. Since it is a decentralized approach, it is suitable for large group of vehicles. By this method, a group of autonomous agents can be guided in unknown and uncertainly changing environment.

9.4.1.2 Disadvantages

The behavior of the formation group cannot be defined explicitly. The mathematical computation and stability analysis are highly difficult. To diminish the computational complexities, the dynamics of individual vehicle is reduced to single integrator dynamics. Kinematic model of vehicle is more complex and has more limitations. It is very difficult to converge the group of vehicles to a desired configuration.

9.4.2 Formation control using leader-follower approach

Here, some vehicles are designated as leaders which follow the desired path, while others are designated as followers to follow the leader vehicles. The representation of the leader-follower approach is schematically presented in Figure 9.8.

Availability of information for each vehicle depends upon the quality and quantity of local sensors assembled with it. Feedback controllers are mainly two types

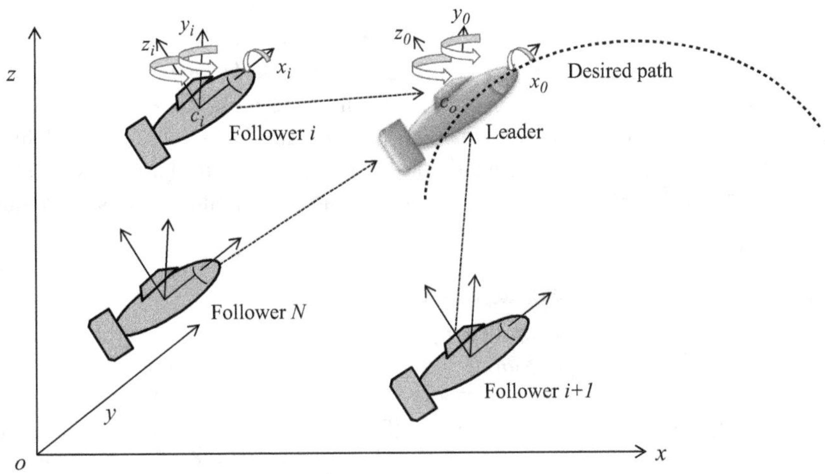

Figure 9.8 Schematic presentation of the leader–follower approach of non-holonomic agents

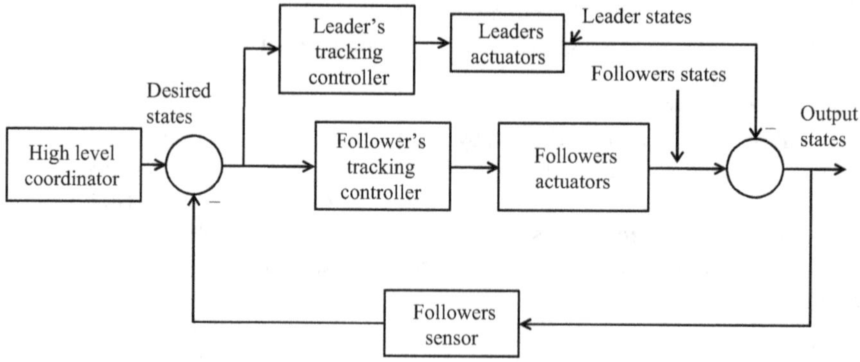

Figure 9.9 Control structure of leader-follower approach

for maintaining leader-follower formations pattern of multiple vehicles [16,17]. The schematic diagram is presented in Figure 9.9.

Beginning from an arbitrary point in space, the leader vehicle/vehicles track the desired trajectory maintaining a particular formation structure i.e. the desired trajectory positions coincide with the positions of leader vehicles. This says that the tracking error should be zero.

It is necessary that, during accomplishing formation, the leader vehicle/vehicles to follower distance remains constant. Some of the techniques of formation control of multiple robotic vehicles by using leader-follower approach are briefly explained below.

9.4.2.1 Distance-angle and distance-distance method

The most used leader-follower methods are *distance-angle, distance-distance* formula, *distance-angle*, or $L-\psi$ methods (Figure 9.10(a,b)). One vehicle follows other two vehicles. Sometimes it is taken as $leader-leader-follower$ [18]. This is equivalent to $L-L$ method of formation. $L-L$ and $L-\psi$ controller may develop for multiple unmanned vessels using sliding mode control with internal stable dynamics [19–24].

9.4.2.2 Distance and bearing angle method

Leader-follower approach using distance and bearing angle between autonomous vehicles is similar to *distance-angle* and *distance-distance* method [25–27]. It is known as distance-bearing (distance-angle) method and is constructed for individual robot by using the motor sensor topology in [28]. Geometry of formations is imposed with distance angle method in [29]. Here, the angle used to localize the vehicle is considered from the follower, not from leader. The formation control required only leader-follower relative distance and bearing angle without any knowledge of leader velocity exactly [26]. The algorithm employs to estimate the velocity profile of the leader vehicle which may be time varying or time invariant. The global problem for

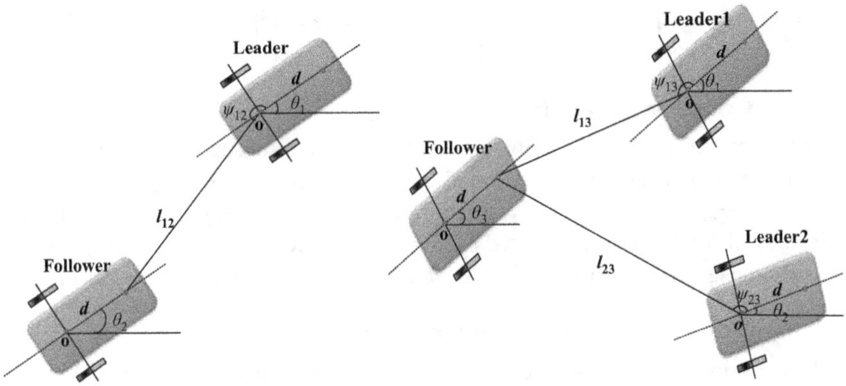

Figure 9.10 (a) Three vehicles in L – ψ formation and (b) L – L formation

formation of a group of multiple robots may resolve to local problems by using distributed control and communication protocol [30]. Here, the group of n robots is distributed into $n-1$ pair of followers and their corresponding leaders. Then positions of followers are maintained using the desired separations and relative bearings. Formation control of multiple unicycles may achieve using distance-angle methods with the unicycles are arranged in hierarchical fashion [31]. The interconnection of elementary unicycle leader-follower units are described by a rooted tree graph.

9.4.2.3 Virtual leader method

The desired path is followed by a virtual vehicle and the other vehicles has to follow the virtual vehicle [32–36]. Followers may follow the leader vehicle by estimating the position and orientation using a fast-color tracking system [34]. This establishes a Beizer trajectory between the current position and the position of the leader vehicle. The virtual concept may be used for objects transportation from one place to another [37]. In this case, the relative orientation between the leaders and each follower are controlled by a fork-lift type constraint. The travelling paths may be generated by using these constraints. The controller may be developed by using virtual leader concept and flatness property of individual vehicles. Here, reformation of the agents happened after splitting from group to avoid collision in time-dependent path [36].

9.4.2.4 Line of sight (LOS) and vision-based method

Line of sight (LOS) guidance method may use to control multiple autonomous vehicles to follow the desired path [38]. The control strategy based on the feedback linearization is derived. When the vehicles track the desired trajectory, the crosstalk errors in communication channels is minimized. Cross-talk based formation control law using LOS method is developed for multiple surface in [39]. Vision-based formation control law is developed in [25,40]. Sometimes, LOS is associated with extended Kalman Filter (EKF) to develop formation control laws [40,41]. Leader-follower problem of formation of multiple vehicles based on vision-based architecture is possible using dual unscented Kalman filter (DUKF) [42].

9.4.2.5 Consensus method

The connectivity is checked among the vehicles in consensus method by taking as a local group. Formation is performed by keeping communication among all the local sub-groups in the group. Decentralized neighbor to neighbor local controller is developed for keeping connected a group of vehicles in formation [43]. Leader-follower consensus controller is developed for a group of multiple spacecraft in [44].

9.4.2.6 Other techniques of leader-follower control strategy

On-board sensors information based leader-follower formation controller is developed in [45]. Leader follower problem is solved by using relative motion theory in [27,46], UAVs in [47]. Formation control of multiple vehicles can achieve using vector field method [48]. Leader-follower approach of multiple vehicles is developed by using the preexisting diagraph approach in [49,50]. Leader-follower formation structure is developed by using phase wave and phase gradient in a similar manner to that of a slime mold (amoeba) in [51]. Here, the controller is designed to control the direction of heading of the follower vehicles according to the heading angles of the leader vehicles. Leader-follower formation control structure is developed considering decentralized master follower formation control of multiple vehicles in [52] and mooring controller in [16].

To overcome the difficulties present in centralized formation control, a distributed formation of multiple mobile robotic vehicles is accomplished by using bond-graph modeling framework in [53]. Formation control of multiple agents is achieved by using multiple co-leaders through minimally persistence co-leader structure in [54]. Electrically driven leader-follower formation controller is developed for controlling a group of multiple mobile robots in limited availability of information environment in [55].

Leader-follower formation control law is achieved by using steady-state error analysis in [56], parametric uncertainties in [57], graphical inequalities in [58], and external disturbances and parametric uncertainties in [59]. Here, the control law may develop using LMI techniques.

Output feedback-based multiple spacecraft leader-follower formation controller is developed in [60], with the utilization of artificial potential functions associated with follower spacecraft in [61,62], network delay along with packet dropout in the communication channel in [63], using relative velocity and relative position of the spacecraft in [64].

9.4.2.7 Advantages

One of the most advantages of this approach is that it can be implemented easily. If the leader is perturbed by any external disturbances, then the formation can be maintained in case of multiple leader groups. The group objective is path following task. The leader-follower approach reduces the total huge task to individual small tasking problem, so that the followers should have states knowledge of the leader vehicles.

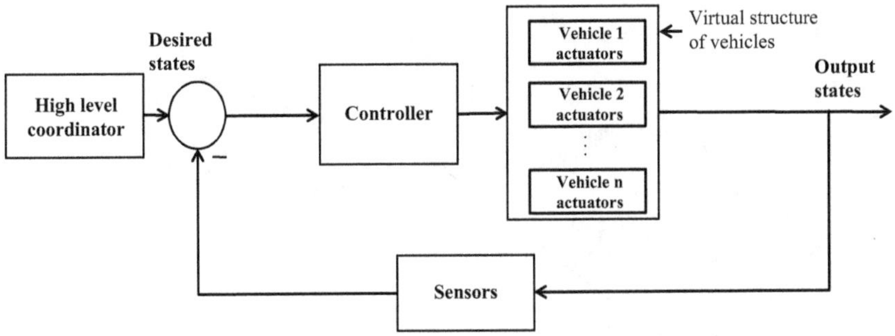

Figure 9.11 Schematic presentation of control structure of formation control of multiple autonomous vehicles via virtual structure approach

9.4.2.8 Disadvantages

In this case, the formation structure does not use any feedback control from the followers to the leader. If the follower vehicle is perturbed from the group, the formation structure cannot be maintained which shows the lack of robustness of this type of approach in case of a follower vehicle is missing. If there is a single leader in the formation group, then this leader vehicle is the point of failure of formation motion. Considering individual leader-follower pair of agent, it obeys local optimal formation control law. But connecting all individual optimal formation control law does not guarantee that it is a global optimal controller.

9.4.3 Formation control using virtual structure approach

The virtual structure concept was developed first in [65]. This is basically used for formation of spacecraft or small satellites flying [66]. The whole formation structure is considered to be a rigid body (Figure 9.12). In this approach, the control structure consists of three steps. These are: (i) virtual structure dynamic analysis, (ii) motion task translation of each agent, and (iii) each agent's controller are derived for tracking. Formation control structure generalization form is shown in Figure 9.11.

Considering there are N numbers of vehicles are in virtual structure. These N number of vehicles will remain correlated if the following equation satisfies:

$$(\eta_v^{1\times N}) - (\eta_d^{1\times N}) = 0^{1\times N}, \tag{9.1}$$

where $\eta_v^{1\times N}$ is the position vector of N robotic vehicles and $\eta_d^{1\times N}$ vector of desired trajectories position.

Following section presents the various applications as well as different method to remain correlated in virtual structure approach.

Formation control of multiple unicycle type robots is achieved by using the kinematics of the individual vehicles with the help of coupling terms among vehicles to keep robustness of the system in case of perturbation of any vehicle from the

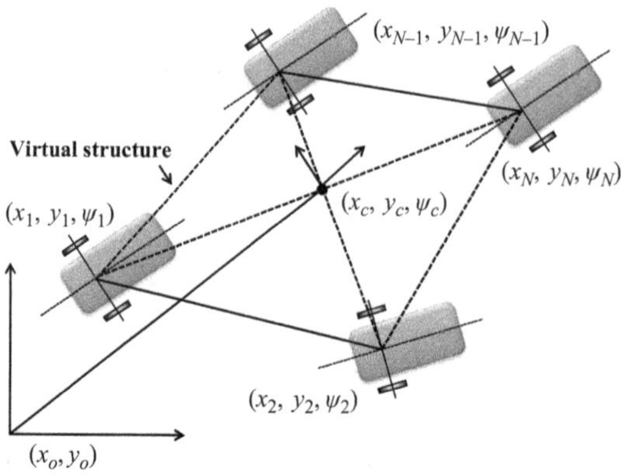

Figure 9.12 Geometric relationships among autonomous vehicles and the imposed virtual structure

group [67]. Formation of multiple AUVs is accomplished by using hierarchical virtual structure in [68,69]. Formation control of a set of aircrafts through virtual structure is achieved by using formation feedback method in [70,71,72], formation feedback method in [73], and behavioral and leader-follower in [4].

9.4.3.1 Advantages

As there is no leader follower concept is implemented in this method, it possesses certain robustness toward perturbation of any vehicle from the group. It is relatively easy to prescribe the behavior for the formation group, and coordination can be maintained perfectly during the maneuvering.

9.4.3.2 Disadvantages

In virtual structure approach, the class of potential applications is limited. This approach is not the optimal choice. This needs to be frequently reconfigured after completion of every formation task. Full states of the internal structure are needed for communication among the agents in this approach.

9.4.4 Formation control using artificial potentials approach

The interaction forces among neighbor vehicles are presented by artificial potential and are designed for keeping inter vehicle spacing according to requirement. These potential functions are of two types, i.e. repulsive potential function (RPF) and attractive potential function (APF). The potential fields as well as potential forces are calculated based on the APFs. The potential fields between the vehicles are used for keeping them with in the formation group. But the RPF is all ways computed between the vehicles as well as the vehicle and obstacle for collision avoidance.

9.4.5 Attractive potential functions

For avoiding splitting of formation group, there exist an APF among the vehicles. This APF is a distance-dependent function between ith and jth vehicles. This value is assumed to be zero when $\eta_i = \eta_j$ and reach maximum value when $\|\eta_i - \eta_j\| \geq d$. d is the safety distance. Generally, the choice of attractive potential function between vehicles grows gradually with the distance between the vehicles. The total attractive potential function is defined as:

$$U_{att}(\eta_i, \eta_j) = \frac{1}{2} k_{att} r_{ij}^2, \tag{9.2}$$

where $r_{ij} = \|\eta_i - \eta_j\|$ is the Euclidean norm of the distance between position of ith and jth vehicle. k_{att} is a scaling factor which is positive. The gradient of this APF which is a vector quantity is presented as:

$$\nabla U_{att}(\eta_i, \eta_i) = k_{att} \left(\eta_i - \eta_j\right). \tag{9.3}$$

The negative gradient of APF between vehicles is the force of attraction and is presented as:

$$F_{att} = -\nabla U_{att}(\eta_i, \eta_i) = -k_{att}(\eta_i - \eta_j). \tag{9.4}$$

Similarly, the APF between vehicle having position η_i and the desired trajectory η_d is presented as:

$$F_{att}^d = -\nabla U_{att}^d(\eta_i, \eta_d) = -k_{att}^d \left(\eta_i - \eta_d\right), \tag{9.5}$$

where k_{att}^d is a positive scaling factor which is a constant.

9.4.6 Repulsive potential functions

To avoid collision among vehicles, there should exist a separation potential functions among the vehicles. This potential function retains maximum value when collision between vehicles will happen, i.e. when $\eta_i = \eta_j$ and gradually converges to zero when $\|\eta_i - \eta_j\| \geq d$. This is maintained for keeping safety distance among the vehicles. This RPF between ith and jth vehicle is given as $U_{rep}(\eta_i, \eta_j)$. This potential function for group of N vehicles may be given as $\sum_{i=1}^{N}\sum_{j=1}^{N} U_{rep}(\eta_i, \eta_j), \quad i \neq j$.

The negative gradient of the repulsive potential function field is given by:

$$F_{rep} = -\nabla \sum_{i=1}^{N}\sum_{j=1}^{N} U_{rep}(\eta_i, \eta_j), \quad i \neq j. \tag{9.6}$$

The total potential function of possessed by the whole system can be computed by taking summation of the repulsive as well as attractive potential functions.

The negative gradient of this potential function is given as:

$$F = F_{att} + F_{att}^d + F_{rep}. \tag{9.7}$$

Some of the techniques of formation control using APF and RPF are explained as follows.

Virtual leader is used for manipulation of group in [74]. Formation controller with collision avoidance is possible by using artificial potential function [12,75,76]. The repulsive as well as attractive potential functions are computed based on the distance between (i) vehicle and vehicle, (ii) vehicle and position of the desired trajectories, and (iii) vehicle and positions of the obstacles [77,78].

Artificial potential function used p-type bump function in [79], consensus protocol [80], and communication delay in [81]. Potential field both in attractive and in repulsive form, and virtual hill method may apply to control a cluster of mobile robots [82]. By this method, the robots are controlled to drive in the direction of decreasing energy.

Multiple unmanned vehicles are swarmed into formation structure using APF with limiting functions in [83]. This approach is efficient from the mathematical computation point and well scalable as per the size of the swarm group.

9.4.6.1 Advantages

There is no need to ordering the vehicles as there is no leader is considered. Any vehicle can interchangeable with any other. There is no much harm in case of missing of a vehicle.

9.4.6.2 Disadvantages

When the repulsive potentials among the vehicles and the vehicle and obstacle are considered, the potential functions are calculated based on the local minima. So when the number of obstacles increases in the travelling environment, this increases with the number of local minima instead of routing toward the global minima. Hence the system formation becomes unstable. One of the major disadvantages of this method is calculation of the total potential function.

When multiple potentials are present, there is chance of equilibrium point may appear due to summation of two potential functions. In this situation, the composite vector vanishes and the chance of trapping the system in undesired equilibrium points increases. The calculation of these equilibrium potential is very difficult.

9.4.7 Formation control using graph-theory approach

Formation control using graph theory is an important and popular approach. Here, a group of multiple vehicles exchange information among themselves according to a pre-specified communication digraph (directed graph) [84]. The whole group is divided in to a number of sub-groups. All the vehicles of a subgroup can interchange information among themselves. The specific communication digraph is shown in Figure 9.12. Different techniques of formation control of multiple vehicles through graph-theory approach are explained below.

Consider a formation is a vector:

$$h = h_p \otimes \begin{pmatrix} 1 \\ 0 \end{pmatrix} \in R^{2N}, \tag{9.8}$$

where $\otimes$ is the Kronecker product. N autonomous vehicles are in formation h at time t if there are vectors $q, w \in R^{2N}$ such that $(x_p)_i(t) - (h_p)_i = q(t)$ and $(x_v)_i(t) = w(t)$,

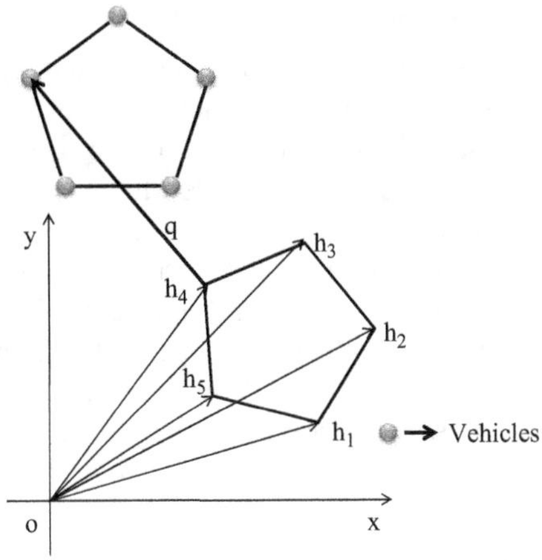

Figure 9.13 Vehicles in formation based on graph theory [84]

for $i = 1 \ldots N$. The robotic vehicles converge to formation h if there exist R^N valued functions $q(\cdot), w(\cdot)$ such that $(x_p)_i(t) - (h_p)_i - q(t) \rightarrow 0$ and $(x_v)_i(t) - w(t) \rightarrow 0$ as $t) \rightarrow \infty$.

where $x_p = ((x_p)_1, \ldots, (x_p)_N)^T$ is the vectors of position like variable and $x_v = ((x_v)_1, \ldots, (x_v)_N)^T$ is the vectors of velocity like variable.

Figure 9.13 shows the interpretation of the vectors connecting vehicles. For upholding communication between a pair of vehicles, there are two types of control structures exist in the literature, i.e. asymmetric and symmetric controls [85]. The various graph theory-based formation control techniques are explained here.

Graph theory-based formation controller for multiple vehicles are developed considering relative information in [86], consensus protocol in [87], communication delay in [88,89,90], dynamic uncertainties in [91], geodesic equations in [92], formation graph rigidity in [85,93,94], game theory approach in [95], and switching topology in [96].

9.4.7.1 Advantages

In most of the cases, rigorous mathematical calculations are not necessary and solutions of formation can be computed using only basic geometry.

9.4.7.2 Disadvantages

When solving the formation problems, the positions of vehicles are considered as the vertices of the graph and for keeping the formation the graphs should be connected. In case of the disconnected graphs, the formation is difficult.

9.4.8 Other control strategies

There are so many types of formation strategies exist in the literature other than the major types of formation strategies explained earlier. These are explained here.

9.4.8.1 Geometrical formation control

A group of multiple vehicles are allotted to maneuver and reach at a desired position without facing any collision using geometrical coordinated path following (CPF) techniques in [97–99] along with communication topology taking into account in [100,101]. Different techniques may use for development of formation controller are: discrete-time periodic communication [102–105], time delay communication [106,107], coalition approach [108], geometric formation control [98], velocity optimization technique [109] or with the adjustment of speed of a virtual target followed the desired path [110], hybrid controller is developed [111], using lumped state errors [112], and separation principle with robust control technique [113]. The formation control problem of multiple AUVs is solved by using the geometric reduction along with Jacobi shape theory in [114].

9.4.8.2 Formation reference point (FRP) method

This type of formation control method uses a virtual formation reference point (FRP). This is particularly applicable to solve the predefined path following and trajectory problems [115–120].

9.4.8.3 Consensus-based formation control method

Consensus formation controllers are developed for autonomous vehicles in [121], group of spacecraft in [122,123].

9.4.8.4 Other methods

Highly efficient formation controllers may be developed by using PD and PID control techniques [12,124], time varying center-based stabilization formation techniques for AUVs are developed [125]. Kuramoto oscillator networks-based controllers are developed in [126]. Other techniques of formation controller are using vehicle speed profiles [127], error of path [128], average position of vehicles with formation variation [129], platoon formation [130], inverse kinematics [131], horizon H^∞ control scheme [132], permutation-invariant formation technique [133], cluster space state method [134], [27], vision-based consensus [135], by coupling of individual agent coordinates [136], expansion and contraction paradigm [137], the properties and area of representation space (RS) [138].

Stable and periodic formation control of a group of nonholonomic vehicles is achieved by arranging the robotic vehicles in a cyclic interconnection topology which is known as 'cyclic pursuit' in mathematics (Figure 9.14) [139].

Multiple spacecraft formation controller of may develop with hierarchical evolutionary algorithm [140]. Here, the genetic algorithm may use for designing spacecraft desired trajectory both in high and low levels of controller. The high level is made for global planning of formation configuration optimization and collision avoidance between spacecraft. The low-level controller is designed for determining optimal trajectories of individual spacecraft which are separate from each other by using Chebyshev polynomials.

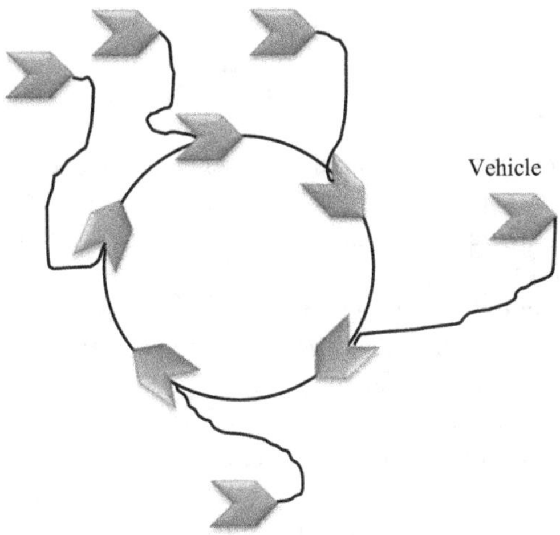

Figure 9.14 Cyclic-pursuit of multiple autonomous vehicles

In general, the leader following and virtual structure approaches are more or less belongs to the centralized coordination strategies but most of the behavioral-based formation control belongs to decentralized coordination approach.

9.5 Communication issues in formation of multiple vehicles

The vehicles should communicate each other during moving in a group. Otherwise, it may communicate to a central hub which is known as mother vehicle/ship. The *communication range* in formation is defined as the maximum distance up to which communication between two vehicles is possible. If the vehicles are communicated directly which are present sufficiently nearby, then it is *local communication*. It is called the *global communication*, if a vehicle communicates with any other vehicle within the communication range. Studies in local communication are more scalable, robust, and easier to build in comparison to global communication, but these are limited in variety and precision in formation. The amount of data or any information is transmitted or received by vehicles in a given period of time can be determined by the *communication bandwidth*. The physical interconnections between vehicles are determined by the *communication topology*. The interconnection structure may be unidirectional or bidirectional. Also the communication topology may be fixed or dynamic, so that the formation structure changes accordingly.

In wireless network, one vehicle uploads the data and another one downloads it. So each communication transaction between sender and acceptor forms a short network packet called a *cycle-initialization*. One cycle consists of a source, destination, and data rate of packet flow. So, the components of a typical cycle-initialization are as follows.

Data Uplink: The central node that sends data or polls one or more vehicles in sequential manner.

Command Downlink: The central node sends command to individual vehicle on the demand of a system operator.

Peer-to-Peer: Any vehicle can transmit data to any other vehicle without any network access control.

Acknowledgement: When providing data to the modem, the ACK flag of the transmitter is set in such a manner that when the receiver receives the data, it immediate sends an acknowledgement in good condition of Cyclic Redundancy Check (CRC).

Special case: communication issues in formation of multiple AUVs

Communication in under water environment is difficult due to the following reasons; time variation in channel of propagation or, multiple propagation chances, small available band width, in travelling medium there is chance of strong attenuation of signal, etc. Acoustic signals are commonly used in AUV communication instead of electromagnetic waves. The rate of data transmission is less in acoustic communication as compared to terrestrial communication. Maintaining formation keeping cooperation through wireless network of multiple AUVs is a difficult task.

To establish appropriate cooperation among the multiple AUVs by overcoming the limitations of underwater environment, a special type of system, i.e. DELPHIS is used in [141].

There exists an architecture in DELPHIS which allows efficient communication among AUVs. The DELPHIS system is independent upon any centralized task allocation unit but total task is uniformly distributed among all the team members. One control/communication method is applied to all AUVs avoiding complex environment structure. The sub-architectures of DELPHIS are mission model, AUV database, and mission controller. Using these components, the total system maintains the communication between AUVs efficiently. A particular filter-based system may develop where AUV localization is possible using acoustic sensor nodes designed for ad-hoc network [142]. Multimode operations based on the control and communication is used to build up a prototype in [143]. An acoustic modem that is flexible enough test different communication algorithms including networking protocols may develop [144]. Due to its high flexible structure, this modem is called as reconfigurable Modem or rModem. The main purpose of this modem is to bring the simulation results and rapid prototyping environment together. An inter-vehicular communication system is used for keeping communication among the fleet robotic vehicles in [145] and with environmental study in [146]. There are two methods for developing communication processes are considered in [147], i.e. information exchange strategy is reduced connectivity loss or communication range is reduced. A cooperative distributed algorithm for the management of ad-hoc underwater network is presented. Autonomous Undersea System Network (AUSNET) for cooperative control of AUVs may be useful [148]. A network is created whose capability is based on the dynamic source routing DSR which includes the application program interface (API) among the network drives and the applications that uses these drives. These are efficient on cooperative motion control of multiple AUVs. Network algorithm, based on time-scheduled operation is developed to locate the AUV under water in [149], a special communication protocol, i.e. Compact Control Language (CCL) is used in [150].

9.6 Formation control sub-problems

For getting more accurate performances, different sub-problems associated with the formations control problems such as collision and obstacle avoidance, path following and trajectory tracking and path generation, formation shape generation, switching between shapes according to situation etc. are necessarily taken into consideration.

9.6.1 Obstacle and collision avoidance

During movement of the formation structure, there are chances of collision among the vehicles themselves as well as the vehicles with the external obstacles. It is necessary to avoid all these types of collisions for performing smooth formation movement. Hence collision avoidance is considered as highly essential sub-problem in the formation of structure movement. The obstacle may be fixed/static i.e. whose position does not change in inertial frame of reference with respect to time or dynamic/moving whose position changes with time [109]. In case of a moving obstacle, the obstacle may appear suddenly with time. This collision avoidance problem can easily be solved by using artificial potential function method. The negative gradient of this artificial potential function is the artificial potential force. This generates the repulsive force among the vehicles or the vehicle and the obstacles to avoid the collision. This is a distance-dependent function, which is inversely proportional to the norm of the distance between two objects involved with collision.

Collision avoidance algorithm is possible by considering the area around the obstacles into different regions [151]. Collision avoidance are possible by the techniques such as Fuzzy logic method [2], Mixed Integer Quadratic Programming (MIQP) optimization method [152], limit cycle process [153–155], dual mode control strategy [156], *line-of-sight obstacle avoidance* principle [157], forming multiple spheres with time (Figure 9.15) [158], and forbidden sphere technique [159].

The problem of collision between vehicles may be solved for a formation group moving in plane with the use of gyroscopic forces [160], null-space-based behavior [15], and virtual leader techniques [18,27,24].

9.6.2 Formation shape generation

Different types of formation shape maintenance and generation are important directions of formation control. The formation of a group of multiple robots is possible in different formation regions [161]. The common shapes that attend by formation structure are square, rectangular, triangular, ring, ellipsoid polygonal, crescent (Figure 9.16).

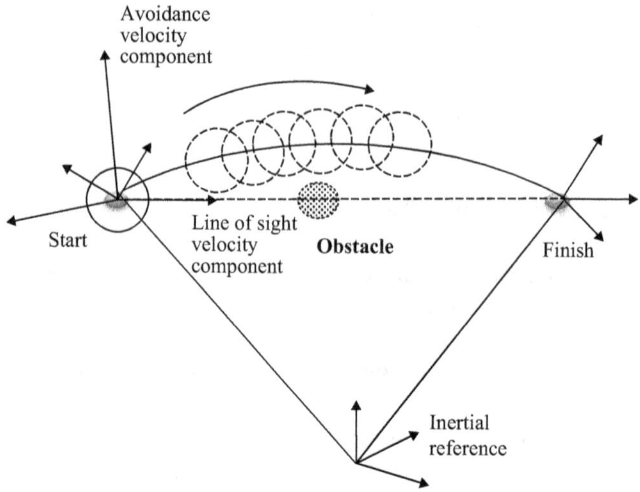

Figure 9.15 Line-of-sight and collision-avoidance path components

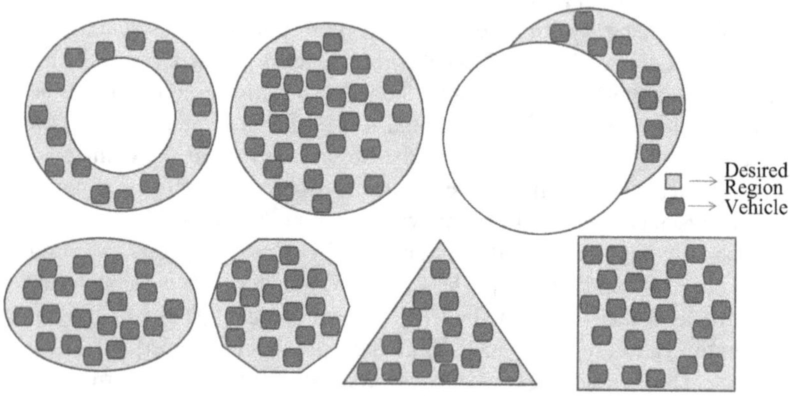

Figure 9.16 Examples of desired regions

Transportation task of objects becomes easier by using the formation control technique [37]. For transporting the objects from one place to another place defines the shape of the formation structure with vertices of this formation structures are presented by the grasping points of the robots/vehicles to the object to be transported. The formation shape generation and alteration are possible easily by using simple geometry of equilateral triangle [162], whereas the communication among the vehicles can be maintained by using the graph theory topology [163].

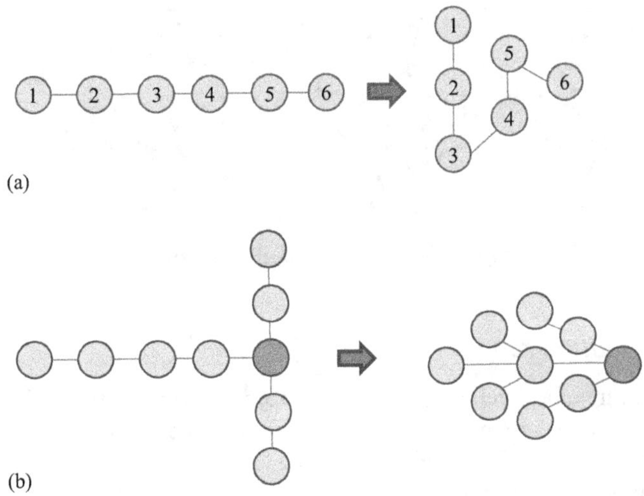

Figure 9.17 *(a) Schematic presentation of six-vehicle formation shape change graphs. (b) Switch from T-formation shape to diamond formation shape*

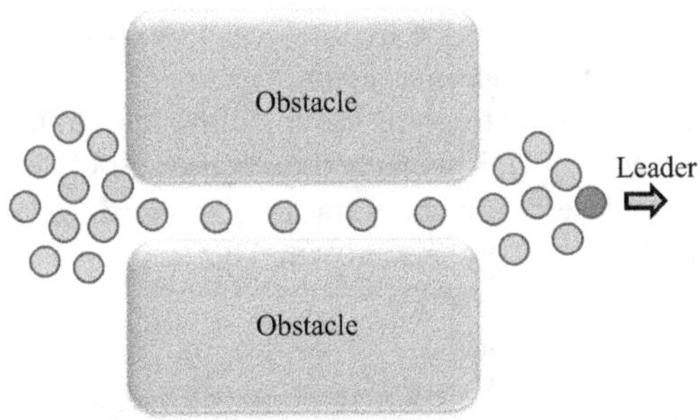

Figure 9.18 *A group of vehicles change their formation shape to pass through a narrow region*

9.6.3 Switching between shapes according to situation

Formation structure shapes can switch from one shape to other, for example switching from straight line shape to wedge shape [164] or the shape change from double platoon to diamond [109], from T-shape to diamond, etc., which are presented in Figure 9.17(a) and (b).

In drastic environmental state the formation structure may attain the shape of an amoeba whose shapes vary accordingly [51]. The shape of the formation structure may vary for passing through narrow region as presented in Figure 9.18 [77].

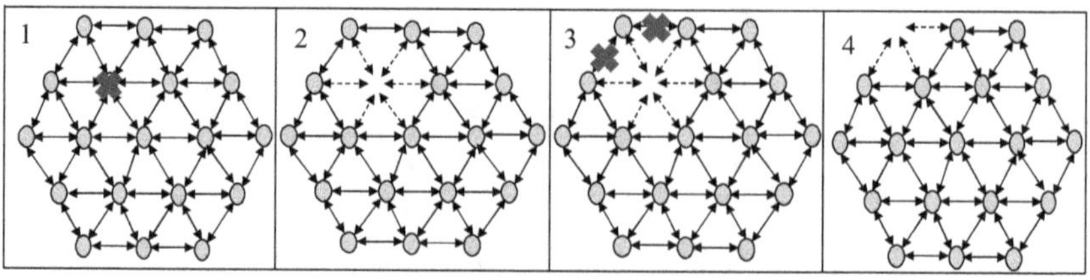

Figure 9.19 Repairing of the formation shape in case of missing of a vehicle

9.6.4 Formation repair

In case of failure of any vehicle present in the formation group, this defective vehicle are escaped from the group and the reconfiguration of the group starts again by repositioning the working vehicle and avoiding the failed one as obstacle. The schematic presentation of the formation repair is presented in Figure 9.19 [165].

9.6.5 Movement of formation structure

On the basis of maneuvering of the formation structure, the movement of formation structure may be three types. These are path generation, path following and trajectory tracking. In case of path generation the suitable way points have to be generated by the controller where to be traveled by the formation structure. This technique is used for movement of the formation structure [133,152]. In case of path following program the whole formation structure has to follow the predefined desired path which does not vary with time. This technique is used in [2] for movement of the formation structure. In case of trajectory tracking problems the formation structure moves along the timedependent predefined path. This technique is used for movement of the formation structure in [68,166].

9.7 Conclusions

In this chapter, an elaborate review on research reported on formation control of a group of multiple AMVs is presented. Taxonomy of formation structure has been included depending on different formation structures, control strategies, and coordination strategies. An elaborate survey on effectiveness of the reported formation control strategies is presented which will motivate the researchers in the area in future control developments of the different autonomous vehicles.

References

[1] C. Reynolds, "Flocks, herds and schools: A distributed behavioral model," *Proc. 14th Annual conf. on Computer graphics and Interactive Techniques* (Association for Computing Machinery), pp. 25–34, 1987.

[2] B. K. Sahu and B. Subudhi, "Flocking Control of Multiple AUVs Based on Fuzzy Potential Functions," *IEEE Trans. Fuzzy Syst.*, vol. 26, no. 5, pp. 2539–2551, 2018.

[3] D. J. Stilwell and B. E. Bishop, "Platoons of Underwater Vehicles," *IEEE Control Syst. Magaz.*, vol. 20, pp. 45–52, 2000.

[4] R. W. Beard, J. Lawton and F. Y. Hadaegh, "A Coordination Architecture for Spacecraft Formation Control," *IEEE Trans. Control Syst. Technol.*, vol. 9, pp. 777–790, 2001.

[5] D. P. Scharf, F. Y. Hadaegh and S. R. Ploen, "A survey of spacecraft formation flying guidance and control (part I): Guidance," *Proc. Am. Control Conf.*, vol. 2, pp. 1733–1739, 2003.

[6] J. G. Bender, "An Overview of Systems Studies of Automated Highway Systems," *IEEE Trans. Vehicular Technol.*, vol. 40, no. 1, pp. 82–99, Feb. 1991.

[7] R. O. Saber, W. B. Dunbar and R. M. Murray, "Cooperative Control of Multi-vehicle Systems Using Cost Graphs and Optimization," *Proc. Am. Cont. Conf.*, Denver, Colorado, pp. 1–15, 2003.

[8] T. Balch and R. Arkin, "Behavior-based Formation Control for Multirobot teams," *IEEE Trans. Robot. Autom.*, vol. 14, no. 6, pp. 926–939, 1998.

[9] H. Chia, H. Hsu and A. Liu, "Applying Various Reference Types to Formation Control of Mobile Robots," *J. Inform. Sci. Eng.*, vol. 23, pp. 1499–1522, 2007.

[10] P. Ogren, M. Egerstedt and X. Hu, "A Control Lyapunov Function Approach to Multiagent Coordination," *IEEE Trans. Robot. Autom.* vol. 18, pp. 847–851, 2002.

[11] G. Ferrari-Trecate, A. Buffa and M. Gati, "Analysis of Coordination in Multi-Agent Systems through Partial Difference Equations," *IEEE Trans. Autom. Control*, vol. 51, no. 6, pp. 1058–1063, 2006.

[12] K. D. Do, "Formation Stabilization and Tracking Control of Mobile Agents Using Local Potential Functions," *Proc. Am. Control Conf.* Minnesota, USA, pp. 2142–2147, Jun. 14–16, 2006.

[13] D. Gu and H. Hu, "Distributed network-based formation control," *Int. J. Syst. Sci.*, vol. 40, no. 5, pp. 539–552, 2009.

[14] X. Kang, H. Xu and X. Feng, "Fuzzy Logic Based Behavior Fusion for Multi-AUV Formation Keeping in Uncertain Ocean Environment," *IEEE Conf.*, China, 2009.

[15] G. Antonelli, F. Arrichiello and S. Chiaverin "Experiments of Formation Control With Multirobot Systems Using the Null-Space-Based Behavioral Control," *IEEE Trans. Control Syst. Technol.*, vol. 17, no. 5, pp. 1173–1182, 2009.

[16] E. Yang and D. Gu, "Nonlinear Formation-Keeping and Mooring Control of Multiple Autonomous Underwater Vehicles," *IEEE/ASME Trans. Mechatronics*, vol. 12, no. 2, pp. 164–178, 2007.

[17] D. B. Nguyen and K. D. Do, "Formation Control of Mobile Robots," *Int. J. Comput. Commun. Control*, vol. 1, no. 3, pp. 41–59, 2006.

[18] M. N. Soorki, H. A. Talebi and S. K. Y. Nikravesh, "A Leader-following Formation Control of Multiple Mobile Robots With Active Obstacle Avoidance," *Iranian Conf. on Elect. Eng.,* Tehran 1–6, 2011.

[19] F. Fahimi, "Sliding-Mode Formation Control for Underactuated Surface Vessels," *IEEE Trans. Robot.*, vol. 23, no. 3, pp. 617–622, 2007.

[20] F. Fahimi, "Sliding Mode Formation Control for Under-Actuated Surface Vessels," *IEEE Trans. Robot.*, vol. 23, no. 3, pp. 617–622, 2007.

[21] F. Fahimi, "Sliding Mode Formation Control for Under-Actuated Autonomous Surface Vehicles," *Proc. Am. Control Conf.* Minnesota, USA, pp. 4255–4260, Jun. 14–16, 2006.

[22] S. A. P. Ramaswamy and S. N. Balakrishnan, "Formation Control of Car-Like Mobile Robots: A Lyapunov Function Based Approach," *Am. Control Conf.*, pp. 657–661, Jun. 11–13, 2008.

[23] J. P. Desai, J. P. Ostrowski and V. Kumar, "Modeling and Control of Formations of Nonholonomic Mobile Robots," *IEEE Trans. Robot. Autom.*, vol. 17, no. 6, pp. 905–908, 2001.

[24] M. N. Soorki, H. A. Talebi and S. Nikravesh, "A Robust Leader-obstacle Formation Control," *IEEE Trans. Syst. Man Cybern. B: Cybern.*, vol. 42, no. 2, pp. 489–494, 2012.

[25] A. P. Dani, N. Gans and W. E. Dixon, "Position-Based Visual Servo Control of Leader-Follower Formation Using Image-Based Relative Pose and Relative Velocity Estimation," *Am. Control Conf.*, MO, USA, pp. 5271–5276, 10–12, 2009.

[26] K. Choi, S. J. Yoo, J. B. Park and Y. H. Choi, "Adaptive Formation Control in Absence of Leader's Velocity Information," *IET Control Theory Appl.*, vol. 4, issue 4, pp. 521–528, 2010.

[27] M. N. Soorki, H. A. Talebi and S. K. Y. Nikravesh, "A Robust Dynamic Leader-Follower Formation Control with Active Obstacle Avoidance," *IEEE Int. Conf. Syst., Man, Cybern.*, pp. 1932–1937, 2011.

[28] G. A. Kaminka, R. Schechter-Glick and V. Sadov, "Using Sensor Morphology for Multirobot Formations," *IEEE Trans. Robot.*, vol. 24, no. 2, pp. 271–282, 2008.

[29] L. Consolini, F. Morbidi, D. Prattichizzo and M. Tosques, "On the Control of a Leader-Follower Formation of Nonholonomic Mobile Robots," *Proc. IEEE Conf. Decision Control*, CA, USA, pp. 5992–5997, 2006.

[30] J. Shao, G. Xie and L. Wang, "Leader-following Formation Control of Multiple Mobile Vehicles," *IET Control Theory Appl.*, vol. 1, issue 2, pp. 545–552, 2007.

[31] L. Consolini, F. Morbidi, D. Prattichizzo and M. Tosques, "Stabilization of a Hierarchical Formation of Unicycle Robots with Velocity and Curvature Constraints," *IEEE Trans. Robot.*, vol. 25, no. 5, pp. 1176–1184, 2009.

[32] R. Cui, S. S. Ge, B. How and Y. S. Choo, "Leader-follower Formation Control of Underactuated AUVs with Leader Position Measurement," *Proc. IEEE Int. Conf. Robot. Autom.*, Japan, pp. 967–984, 2009.

[33] J. Almeida, C. Silvestre and A. Pascoal, "Cooperative Control of Multiple Surface Vessels in the Presence of Ocean Currents and Parametric Model Uncertainty," *Int. J. Robust Nonlinear Control*, vol. 20, no. 14, pp. 1549–1565, 2010.

[34] W. Zhao, Y. Hu and L. Wang, "Leader-following Formation Control of Multiple Vision-based Autonomous Robotic Fish," *Joint 48th IEEE Conf. on Decision and Control and 28th Chinese Control Conf.* Shanghai, PR China, pp. 579–584, Dec. 16–18, 2009.

[35] G. W. Gamage, G. K. I. Mann and R. G. Gosine, "Leader Follower Based Formation Control Strategies for Nonholonomic Mobile Robots: Design, Implementation and Experimental Validation," *Am. Control Conf.*, MD, USA, pp. 224–229, Jun. 30–Jul. 02, 2010.

[36] A. Ailon and I. Zohar, "Control Strategies for Driving a Group of Nonholonomic Kinematic Mobile Robots in Formation Along a Time-Parameterized Path," *IEEE/ASME Trans. Mechatronics*, vol. 17, no. 2, pp. 326–336, 2012.

[37] A. Yufka, O. Parlaktuna and M. Ozkan, "Formation-Based Cooperative Transportation by a Group of Non-holonomic Mobile Robots," *IEEE Int. Conf. Syst. Man Cybern.*, Istanbul, Turkey. pp. 3300–3307, Oct. 2010.

[38] Y. Wang, W. Yan and W. Yan, "A Leader-Follower Formation Control Strategy for AUVs Based on Line-of-Sight Guidance," *Proc. IEEE Int. Conf. on Mechatr. and Automation*, Changchun, China, pp. 4863–4867, Aug. 2009.

[39] E. Borhaug, A. Pavlov and K. Y. Pettersen, "Cross-track Formation Control of Underactuated Surface Vessels," *Proc. IEEE Conf. on Decision and Control*, CA, USA, pp. 5955–5961, Dec.13–15, 2006.

[40] G. Luca, F. Morbidi and D. Prattichizzo, *et al.*, "Vision-Based Localization for Leader–Follower Formation Control," *IEEE Trans. Robot.*, vol. 25, no. 6, pp. 1431–1438, 2009.

[41] P. Vela, A. Betser, J. Malcolm and A. Tannenbaum, "Vision-Based Range Regulation of a Leader-Follower Formation," *IEEE Trans. Control Syst. Technol.*, vol. 17, no. 2, pp. 442–448, 2009.

[42] O. A. A. Orqueda and R. Fierro, "Visual Tracking of Mobile Robots in Formation," *Proc. Am. Control Conf.* USA, pp. 5940–5945, Jul. 11–13, 2007.

[43] N. Sorensen and W. Ren, "A Unified Formation Control Scheme with a Single or Multiple Leaders," *Proc. Am. Control Conf.*, USA, pp. 5412–5418, Jul. 11–13, 2007.

[44] A. Abdessameud and A. Tayebi, "Attitude Synchronization of a Spacecraft Formation Without Velocity Measurement," *Proc. of IEEE Conf. on Decision and Control*, Cancun, Mexico, pp. 3719–3724, Dec. 9–11, 2008.

[45] T. Gustavi and X. Hu, "Observer-Based Leader-Following Formation Control Using Onboard Sensor Information," *IEEE Trans. Robot.*, vol. 24, no. 6, pp. 1457–1462, 2008.

[46] M. Defoort, T. Floquet, A. Kokosy and W. Perruquetti, "Sliding-Mode Formation Control for Cooperative Autonomous Mobile Robots," *IEEE Trans. Indus. Electron.*, vol. 55, no. 11, pp. 3944–3953, 2008.

[47] D. Walle, B. Fidan, A. Sutton, C. Yu and B. D. O. Anderson, "Non-hierarchical UAV Formation Control for Surveillance Tasks," *Am. Control Conf.*, pp. 777–782, Jun. 11–13, 2008.

[48] J. Kwon, C. Kim, H. Lee, D. Chwa and S. Hong, "Nonlinear Control for Column Formation of Wheeled Mobile Robots Based on Vector Field Method," *ICROS-SICE Int. Joint Conf.*, Japan, pp. 4087–4091, Aug. 18–21, 2009.

[49] J. Guo Z. Lin, M. Cao and G. Yan, "Adaptive control Schemes for Mobile Robot Formations with Triangularised Structures," *IET Control Theory Appl.*, vol. 4, no. 9, pp. 1817–1827, 2010.

[50] J. Guo, Z. Lin, Ming Cao and G. Yan, "Adaptive Leader-Follower Formation Control for Autonomous Mobile Robots," *Am. Control Conf.*, MD, USA, pp. 6822–6827, Jun. 30–Jul. 02, 2010.

[51] D. Kurabayashi, T. Choh, J. Cheng and T. Funato, "Adaptive Formation Transition among a Mobile Robot Group based on Phase Gradient," *Proc. IEEE Int. Conf. Robotics and Biomimetics*, Bangkok, Thailand, pp. 2001–2006, Feb. 21–26, 2009.

[52] A. K. Ray, P. Benavidez, L. Behera and M. M. Jamshidi, "Decentralized Motion Coordination for a Formation of Rovers," *IEEE Syst. J.*, vol. 3, no. 3, pp. 369–381, 2009.

[53] M. J. Daigle, X. D. Koutsoukos and G. Biswas, "Distributed Diagnosis in Formations of Mobile Robots," *IEEE Trans. Robot.*, vol. 23, no. 2, pp. 553–569, 2007.

[54] T. H. Summers, C. Yu, B. D. O. Anderson and S. Dasgupta, "Control of Coleader Formations in the Plane," *Joint 48thIEEE Conf. on Decision and Control and 28th Chinese Control Conf.* Shanghai, PR China, pp. 8334–8339, Dec. 16–18, 2009.

[55] B. S. Park, J. B. Park and Y. H. Choi, "Adaptive Formation Control of Electrically Driven Nonholonomic Mobile Robots with Limited Information," *IEEE Trans. Syst. Man Cybern. B: Cybern*, vol. 41, no. 4, pp. 1061–1075, 2011.

[56] M. A. Golkar, A. Momeni, K. Moezzi, A. G. Aghdam and I. Mantegh, "Convergence Analysis and Controller Design for a Team of Mobile Robots Subject to Measurement Error," *Am. Control Conf.,* San Francisco, CA, USA, pp. 3350–3356, Jun.29–Jul.01, 2011.

[57] M. S. Queiroz, Q. Yan, G. Yang and V. Kapila, "Global Output Feedback Tracking Control of Spacecraft Formation Flying with Parametric Uncertainty," Proc. *Conf. Decision & Control*, Arizona, USA, pp. 584–589, Dec. 1999.

[58] M. Mesbahi and F. Y. Hadaegh, "Graphs, Matrix Inequalities, and Switching for the Formation Flying Control of Multiple Spacecraft," *Proc. Am Control Conf.,* San Diego, California, pp. 4148–4152, Jun. 1999.

[59] B. S. Park, J. B. Park and Y. H. Choi, "Robust adaptive formation control and collision avoidance for electrically driven non-holonomic mobile robots," *IET Control Theory Appl.*, vol. 5, no. 3, pp. 514–522, 2011.

[60] Q. Yan, G. Yang, V. Kapila and M. S. Queiroz, "Nonlinear Dynamics and Output Feedback Control of Multiple Spacecraft in Elliptical Orbits," *Proc. American Control Conf.*, Chicago, Illinois, pp. 839–843, Jun. 2000.

[61] T. D. Le, K. Furusawa and T. Hayakawa, "Orbital Formation Control of Multiple Spacecraft," *Proc. American Control Conf.*, Baltimore, USA, pp. 3636–3641, Jun. 30–Jul. 02, 2010.

[62] H. Wong, V. Kapila and A. Sparks, "Adaptive Output Feedback Tracking Control of Multiple Spacecraft," *Proc. Am. Control Conf.*, Arlington, VA, pp. 698–703, Jun. 25–27, 2001.

[63] X. Liu and K. D. Kumar, "Formation Control of Spacecraft Flying with Network-induced Delays and Packet Dropouts," *IEEE Int. Conf. Control and Automation*, Xiamen, China, pp. 480–485, Jun. 9–11, 2010.

[64] L. Hui and J. Li, "Terminal Sliding Mode Control for Spacecraft Formation Flying," *IEEE Trans. Aerospace Electron. Syst.*, vol. 45, no. 3, pp. 835–846, 2009.

[65] K. H. Tan and M. A. Lewis, "Virtual Structures for High-precision Cooperative Mobile Robotic Control," *Proc. IEEE/RSJ Int. Conf. Intelligent Robots Syst.*, vol. 1, pp. 132–139, 1996.

[66] R. W. Beard, J. Lawton and F. Y. Hadaegh, "A Feedback Architecture for Formation Control," *Proc. Am. Control Conf*, vol. 6, pp. 4087–4091, 2000.

[67] T. H. A. van den Broek, N. van de Wouw and H. Nijmeijer, "Formation Control of Unicycle Mobile Robots: a Virtual Structure Approach," *Joint 48thIEEE Conf. Decision and Control and 28th Chinese Control Conf.* Shanghai, PR China, pp. 8328–8333, Dec. 16–18, 2009.

[68] J. Yuan and G. Y. Tang, "Formation Control for Mobile Multiple Robots Based on Hierarchical Virtual Structures," *IEEE Int. Conf. Control Automation*, Xiamen, China, pp. 393–398, Jun. 9–11, 2010.

[69] S. J. Yoo, J. B. Park and Y. H. Choi, "Adaptive Formation Tracking Control of Electrically Driven Multiple Mobile Robots," *IET Control Theory Appl.*, vol. 4, no. 8, pp. 1489–1500, 2010.

[70] R. W. Beard, J. Lawton and F. Y. Hadaegh, "A Feedback Architecture for Formation Control," *Proc. Am. Control Conf.*, Chicago, Illinois, pp. 4087–4091, Jun. 2000.

[71] R. W. Beard, J. Lawton and F. Y. Hadaegh, "A Coordination Architecture for Spacecraft Formation Control," *IEEE Trans. Control Syst. Technol.*, vol. 9, no. 6, pp. 777–790, 2001.

[72] M. Xin, S. N. Balakrishnan and H. J. Pernicka, "Multiple Spacecraft Formation Control with θ–D Method," *IET Control Theory Appl.*, 2007, vol. 1, no. 2, pp. 485–493, 2007.

[73] S. Wang and H. Schaub, "Nonlinear Feedback Control of a Spinning Two-Spacecraft Coulomb Virtual Structure," *IEEE Trans. Aerospace Electronic Syst.*, vol. 47, no. 3, pp. 2055–2067, 2011.

[74] N. E. Leonard and E. Fioroly, "Virtual Leader, Artificial Potentials and Coordinated Control Groups," *Proc. IEEE Conf. Decision and Control*, pp. 2968–2973, 2001.

[75] K. D. Do, "Formation Control of Mobile Agents Using Local Potential Functions," *Proc. Am. Control Conf.* Minnesota, USA, pp. 2148–2153, Jun. 14–16, 2006.

[76] J. A. Gouvea, A. R. Pereira, L. Hsu and F. Lizarralde, "Adaptive Formation Control of Dynamic Nonholonomic Systems Using Potential Functions," *Am. Control Conf.*, MD, USA, pp. 230–235, Jun. 30–Jul. 02, 2010.

[77] J. Wang, X. Wu and Z. Xu, "Decentralized Formation Control and Obstacles Avoidance based on Potential Field method," *Proc. Int. Conf. on Machine Learning and Cybernetics*, Dalian, pp. 803–808, Aug. 13–16, 2006.

[78] A. R. Pereira, L. Hsu and R. Ortega, "Globally Stable Adaptive Formation Control of Euler-Lagrange Agents via Potential Functions," *Am. Control Conf.*, MO, USA, pp. 2606–2611, Jun. 10–12, 2009.

[79] K. D. Do, "Formation Tracking Control of Unicycle-Type Mobile Robots With Limited Sensing Ranges," *IEEE Trans. Control Syst. Technol.*, vol. 16, no. 3, pp. 527–538, 2008.

[80] D. Xue, J. Yao, G. Chen and Y. L. Yu, "Formation Control of Networked Multi-agent Systems," *IET Control Theory Appl.*, vol. 4, no. 10, pp. 2168–2176, 2010.

[81] L. M. Wachter and L. E. Ray, "Stability of Potential Function Formation Control with Communication and Processing Delay," *Am. Control Conf.*, MO, USA, pp. 2997–3004, Jun. 10–12, 2009.

[82] Y. Kang, M. Lee, C. Kim, S. Yoon and C. Noh, "A Study of Cluster Robots Line Formatted Navigation Using Potential Field Method," *IEEE Int. Conf. on Mechatronics and Automation*, Beijing, China, pp. 1723–1728, Aug. 7–10, 2011.

[83] L. E. Barnes, M. A. Fields and K. P. Valavanis, "Swarm Formation Control Utilizing Elliptical Surfaces and Limiting Functions," *IEEE Trans. Syst. Man Cybern. B: Cybern.*, vol. 39, no. 6, pp. 1434–1445, 2009.

[84] G. Lafferriere, A. Williams, J. Caughman and J. Veerman, "Decentralized Control of Vehicle Formations," *Syst. Control Letters*, vol. 54, pp. 899–910, 2005.

[85] B. Fidan, C. Yu and B. D. O. Anderson, "Acquiring and Maintaining Persistence of Autonomous Multi-vehicle Formations," *IET Control Theory Appl.*, vol. 1, no. 2, pp. 452–460, 2007.

[86] W. Dong and J. A. Farrell, "Formation Control of Multiple Underactuated Surface Vessels," *IET Control Theory Appl.*, vol. 2, no. 12, pp. 1077–1085, 2008.

[87] Y. H. Chang, C. Chang, C. L. Chen and C. W. Tao, "Fuzzy Sliding Mode Formation Control for Multirobot Systems: Design and Implementation," *IEEE Trans. Syst. Man, Cybern. B: Cybern.*, vol. 42, no. 2, pp. 444–457, 2012.

[88] W. Dong and J. A. Farrell, "Cooperative Control of Multiple Nonholonomic Mobile Agents," *IEEE Trans. Automat. Control*, vol. 53, no. 6, pp. 1434–1448, 2008.

[89] W. Dong and J. A. Farrell, "Decentralized Cooperative Control of Multiple Nonholonomic Systems," *Proc. IEEE Conf. on Decision and Control*, pp. 1486–1491, Dec. 12–14, 2007.

[90] J. A. Marshall and M. E. Broucke, "Symmetry Invariance of Multiagent Formations in Self-Pursuit," *IEEE Trans. Automat. Control*, vol. 53, no. 9, pp. 2022–2032, 2008.

[91] W. Dong, "Formation Control of Multiple Wheeled Mobile Robots with Uncertainty," *IEEE Conf. on Decision and Control*, Atlanta, GA, USA, pp. 4492–4497, Dec. 15–17, 2010.

[92] W. Zhang and J. Hu, "Optimal Multi-Agent Coordination under Tree Formation Constraints," *IEEE Trans. Autom. Control*, vol. 53, no. 3, pp. 692–705, 2008.

[93] J. M. Hendrickx, B. Fidan, C. Yu, B. D. O. Anderson and V. D. Blondel, "Formation Reorganization by Primitive Operations on Directed Graphs," *IEEE Trans. Automat. Control*, vol. 53, no. 4, pp. 968–979, 2008.

[94] W. Dong and J. Farrell, "Formation Control of Multiple Mobile Robots with Uncertainty," *Am. Control Conf.*, pp. 1412–1417, Jun. 11–13, 2008.

[95] D. Gu, "A Differential Game Approach to Formation Control," *IEEE Trans. Control Syst. Technol.*, vol. 16, no. 1, pp. 85–93, 2008.

[96] H. Min, S. Wang F. Sun Z. Gao and Y. Wang, "Distributed Six Degree-of-freedom Spacecraft Formation Control with Possible Switching Topology," *IET Control Theory Appl.*, vol. 5, no. 9, pp. 1120–1130, 2011.

[97] R. Ghabcheloo, I. Kaminer, A. P. Aguiar and A. Pascoal, "A General Framework for Multiple Vehicle Time-Coordinated Path Following Control," *Am. Control Conf.* MO, USA, pp. 3071–3076, Jun. 10–12, 2009.

[98] F. Zhang, "Geometric Cooperative Control of Particle Formations," *IEEE Trans. Autom. Control*, vol. 55, no. 3, pp. 800–804, 2010.

[99] F. Dorfler and B. Francis, "Geometric Analysis of the Formation Problem for Autonomous Robots," *IEEE Trans. Autom. Control*, vol. 55, no. 10, pp. 2379–2384, 2010.

[100] J. Almeida, C. Silvestre and A. Pascoal, "Cooperative Control of Multiple Surface Vessels in the Presence of Ocean Currents and Parametric Model Uncertainty," *Int. J. Robust Nonlinear Control*, vol. 20, pp. 1549–1565, 2010.

[101] J. Almeida, C. Silvestre and A. Pascoal, "Cooperative Path-Following of Multiple Surface Vessels with Parametric Model Uncertainty and in the Presence of Ocean Currents,"*6th IFAC Symp. Intelligent Autonomous Vehicles*, vol. 40, no. 15, pp. 84–89, 2007.

[102] J. Almeida, C. Silvestre and A. M. Pascoal, "Cooperative Control of Multiple Surface Vessels with Discrete-time Periodic Communications," *Int. J. Robust Nonlinear Control*, vol., 22, no. 4, pp. 398–419, 2012.

[103] J. Almeida, C. Silvestre and A. Pascoal, "Coordinated Control of Multiple Vehicles with Discrete-time Periodic Communications," *Proc. 46th Conf. Decision and Control*, New Orleans, LA, USA, pp. 2888–2893, 2007.

[104] R. Ghabcheloo, A. Pascoal, C. Silvestre and I. Kaminer, "Coordinated Path Following Control of Multiple Vehicles subject to Bidirectional Communication Constraints," *Springer-Verlag, chapt. Group Coordination and Cooperative Control*, pp. 93–111, 2006.

[105] A. P. Aguiar and A. M. Pascoal, "Coordinated Path-Following Control for Nonlinear Systems with Logic-Based Communication," *Proc. 46th IEEE Conf. on Decision and Contr*. New Orleans, LO, USA, pp. 1473–1479, 2007.

[106] R. Ghabcheloo, A. P. Aguiar, A. Pascoal and C. Silvestre, "Coordinated Path-following Control of Multiple AUVs in the Presence of Communication Failures and Time Delays," *7th IFAC Conf. on Maneuvering and Control of Marine Craft*, Lisbon, Portugal, pp. 1–7, 2006.

[107] R. Ghabcheloo, A. Aguiar, A. Pascoal, C. Silvestre, I. Kaminer and J. Hespanha, "Coordinated Path-following Control of Multiple Underactuated Autonomous Vehicles in the Presence of Communication Failures," *Proc. IEEE Conf. on Decision and Control*, pp. 4345–4350, 2006.

[108] L. Vig and J. A. Adams, "Multi-Robot Coalition Formation," *IEEE Trans. Robot*, vol. 22, no. 4, pp. 637–649, 2006.

[109] S. Liu, D. Sun and C. Zhu, "Coordinated Motion Planning for Multiple Mobile Robots Along Designed Paths with Formation Requirement," *IEEE/ASME Trans. Mechatron.*, vol. 16, no. 6, pp. 1021–1031, 2011.

[110] F. Vanni, A. P. Aguiar and A. M. Pascoal, "Cooperative Path-Following of Underactuated Autonomous Marine Vehicles with Logic-based Communication," *IFAC Proceedings Volumes*, vol. 41, no. 1, pp. 107–112, 2008.

[111] Y. Lan, G. Yan and Z. Lin, "Synthesis of Distributed Control of Coordinated Path Following Based on Hybrid Approach," *IEEE Trans. Automat. Control*, vol. 56, no. 5, pp. 1170–1175, 2011.

[112] A. Zou and K. D. Kumar, "Distributed Attitude Coordination Control for Spacecraft Formation Flying," *IEEE Trans. Aerospace Electron Syst.*, vol. 48, no. 2, pp. 1329–1346, 2012.

[113] A. Zou and K. D. Kumar, "Neural Network-Based Distributed Attitude Coordination Control for Spacecraft Formation Flying With Input Saturation," *IEEE Trans. Neural Networks Learning Syst.*, vol. 23, no. 7, pp. 1155–1162, 2012.

[114] H. Yang and F. Zhang, "Geometric Formation Control for Autonomous Underwater Vehicles," *IEEE Int. Conf. Robot. Autom.*, Alaska, USA, pp. 4288–4293, May 2010.

[115] R. Skjetne, S. Moi and T. I. Fossen, "Nonlinear formation control of marine craft," *Proc. 41st IEEE Conf. Decision and Control*, NV, USA, pp. 1699–1704, Dec. 2002.

[116] R. Skjetne, I. A. F. Ihle and T. I. Fossen, "Formation Control by Synchronizing Multiple Maneuvering Systems," *Proc. 6th IFAC Conf. Manoeuvring Control Marine Crafts*, Girona, Spain, pp. 280–285, Sep. 2003.

[117] I. A. F. Ihle, R. Skjetne and T. I. Fossen, "Nonlinear Formation Control of Marine Craft with Experimental Results," *Proc. 43rd IEEE Conf. Decision Control*, Paradise Island, pp. 680–685, 2004.

[118] R. Cui "Formation Control of Autonomous Underwater Vehicles Under Fixed Topology," *IEEE Int. Conf. Control Autom.*, pp. 2913–2918, May 2007.

[119] W. Van, "Formation Control of Underactuated Autonomous Underwater Vehicles in Horizontal Plane," *Proc. IEEE Int. Conf. Autom. Logistics*, pp. 822–827, Sep. 2008.

[120] J. Ghommam and F. Mnif, "Coordinated Path Following Control for a Group of Underactuated Surface Vessels," *IEEE Trans. Indus. Electron.*, vol. 56, no. 10, pp. 3951–3963, 2009.

[121] W. Ren, "Consensus Strategies for Cooperative Control of Vehicle Formations," *IET Control Theory Appl.*, vol. 1, no. 2, pp. 505–512, 2007.

[122] I. Okoloko, "Path Planning for Multiple Spacecraft Using Consensus with LMI Avoidance Constraints," *IEEE Aerospace Conf. Proc.*, pp. 1–8, 2012.

[123] T. Hayakawat, T. Matsuzawat and S. Harat, "Formation Control of Multi-Agent Systems with Sampled Information – Relationship Between Information Exchange Structure and Control Performance," *Proc. IEEE Conf. on Decision and Control*, CA, USA, Dec. 13–15, pp. 4333–4338, 2006.

[124] S. P. Hou and C. C. Cheah, "PD Control Scheme for Formation Control of Multiple Autonomous Underwater Vehicles," *IEEE/ASME Int. Conf. Adv. Intell. Mechatron.*, Singapore, pp. 356–361, 2009.

[125] L. Brinon, A. Seuret and C. Canudas, "Translation Control of a Fleet Circular Formation of AUVS Under Finite Communication Range," *48th IEEE Conf. Decision Control 28th Chinese Control Conf.*, Shanghai, China, pp. 8354–8350, Dec. 2009.

[126] R. Ghabcheloo, A. P. Aguiar, A. Pascoal and C. Silvestre, "Synchronization in Multi-agent Systems with Switching Topologies and Non-homogeneous Communication Delays," *Proc. 46th IEEE Conf. on Decision and Control,* New Orleans, LA, pp. 2327–2332, 2007.

[127] E. Borhaug, A. Pavlov and K. Y. Pettersen, "Straight Line Path Following for Formations of Underactuated Underwater Vehicles," *Proc. Decision and Control*, pp. 2905–2912, Dec. 12–14, 2007.

[128] W. J. Dong, Yi Guo and J. A. Farrell, "Formation Control of Nonholonomic Mobile Robots," *Proc. Am. Control Conf.* Minnesota, USA, pp. 5602–5607, Jun. 14–16, 2006.

[129] I. A. F. Ihle, J. Jouffroy and T. I. Fossen, "Formation Control of Marine Surface Craft: A Lagrangian Approach," *IEEE J. Oceanic Eng.*, vol. 31, no. 4, pp. 922–934, 2006.

[130] D. Gray, R. Daniel and J. Stilwell, "Decentralized Control and Estimation for a Platoon of Autonomous Vehicles with a Circulant Communication Network," *Proc. Am. Control Conf.* Minnesota, USA, pp. 743–748, Jun. 14–16, 2006.

[131] G. Antonelli and S. Chiaverini, "Kinematic Control of Platoons of Autonomous Vehicles," *IEEE Trans. Robot.*, vol. 22, no. 6, pp. 1285–1292, 2006.

[132] T. Yang, Z. Liu, H. Chen and R. Pei, "Distributed Robust Control of Multiple Mobile Robots Formations via Moving Horizon Strategy," *Proc. Am. Control Conf.* Minnesota, USA, pp. 2838–2843, June 14–16, 2006.

[133] S. Kloder and S. Hutchinson, "Path Planning for Permutation-Invariant Multirobot Formations," *IEEE Trans. Robot.*, vol. 22, no. 4, pp. 650–665, 2006.

[134] C. A. Kitts and I. Mas, "Cluster Space Specification and Control of Mobile Multirobot Systems," *IEEE/ASME Trans. Mechatronics*, vol. 14, no. 2, pp. 207–218, 2009.

[135] N. Moshtagh, N. Michael, A. Jadbabaie and K. Daniilidis, "Vision-Based, Distributed Control Laws for Motion Coordination of Nonholonomic Robots," *IEEE Trans. Robot.*, vol. 25, no. 4, pp. 851–860, 2009.

[136] D. Pais, M. Cao and N. E. Leonard, "Formation Shape and Orientation Control Using Projected Collinear Tensegrity Structures," *Am. Control Conf.* MO, USA, pp. 610–615, Jun. 10–12, 2009.

[137] S. S. Stankovic, D. M. Stipanovic and M. S. Stankovi, "Decentralized Overlapping Tracking Control of a Formation of Autonomous Unmanned Vehicles," *Am. Cont. Conf.* MO, USA, pp. 3878–3883, Jun. 10–12, 2009.

[138] J. Su and W. Xie, "Motion Planning and Coordination for Robot Systems Based on Representation Space," *IEEE Trans. Syst., Man, Cybern. B: Cybern.*, vol. 41, no. 1, pp. 248–259, 2011.

[139] J. A. Marshall and D. Tsai, "Periodic Formations of Multivehicle Systems," *IET Control Theory Appl.*, vol. 5, no. 2, pp. 389–396, 2011.

[140] S. Wang and C. Zheng, "A Hierarchical Evolutionary Trajectory Planner for Spacecraft Formation Reconfiguration," *IEEE Trans. Aerospace Electron. Syst.*, vol. 48, no. 1, pp. 279–289, 2012.

[141] C. Sotzing and D. M. Lane, "Improving the Coordination Efficiency of Limited Communication Multi-AUV Operations using a Multi-Agent Architecture," *J. Field Robot.*, vol. 27, no. 4, pp. 412–429, 2010.

[142] D. Prasser and M. Dunbabin, "Sensor Network Based AUV Localisation," *7th Int. Conf. Field and Service Robots*. Berlin, Heidelberg: Springer-Verlag, pp. 285–294, 2010.

[143] T. Cuff and R. Wall, "Support Platform and Communications to Manage Cooperative AUV Operations," *Proc. Oceans'06 Asia Pacific IEEE, Singapore*, pp. 1–8, May 2006.

[144] E. Sozer and M. Stojanovic, "Simulation and Rapid Prototyping Environment for AUV Networks," *Proc. Int. Symp. Unmanned Untethered Submersible Technology*, pp. 1–7, 2005.

[145] L. Brignone, J. Alves and J. Opderbecke, "GREX Sea Trials: First Experiences in Multiple Underwater Vehicle Coordination Based on Acoustic Communication," *Proc. OCEANS 2009-EUROPE. OCEANS'09*, Bremen, pp. 1–6, May 2009.

[146] M. Porfiri, D. G. Roberson and D. J. Stilwell, "Environmental Tracking and Formation Control of a Platoon of Autonomous Vehicles Subject to Limited Communication," *ICRA*, pp. 595–600, 2006.

[147] A. Caiti and A. Munafo, "Cooperative Distributed Behaviours of An AUV Network for Asset Protection with Communication Constraints," *Proc. IEEE OCEANS'11,* Santander, Spain, pp. 1–6, Jun. 2011.

[148] C. Benton, J. Kenney, R. Nitzel, R. Blidberg, S. Chappell and S. Mupparapu, "Autonomous Undersea Systems Network (AUSNET) – Protocols to Support Ad-Hoc AUV Communications," *Proc. IEEE/OES Works. Multiple Autonomous Underwater Vehicle Operations,* S. E., ME, USA, pp. 83–87, Jun. 2004.

[149] M. Stojanovic, L. Freitag, J. Leonard and P. Newman, "A Network Protocol for Multiple AUV Localization," in *Proc. IEEE OCEANS02 Conf.*, pp. 604–611, 2002.

[150] R. Stokey, L. Freitag and M. Grund, "A Compact Control Language for AUV Acoustic Communication," *OCEAN-Europe*, vol. 2, pp. 1133–1137, 2005.

[151] Q. Jia and G. Li "Formation Control and Obstacle Avoidance Algorithm of Multiple Autonomous Underwater Vehicles (AUVs) Based on Potential Function and Behaviour Rules," *IEEE Int. Conf. on Automation and Logistics*, Jinan, China, pp. 569–573, Aug. 2007.

[152] T. Kopfstedt, M. Mukai, M. Fujita and O. Sawodny, "Formation Control for Mobile Robots in Partially Known Environments Using Mixed Integer Programming and Fuzzy Systems," *SICE-ICASE Int. Joint Conf.,* Busan, Korea, pp. 1832–1837, Oct. 18–21, 2006.

[153] P. Ghorbanian, S. G. Nersesov and H. Ashrafiuon, "Obstacle Avoidance in Multi-Vehicle Coordinated Motion via Stabilization of Time Varying Sets," *Am. Control Conf.,* San Francisco, CA, USA, pp. 3381–3386, Jun. 29–Jul. 01, 2011.

[154] L. P. Ellekilde and J. W. Perram, "Tool Center Trajectory Planning for Industrial Robot Manipulators Using Dynamical Systems," *Int. J. Robot. Res.*, vol. 24, no. 5, pp. 385–396, 2005.

[155] S. G. Nersesov, P. Ghorbanian and A. G. Aghdam, "Stabilization of Sets with Application to Multi-vehicle Coordinated Motion," *Automatica*, vol. 46, pp. 1419–1427, 2010.

[156] X. Wang, V. Yadav and S. N. Balakrishnan, "Cooperative UAV Formation Flying with Obstacle/Collision Avoidance," *IEEE Trans. Control Syst. Technol.*, vol. 15, no. 4, pp. 672–679, 2007.

[157] S. Zhai, B. Fidan, S. C. Ozturk and V. Gazi, "Single View Depth Estimation Based Formation Control of Robotic Swarms: Obstacle Avoidance, Simulation, and Practical Issues," *16th Mediterranean Conf. Control Autom.*, France, pp. 1162–1167, Jun. 25–27, 2008.

[158] L. Seereeram, M. Ravichandran, R. Smith and R. Beard, "Multi-spacecraft Trajectory Optimization and Control Using Genetic Algorithm Techniques," *IEEE Aerospace Conf. Proc.*, vol. 7, pp. 99–108, 2000.

[159] C. Sultan, S. Seereeram, R. K. Mehra and F. Y. Hadaegh, "Energy Optimal Reconfiguration for Large Scale Formation Flying," *Proc. Am. Control Conf.*, Boston, Massachusetts, pp. 2986–2991, Jun. 30-Jul. 2, 2004.

[160] D. E. Chang, S. C. Shadden, J. E. Marsden and R. O. Saber, "Collision avoidance for multiple agent systems," *Proc. 42nd IEEEConf. Decision Control*, Hawaii, USA, Dec. 2003.

[161] C. C. Cheah, S. P. Hou and J. J. E. Slotine, "Region Following Formation Control for Multi-Robot Systems," *IEEE Int. Conf. Robotics and Automation CA*, USA, pp. 3796–3801, May 19–23, 2008.

[162] T. Kato, K. Watanabe and S. Maeyama, "A Formation Method for Heterogeneous Multiple Robots by Specifying the Relative Position of Each Robot," *SICE Annual Conf.*, Taipei, Taiwan, pp. 3274–3277, Aug. 18–21, 2010.

[163] R. O. Saber, J. A. Fax and R. M. Murray. "Consensus and Cooperation in Networked Multi-Agent Systems," *Proc. IEEE*, vol. 95, no. 1, pp. 215–233, 2007.

[164] T. Keviczky, F. Borrelli, K. Fregene, D. Godbole and G. J. Balas, "Decentralized Receding Horizon Control and Coordination of Autonomous Vehicle Formations," *IEEE Trans. Control Syst. Technol.*, vol. 16, no. 1, pp. 19–33, 2008.

[165] R. Mead, R. Long and J. B. Weinberg, "Fault-Tolerant Formations of Mobile Robots," *IEEE/RSJ Int. Conf. Intelligent Robots Syst.*, St. Louis, USA, pp. 4805–4810, Oct. 11–15, 2009.

[166] R. W. Beard and T. W. McLain, "Multiple UAV Cooperative Search Under Collision Avoidance and Limited Range Communication Constraints," *Proc. 42nd IEEE Conf. Decision Control*, Hawaii, USA, Dec. 2003.

Chapter 10

Hydro-acoustic communications and networking in contemporary underwater robotics: instruments and case studies

Konstantin Kebkal[1], Oleksiy Kebkal[1], Veronika Kebkal[1], Ievgenii Glushko[1], Antonio Pascoal[2], Miguel Ribeiro[2], Manuel Rufino[2], Luis Sebastião[2], Giovanni Indiveri[3], Lorenzo Pollini[4], and Enrico Simetti[5]

Abstract

This chapter describes the results of an easy-to-achieve implementation of an underwater acoustic (UWA) sensor network capable of operating in different modes for the exchange of UWA data among mobile and/or stationary network nodes that may include autonomous surface vehicles, autonomous underwater vehicles (AUVs), remotely operated vehicles, and benthic stations. The data exchanged are crucial in a number of applications that include cooperative multiple vehicle navigation and control, mission status assessment, and environmental sensing. A key novelty of the implementation described is the use of UWA modems incorporating the EviNS dedicated Networking Software Framework. The latter is a compact open-source, open-architecture software that is undemanding in terms of computer resources and can be installed directly on a UWA modem platform as part of its standard software. The practical usefulness and advantage of this solution stems from the fact that it significantly reduces the costs involved in setting up a UWA network while retaining small dimensions, small weight, and high energy efficiency of each hydro-acoustic network node (e.g. modems performing the functions of network communication devices). Several case studies with experimental networks based on UWA modems incorporating the EviNS Framework, with pre-installed but freely changeable media-access and routing protocols, are presented in this chapter. Experimental results aimed at assessing the performance of the UWA network herein described are presented in relation to the combination of simple medium-access control protocols

[1]Evologics GmbH, Germany
[2]Laboratory of Robotics and Engineering, Systems (LARSyS), ISR/Instituto Superior Tecnico (IST), University of Lisbon, Portugal
[3]University of Salento (ISME node), Lecce, Italy
[4]University of Pisa (ISME node), Pisa, Italy
[5]University of Genova (ISME node), Genova, Italy

(providing uncoordinated access), wide-spread flooding-based routing protocols (based on so-called sequence number control), and polling protocols (implementing sequential and broadcast communications in networks with a centralized topology). Apart from providing details on the application of the EviNS Framework, the chapter offers a detailed analysis of the performance of an ad-hoc underwater sensor network operating in a shallow water area, i.e. in a hydro-acoustic environment of large practical interest. The chapter provides also a thorough analysis of the communication performance achievable with multiple nodes in a network with a centralized topology and affords the reader details of the practical estimation of the data rates achievable in communications involving multiple underwater modems performing simultaneous and asynchronous communications as elements of a mobile network. Even though the experiments reported were limited to specific cases, the number of network nodes used, the network geometries, and the fact that the underwater networks involved operated in complicated hydro-acoustic environments capture frequently occurring combinations of circumstances that occur in real off-shore practice.

10.1 Introduction

The technology at the core of underwater digital information exchange belongs to the topics of rapidly growing interest and is witnessing increasingly higher demand. In this context, one of the most common tasks is sensor data exchange in a network between source(s) and receiver(s) through a series of intermediate nodes – underwater acoustic (UWA) modems performing relay functions (Figure 10.1(a) and (c)). Another task of considerable practical interest is underwater positioning of mobile or stationary underwater objects equipped with UWA modems (Figure 10.1(a) and (b)). The task of long-baseline positioning, that is, positioning of an object based on the measurements of ranges to a number of underwater transponders at known locations, is usually solved, in essence, by means of a UWA network whereby data packets are exchanged between the object being positioned and the geo-referenced nodes of the UWA network (stationary or mobile). Since UWA modems involved in solving the task have to be equipped with media access control protocols (and in some cases also with routing protocols), long-baseline positioning is usually considered as one of the UWA networking tasks.

The problem of underwater object positioning can also be solved by resorting to ultra-short baseline (USBL) solutions, consisting of an antenna with multiple receivers in close proximity of each other. In a possible configuration, this system allows measuring the ranges and angles of arrival of the incoming sound waves emitted by the object to be positioned. In this set-up, multiple objects can be positioned and the USBL antennas can also be multiple (e.g. two USBL systems are often used on large vessels supporting underwater technical work, such as cable layers or dredging vessels). Thus, to resolve collisions during the transmission and reception of data packets between the objects being positioned and the antennas, at least media access control algorithms must be used, thus reinforcing the idea that ultrashort-baseline positioning is one of the UWA network tasks.

Remote supervisory control of groups of underwater robots by means of UWA links are also directly related to the tasks of UWA network. Controlled objects can go

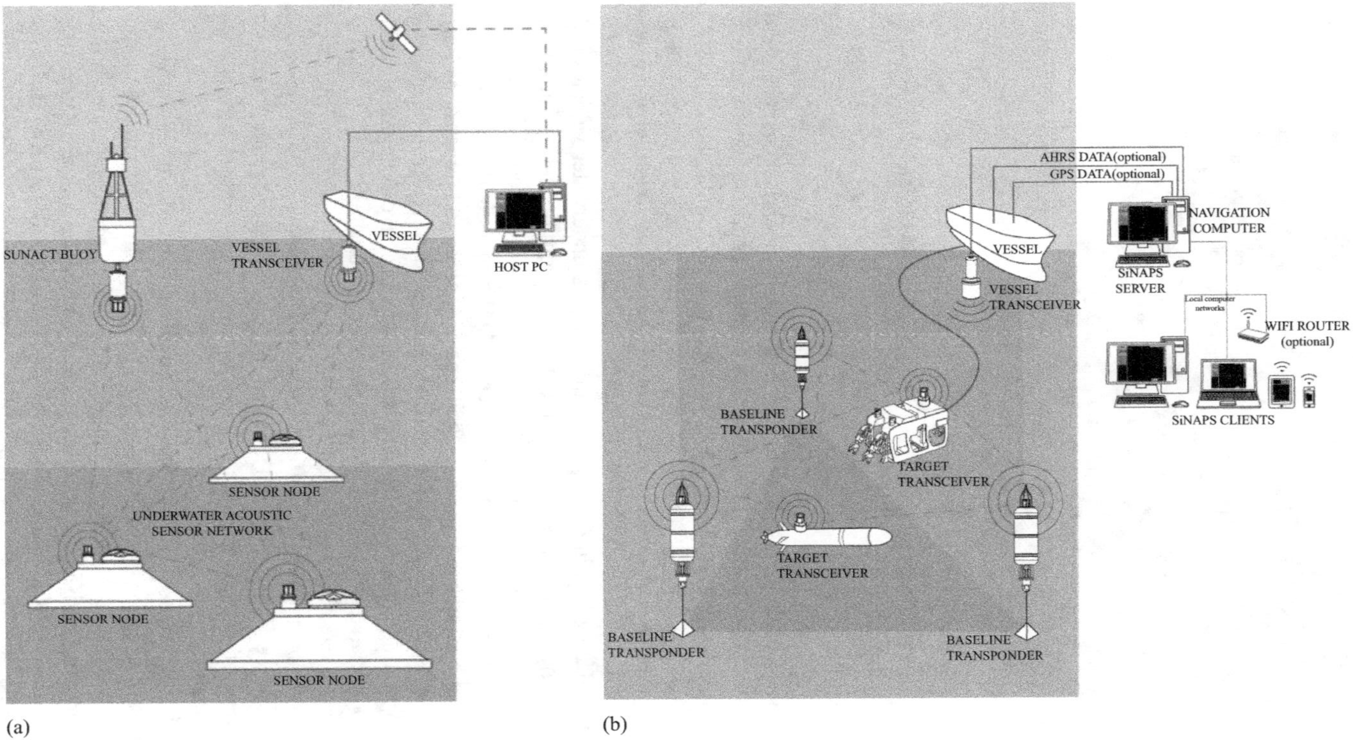

Figure 10.1 Examples of UWA networks

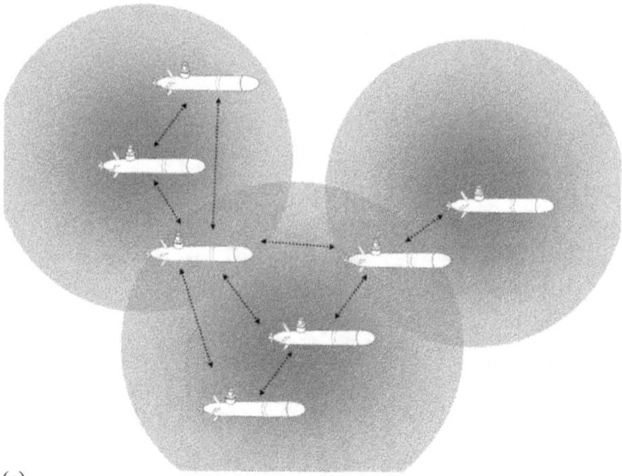

(c)

Figure 10.1 Continued

beyond the range of the UWA modem installed on the supervisory control platform (or fall into the shadow zone). The only way to establish connection with such objects is to request their neighbouring nodes that they search for a communication path and use them as relay nodes. Obviously, except for media access protocols, one of the requirements for UWA modems in such tasks is their ability to implement a network routing protocol.

The task of coordinating a group of autonomous underwater vehicles (AUVs) is of utmost importance in a variety of scientific and commercial operations at sea. There are many concepts related to the applications of coordinated AUVs underwater. One thing is common to them: the number of AUVs in a coordinated group tends to grow, and substantial reduction of their costs, weight, and dimensions becomes absolutely necessary. The cheaper the AUVs and the smaller their dimensions and associated operating costs, the more AUVs the user can afford to deploy and thus the larger the coordinated group of AUVs that the user can create.

Usually, the size of a coordinated AUV group is directly related to the effectiveness of the user's operation at sea (effectiveness in what concerns the usage of such an expensive resource as vessel time). However, the requirement to reduce the costs and the dimensions of an AUV imposes strict requirements on the cost, size and weight of its components.

The classic design of a UWA network involves a set of subsea nodes, which include at least a sensor, a modem and, separately, an auxiliary processor module for integrating the sensor and the modem. The auxiliary processor module requires a pressure housing with underwater connectors and cables and a battery for powering the processor. As a consequence, the device may end up being rather costly. In addition, the development and production of such an auxiliary module requires time-consuming work of well-qualified personnel, thus increasing the cost of the device further. Depending on the requirements for its operation depth and autonomy, the dimensions of such a module can be relatively large and heavy. Obviously, any alternative design that eliminates the need for such an auxiliary module involves a reduction in weight, size and cost of each of the nodes of the UWA sensor network. It is with this objective in mind that in this chapter, we describe an alternative design, together with the hardware and software implementation, of a UWA network node. This chapter is organized as follows: Section 10.1 describes the UWA modem built upon the S2C technology of Evologics as a platform for specialized user applications. Section 10.2 presents a case study on the operation of the UWA modems in an ad-hoc network. Section 10.3 describes another case study on the operation of the UWA modems in a bi-modal network with a centralized topology, and Section 10.4 gives the details of a case study on the operation of the UWA modems with integrated atomic clocks for positioning of a group of AUVs.

10.2 The S2C modem of Evologics as a platform for specialized user applications

From the point of view of the hardware architecture, the S2C modem of Evologics [1,2] is a stack of three processor boards. The field-programmable gate array (FPGA) and digital signal processor (DSP) boards execute in real-time the basic functions of the modem: namely, modulation, demodulation, synchronization, functions of error correction coding, delivery acknowledging, data queuing, and prioritization of data streams. The third modem's board, the universal processor unit, based on an ARM architecture, executes the functions of servicing the physical layer protocols of the modem, as well as the protocols for medium access control layer and data routing. To save battery energy, the ARM board has a relatively low performance. Nevertheless, the peculiarity of this board is the existence of a Linux operating system and a tool-chain, open to users for their external applications [3], i.e. to accommodate their additional executable modules directly on the hardware platform of the modem.

Because of the architecture, operating system, tool-chain and a set of specialized software adopted, the hardware platform of the S2C modem affords the user the capabilities to develop and run his own applications created instead of or in addition to the factory functions of the modem.

Currently, there are a variety of software frameworks which provide the user with the ability to design and operate his own networking protocols. These are specialized software frameworks with open source code and open architectures, for example, SUNSET [4] and DESERT [5]. The use of such frameworks, however, requires a high-performance processor module to run. This fact makes it impossible to execute these frameworks on-board of a low-power (low-performance) processor board of a UWA modems (UWA modems have always low-performance processors; otherwise, they would require heavy and large batteries, or else their period of autonomous operation would be very short).

For UWA network communications and many other tasks, Evologics has developed a software framework that does not require a high-performance processor and, nevertheless, provides the user with the technical means to create UWA networks with required topology and properties, and to integrate with the UWA modems a variety of sensors and other equipment. The peculiarity of this framework also consists of a low-entry barrier (for the user) into the process of creating custom applications (i.e. to modify or implement new network protocols for running them on-board the UWA modem, there is no need to employ highly qualified programmers of embedded systems). This framework is titled EviNS (Evologics intelligent Networking Software Framework) [6,7] and is briefly described below. Also, examples of its application for the organization of a network data exchange among a plurality of UWA modems, as well as for the organization of underwater positioning processes, are presented in the following sections.

10.3 Architecture of the software framework EviNS

The EviNS framework architecture is shown in Figure 10.2. The framework contains a permanent (frozen) part shown in grey colour. The permanent part does not need to be changed by the user from one configuration to another one, and does not require any changes for adaptation to a variety of user tasks. To adapt the software framework to a user's task, so-called hot spots (or extension points) are provided, in which frequently changing users' programs can be placed. User programs are indicated by white rectangular windows, uniting software modules related to similar tasks. For example, the MAC window contains media access protocols, while the routing window consists of routing protocols. The motion window includes drivers of different

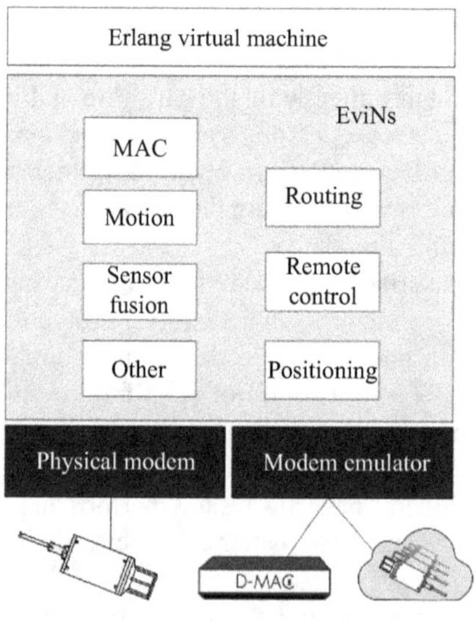

Figure 10.2 Architecture of the software framework EviNS

motion sensors, the positioning window refers to modules for the implementations of several positioning algorithms, and the remote control window consists of so-called back-seat drivers. The framework itself operates under an Erlang Virtual Machine and represents a compact software. This enables its operation on-board a non-powerful hardware platform (or on board other posix-compatible hardware, representing in this case a so-called modem emulator). The framework software is an open-source product (the source code is distributed via Github [6]).

To simplify the development of an application by an end-user (customization of an existing one or adaptation of the modem's factory application to his own tasks), the EviNS framework contains a large number of different software modules and protocols implemented and pre-installed by the modem manufacturer. In addition, the EviNS software framework allows for the inclusion of new user modules, in particular through abstract classes (the extension points), for which no specific implementations are provided by the manufacturer. For the implementation of a specific behaviour of the framework by the user, the manufacturer provides the latter with detailed rules and requirements.

As shown in the architecture depicted in Figure 10.2, in addition to the user tasks inherent to the communication device, the modem with EviNS framework is also capable of providing solutions for a number of additional tasks, especially for the integration of the modem with other devices such as third-party equipment drivers, sensors protocols and different customized communications applications. The approach implemented in the framework allows to flexibly use or quickly adapt the UWA modem to a wide range of user tasks.

The EviNS architecture implemented in the S2C modem not only simplifies the user's effort for creating customized network nodes, but also eliminates the need to develop and include intermediate hardware in the network node, in particular, an auxiliary processor module for integrating sensors with the modem. These circumstances – the capabilities to run the integration software directly on the hardware platform of the UWA modem – contribute to reduce the cost, weight and dimensions of the UWA network hardware, and thus to reduce the cost, weight and dimensions of their eventual carriers, the AUVs. Below are presented three case studies on the practical use of UWA modems with the EviNS framework as UWA network nodes for data exchange and multiple AUV positioning purposes.

10.4 Case studies

The three case studies presented below were conducted in the scope of the European Union's WiMUST (Widely scalable Mobile Underwater Sonar Technology) project, Horizon 2020-ICT-2014-1 research and innovation programme, grant agreement No. 645141 [8].

The WiMUST project aimed at conceiving and designing an intelligent team of cooperative autonomous marine robots, acting as intelligent sensing and communicating nodes of a reconfigurable moving acoustic network that could drastically improve the efficiency of the methodologies used to perform geophysical and geotechnical acoustic surveys at sea.

Figure 10.3 WiMUST – cooperative and networked control/navigation system

The novel key feature of the WiMUST system consisted of the use of a team of cooperative autonomous marine robots, acting as intelligent sensing and communicating nodes of a reconfigurable moving acoustic network (the main idea is depicted in Figure 10.3).

Recent developments have shown that there is vast potential for groups of marine robots acting in cooperation to drastically improve the methods available for ocean exploration and exploitation. Traditionally, seismic reflection surveying is performed by vessel towed streamers of hydrophones acquiring reflected acoustic signals generated by acoustic sources (either towed or on-board a vessel). In this context, geotechnical surveying for civil and commercial applications (e.g. underwater construction, infrastructure monitoring, mapping for natural hazard assessment, environmental mapping) aims at seafloor and sub-bottom characterization using towed streamers of fixed length that are extremely cumbersome to operate. The vision underlying the WiMUST project was that of developing advanced cooperative and networked control/navigation systems to enable a large number (tens) of marine robots (both on the surface and submerged) to interact by sharing information as a coordinated team (not only in pairs) (see [9,10] for further details).

10.4.1 Case study – operation of the UWA modems in an ad-hoc network

10.4.1.1 The tasks

A group of AUVs of the WiMUST system can be deployed in a water column and be distributed over a large area. The AUVs require coordinated operation in variety of applications and thus require data exchange among themselves and their support vessel at the surface.

Organization of real-time access to underwater collected data requires the sensors to be fitted out with the means for digital communications. In particular, AUVs operating in group require intensive exchange of navigation data when engaged in cooperative manoeuvring or performing coordinated measurements. Additionally, the mobile assets of the WiMUST system must be able to play the role of network nodes, also for many other tasks, e.g. when one or multiple AUVs fall behind or deviate from the formation, or drift to a large distance from their support ship during possibly long AUV deployment or retrieving phases. When the AUVs are distributed in a large shallow water area they need to be accessible via the network (via their neighbours) and obtain necessary coordinating information – to join formation, to start a mission, etc.

10.4.1.2 UWA modems used

From a software viewpoint, the UWA modem of the S2C technology [1] with the installed EviNS framework is a UWA communication device which, in addition to the mandatory set of physical layer protocols, contains also many protocols of the media access control layer and network routing protocols. Moreover, when the mandatory protocols operate as proprietary software, then the set of EviNS framework protocols is the software with open source code, which allows changes, improvement, or substitution of software modules defined by the user.

This EviNS feature was applied in the scope of the EU project WiMUST. Namely, to create a bi-modal (dual-frequency) UWA network carried by a group of AUVs.

In a UWA network of this type, the S2C modems of the upper frequency range (HF-modems) ensured the operation of the network of simple centralized topology based on the network protocol called "polling" [11]. HF modems were used to deliver payload data from seismic streamers towed by the AUVs to the support vessel at the surface (to the operator's terminal).

In a complementary manner, the UWA modems of the middle frequency range (MF-modems) provided the network functionality using both centralized and decentralized ad-hoc topologies. In the case where the modems were used for operation in a network with a decentralized topology (for example, in the so-called neighbourhood discovery and ad-hoc data exchange modes), the combination of the protocols – Sequence Number Control Flooding (SNCF) on the network layer [12] and the Carrier Sense Media Access "Aloha" (CSMA-Aloha) [13] on the MAC layer – was used. In the case of using the UWA modems for operation in a network with a centralized topology, a polling protocol was applied (for example, for the transmission of navigation data to the AUVs from the support vessel, as well as delivery of payload data from the AUVs to the vessel).

In the scope of the WiMUST project, a series of sea tests were executed for validation of media access and routing protocols for long range communication of the WiMUST underwater assets. The combination of the MAC protocol CSMA-Aloha and the routing protocol SNCF was selected as the most prospective solution. Some of the results obtained are presented in the next section. In general, the network was able to select different routes under different conditions for data delivery from a source to a destination. In the case of no direct link between the source and the destination, data delivery occurred successfully in a large majority of the delivery attempts. Even in case of very poor connectivity between source and destination, end-to-end data delivery was possible (e.g. after a second or third attempt) usually via several neighbouring network nodes.

10.4.1.3 The solution – data exchange in an ad-hoc network

In the case of several UWA modems operating simultaneously in the same area, there may occur collisions of data packets transmitted to data recipient by several sources, bringing about partial or entire loss of data packets. With the growing number of UWA modems operating simultaneously, the probability of such collisions can increase dramatically, making modem's operation ineffective or even impossible. Also, with growing distances between network nodes and/or adverse hydrology, a set of the UWA modems, e.g. those operating as data recipients, may happen to be out of the range of the UWA modems operating as data sources. Thus, data exchange among the individual network nodes or network parts that have no sufficient connectivity to each other may not be possible.

In radio communications, there are numerous well-known methods for resolving data collisions during modem operations in a network, as well as methods for data routing in a network via relay nodes between its directly non-connected parts [14,15]. In UWAs, however, their application is limited, or they can even be inapplicable at all [16–18].

A number of recent publications, e.g. [16,19,20] has been devoted to the elaboration and development of new methods for UWA modems for providing resolution of data collisions, as well as effective data routing. Several of these methods have been implemented by Evologics and incorporated in the EviNS Framework [6,7] as part of the modems standard software. The set of implemented MAC protocols currently pre-installed in the EviNS currently includes:

1. Control protocol with uncoordinated access to UWA medium – Carrier-Sense Multiple Access-Aloha (CSMA-Aloha) [13];
2. Control protocol T-Lohi [21] based on contention rounds between network nodes to conquer an access to the UWA medium;
3. Non-synchronous control protocol based on minimizing the duration of a handshake by taking advantage of the receiver's tolerance to interference when two nodes are closer than the maximal transmission range (Distance Aware Collision Avoidance Protocol – DACAP) [22].

The pre-installed set of routing programs in the EviNS framework includes:

1. Protocol based on a flooding algorithm: Sequence Number Controlled Flooding – SNCF [12],
2. Protocol based on a dynamic probabilistic flooding algorithm: Dynamic Probabilistic Flooding – DPFlood [23],
3. Protocol based on the exchange of control packets combining service data and payload data (the control packet of establishing path is carried by the data packet) – Information Carrying Based Routing Protocol – ICRP [24,25].

Routing protocols here imply data exchange capability with and without acknowledgement.

As mentioned above, all the pre-installed EviNS protocols are open-source programs, which Evologics distributes with GPL/MIT licence; they can be freely modified by the user as required.

As representative examples, this chapter describes the results of experiments on a UWA network capable of accomplishing tasks such as the exchange of small data packets via the network; e.g. AUV commands, navigation data, sensor data, and current information on the status of AUV modules.

It is important to remark that UWA networks of mobile nodes (or stationary nodes in an environment with fast-changing hydrology) may face the problem of alternating acoustically illuminated/unilluminated areas relatively to each other and, due to this fact, they may have difficulties in exchanging data (up to full loss of the connectivity between source and recipient). In this connection, a common requirement for a UWA network is its self-organizing ability as well as its capability to operate under conditions where the network changes its structure and geometry (dynamic ad-hoc network). In this case study, the organization of the example network and of the source-recipient message delivery is based on the flooding algorithm – SNCF [12]. The order of uncoordinated access to the medium is built upon the Carrier-Sense Multiple Access-Aloha (CSMA-Aloha) algorithm modified for the UWA medium [13].

10.4.1.4 Organization of data exchange in the network

According to the CSMA-Aloha algorithm, a modem initially listens to the activity in the network neighbouring nodes and transmits data packets to the UWA medium in the absence of such activity, or switches to the back-off mode during a short pseudo-random time interval (the order of several seconds) if any "neighbours" activity is detected.

The CSMA-Aloha adopted is an improved version of the generic Aloha protocol widely used in the radio communications. The improvement was proposed in [13]. Before data are transmitted to the UWA propagation medium, each network node tries to determine, within a short pseudo-random time interval, if the signal propagation medium is free from UWA activity of the neighbouring nodes. If the medium is occupied, the data packet transmission is postponed during a pseudo-random time (back-off time), repeatedly if required, until the medium proved to be unoccupied. The data packet sending the time slot is selected pseudo-randomly to prevent synchronized attempts from accessing the propagation medium from several nodes.

The routing protocol – SNCF – is based on the data re-directing principle in the form of "take and pass further". Data transmission can be initiated by any UWA network node to send data to any other node. In the process of data transmission, all the relay nodes that happened to be within the transmitting node's "hearing" zone and capable of receiving these data, pass the data further until it reaches the sink (recipient's address). To rule out fatal network overfilling with an avalanche of repeatedly received data packets, the re-directed packets are provided with the transmitting network node's number and the sink node address [12]. As a packet is received, the receiving node compares its number with a table containing numbers of the already re-directed packets. Only packets with new numbers are subject to re-directing.

10.4.1.5 Network parameters, a study

In many applications, when planning a mission that requires the delivery of control commands or current sensor data, the user is interested predominantly in short data delivery latencies and high probabilities of delivering data to the end node (data recipient) of the UWA network. The setting up of the experiment and the analysis of its results described in this chapter took into account primarily the necessity for doing practical evaluation of these network parameters (data delivery latencies and delivery probabilities).

In studying the probabilities of data delivery from the source to the recipient, one of the requirements for the setting up the experiment is, on the one hand, the capability to vary the packet size (for different packet sizes their delivery probability may vary). On the other hand, when studying data delivery latencies in differently organized networks, one of the requirements for setting up the experiment is combining MAC and routing protocols with the size of transmitted data packets remaining unchanged. To overcome this contradiction, the data packet size varied in the experiment. In order to exclude the data packet size effect on the data delivery latency in a network of a given organization, the data packet duration was subtracted from the time it took the packet to pass every network hop. With this procedure, it became possible to evaluate the latencies caused by the algorithm of data propagation in a network rather than data size (an advantage of this approach is the fact that in future work it will be possible to do a comparison of overheads in networks using all kinds of combinations of MAC and routing protocols).

Such an approach does not exclude the possibility of evaluating the entire latency of the targeted size data propagating in a network. To obtain this time, it is sufficient to add to the estimated delays (connected with certain combinations of network protocols) the data packet length of practical interest (or the packet size with the known communication rate) multiplied by the number of network hops covered by data on the way from the data source to the data sink.

10.4.1.6 Experiment description

Experiments were conducted in shallow waters two times – in the fresh water Lake Werbellinsee, Germany, and in the slightly salty Lake of Tunaicha, Sakhalin Island (experiments jointly with another project running in the same time). The most representative results were obtained in Tunaicha Lake and are discussed next.

The topology of the network is presented in Figure 10.4. The S2C series UWA modems with the pre-installed EviNS software framework were used. The payload data were transmitted using so-called instant messages [26], that is, short messages (up to 512 bits) intended for the transmission of small size data: control commands, navigation strings, and status information of the exchange participants. The data packet duration varied within 200–300 ms.

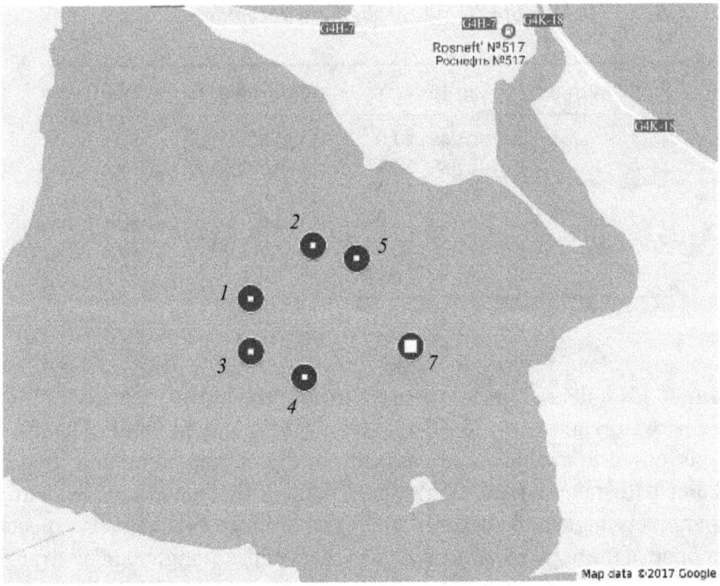

Figure 10.4 Network deployment site (network nodes are numbered)

The hydro-acoustic network consisted of six nodes. Each node represented a multi-functional device comprising the UWA modem S2CR7/17, a 10-Ah accumulator battery, a deep-water syntactic float, and a releaser. One of the network nodes always acted as a data source generating every 15-s data packets to be networked in succession to four (of the six) most remote network nodes in a pseudo-random order.

In each case of data exchange, both the source node and the recipient node recorded in the modem hardware memory all the packet information necessary for the subsequent analysis: timestamp of data packet transmission/reception, its information content and detailed characteristics of UWA communication channels between all the pairs of network nodes (network hops) involved in the data propagation from the source to the recipient.

The network nodes were located at a distance of 0.52–2.01 km from each other. Depths at the network deployment site were 12–15 m. The coordinates of each of the network nodes are given in Table 10.1.

The sound velocity was 1,406.6 m/s down to the depth of 8.1 m, whereupon it increased linearly to 1,413.9 m/s down to the depth of 13.9 m. With such a profile, the acoustic energy emitted by node 7 from the depth of 12 m concentrated in the direction of smaller depths, and network nodes 2, 3, 4, 5 located in the water column higher than node 7 were well within an acoustically illuminated zone. Node 1, however, being deeper than node 7, was in the acoustic shadow zone.

The different depths allowed studying situations where the quality of the received signals was noticeably different. To rule out situations where any network node could deliver data to any other node directly, the UWA modems' source level was set at

Table 10.1 Coordinates of the network nodes

Node number	Latitude, N	Longitude, E	Depth, m
1	46° 49′ 40,0″	143° 07′ 00,0″	13
2	46° 49′ 59,9″	143° 07′ 35,2″	10
3	46° 49′ 20,0″	143° 07′ 00,0″	10
4	46° 49′ 10.0″	143° 07′ 30,0″	10
5	46° 49′ 55.0″	143° 08′ 00,0″	10
7	46° 49′ 21.5″	143° 08′ 30,6″	12

the minimum (168 dB re 1 μPa at 1 m). Besides, the receive sensitivity for each of the modems was reduced by 20 dB relative to the standard value. The use of such settings permitted to exclude the possibility of direct data communication between nodes 7 and 1 (distance of 2.01 km). Other pairs in the network were characterized by comparatively high probability (0.5–0.9) of a direct data delivery to each other. However none of the network nodes could connect with a "neighbour" with probability equal to 1.

Operations in such conditions forced the network protocol to involve different data delivery routes from the source to the recipient. This permitted evaluation of the network basic characteristics in terms of simplifying digital communication between a data source and a data recipient under different connectivity conditions.

The network basic characteristics are assessed in this work in terms of: (1) entire latency of data packet propagation in the network, (2) average latency of data packet propagation per network hop, and (3) experimental probability of data delivery to the recipient on the first attempt. The most important results of the experiments are shown below in the figures with the accompanying explanations.

10.4.1.7 Experiment description

In total, 249 data packets were generated by the source (node 7) and networked to address nodes 1, 2, 4, 5 (addressed arbitrarily). Node 3 that had the best connectivity to node 7, i.e. the highest probability of data delivery from node 7 on the first attempt, was used as a relay node only (as a reception node it was of no special interest, since it would have very often received data directly from the source; the usefulness of the rest of the network nodes would have consisted primarily in the delivery of duplicating data).

The number of packets addressed in the network to node 1 was 49, to node 2 – 72, to node 4 – 55, and to node 5 – 72. In Figure 10.5(a)–(d), the black round marker designates the data source (always node 7), the light green marker indicates the data recipient (1, 2, 4 or 5), and the ring-shaped blue marker shows the network nodes involved in the exchange as relays.

Figure 10.5(a) shows the data packet delivery routes between data source node 7 and data packet recipient node 1. Each network hop is marked with the number of the messages it forwarded.

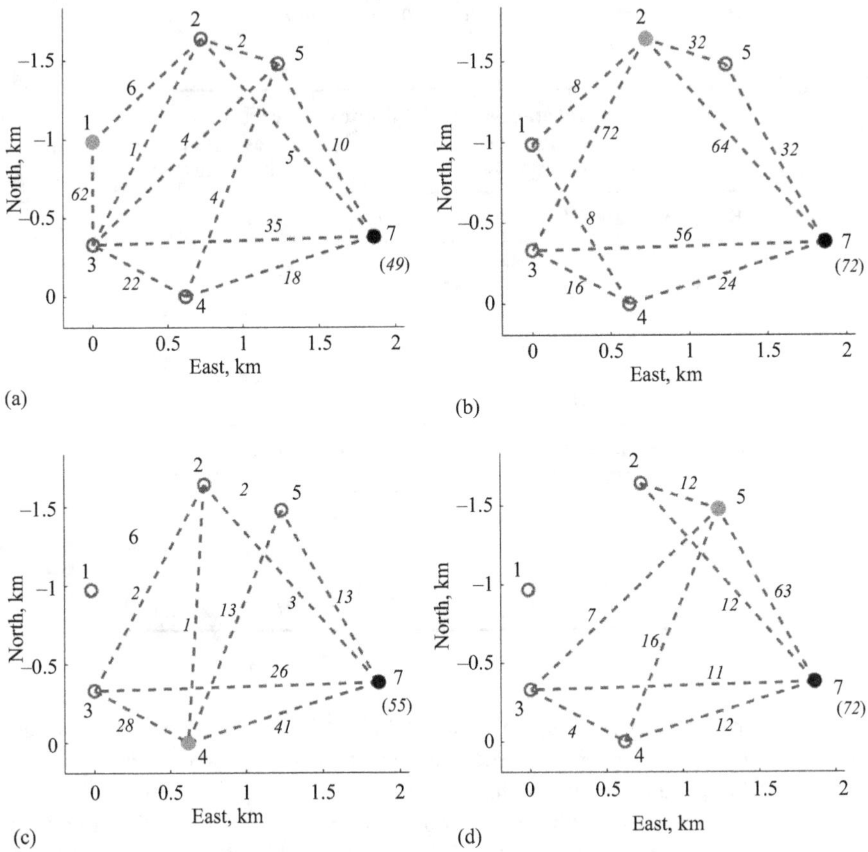

Figure 10.5 Propagation routes between the data source and recipient: (a) recipient – node 1, (b) recipient – node 2, (c) recipient – node 4 and (d) recipient – node 5

Since the direct link between the source and recipient (nodes 1 and 7) was excluded, all the packets were delivered via the intermediate relay nodes. Table 10.2 shows that the number of hops varied from 2 to 4. The shortest route passed through nodes 7-3-1 and 7-2-1. The longest route is through nodes 7-5-4-3-1. Statistics on the routes of signal propagation in the network for different data recipients is also provided in Table 10.2.

The specific features of the SNCF network protocol dictate that the data packets can reach the recipient along several routes. In this connection, instead of 49 networked packets, 68 packets were delivered to the receiver (136%).

Since in practice in the context of transmitting control commands and navigation data the fastest data delivery paths are of the greatest interest, the duplicated packets (packets with the same number, but longer packet delivery time) were discarded by the recipient node. Thus, in Figure 10.5(a) for each of the 49 packets only the shortest

Table 10.2 Statistics on data transmission routes in the network

Route	Percentage of messages passed the route (%)	Route	Percentage of messages passed the route (%)
Recipient – node 1		**Recipient – node 2**	
7-2-1	6.1	7-2	36.4
7-2-3-1	2.0	7-3-2	31.2
7-3-1	53.1	7-4-1-2	4.5
7-4-3-1	26.5	7-4-3-2	9.1
7-5-2-1	8.2	7-5-2	18.2
7-5-4-3-1	4.1		
Recipient – node 4		**Recipient – node 5**	
7-2-3-4	1.3	7-2-5	8.8
7-2-4	1.3	7-3-4-5	2.9
7-3-4	35.1	7-3-5	4.4
7-4	53.2	7-4-5	11.8
7-5-4	9.1	7-5	72.1

route is taken into account. Also, in Table 10.2 the column "Percentage of messages passed the route" is related only to messages having the smallest delivery latency in the network (duplicated messages that reached the recipient are not counted).

In the southern part of the network, nodes were characterized by better connectivity than in the northern part: the probability of data packet delivery directly to node 3 reached 0.9, to nodes 4 and 5 – approximately 0.65, and to node 2 – approximately 0.55. This connectivity accounted for the data delivery primarily along the routes including nodes 3 and 4. This does not mean that the rest of network nodes were useless in such connectedness conditions. In more than 14% of cases alternative data delivery routes proved to be faster.

Data delivery latencies

Figure 10.6(a) shows the time intervals in which, during the experiment, data were passing through the network from the source to the corresponding recipient (horizontal axis is marked with indexes corresponding to the data recipient nodes). Also, Figure 10.6(b) shows the time intervals in which the average delivery latency per network hop was estimated on the propagation route between the data source and the corresponding data recipients.

In the experiment, the total data packet delivery time from source 7 to recipient 1 was between 2.20 and 12.99 s. Considerable variations of this value are due to the specific features of the CSMA-Aloha protocol. After data packet reception each relay node waited for a short time interval (random, uniformly distributed between 0 and 3 s) and if the channel was unoccupied, it re-directed the data packet to the UWA

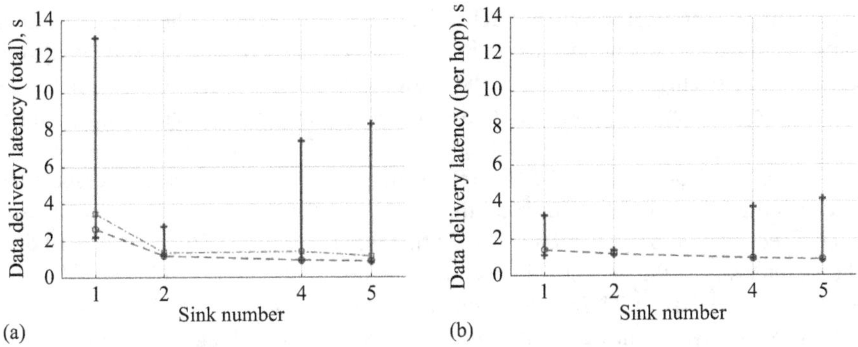

Figure 10.6 Time of data packet propagation between node 7 and receiving nodes 1, 2, 4, 5: (a) packet total propagation time when passing network hops from the source to the recipient, and (b) average time of passing one network hop

medium as a broadcast message. If the channel was occupied, the MAC protocol switched to the back-off mode for the duration of 2^n c, where n, a positive integer, is the number of an attempt to listen to the channel with repeatedly detected acoustic activity of the neighbouring nodes.

Information on the data packet delivery latency, important in practice, is depicted as the median of the distribution, and is indicated with a ring-shaped marker. When data were sent to node 1, the distribution median was 2.64 s (with an average value of 3.46 s). It can be seen that the median value is shifted markedly in the direction of smaller data delivery duration values. This is due to the fact that when the data propagated in the network via the SNCF protocol, there were normally several routes, one of them significantly shorter than the others (duplicated data delivered along longer routes were discarded).

Similarly, Figure 10.5(a) shows the results of the experiments (packet delivery routes) involving data transfer between the data generating node 7 and node 2. In 36% of the cases the delivery was direct. The rest of the data was delivered by the network along alternative routes – i.e. with the relatively weak source–receiver connectivity the main contribution of data delivery to the recipient was provided by the relay nodes. Due to delivery via multiple routes, the addressee node 2 received 176 rather than 72 packets sent by the source (244%). The shortest route was the direct one, i.e. 7-2. The longest routes passed via nodes 7-4-1-2 and 7-4-3-2. Figure 10.6(a) shows an interval of the packet delivery latencies (index 2 on the horizontal axis). In general, the delivery routes were shorter than in the case of delivery between nodes 7 and 1, and the data delivery latency was shorter staying within 1.18–2.79 s. The distribution median was 1.19 s (when the average was 1.36 s). In much the same way as in the first case, one of the multiple routes was shorter than the others (duplicated data delivered on longer routes were discarded). As a consequence, the median value was also displaced in the direction of smaller data delivery latencies.

Figure 10.5(c) and (d) shows packet delivery routes between the data generating node 7 and data recipient nodes 4 and 5. In 53% and 69% of cases, respectively, the data delivery occurred directly from the source to recipients. The rest of the data was delivered by the network along alternative routes. In the exchange with nodes 4 and 5, the shortest routes were correspondingly 7-4 and 7-5, and the longest routes were, respectively, 7-2-3-4 and 7-3-4-5. Due to the multi-route delivery, the addressee node 4 received 83 packets rather than 55 that were sent (151%), whereas the addressee node 5 received 98 packets rather than 72 that were sent to it (136%). Figure 10.6(a) shows intervals of the packet delivery full time distribution (next to indexes 4 and 5 on the horizontal axis). They have comparatively broad limits (0.94–7.40 s and 0.87–8.32 s, respectively), however, in much the same manner, long delivery durations were comparatively rare, and since the delivery routes were even shorter than in the case of data delivery between nodes 7 and 2, the distribution median for the packet delivery latency to nodes 4 and 5 assumed even smaller values of 0.95 and 0.88 s, respectively.

Statistical data delivery probabilities

Because of the data delivery route duplication permitted by the network layer protocol, the addressee node received more than one packet per each sent packet, in particular, from 136% to 252% of the sent numbers. In this connection, if the data delivery delay, in which the packet is considered to be valid (unexpired), was not limited, the experimental probability of data delivery from the source to any receiver in the network was equal to 1.

If, however, the data expiration time is limited to e.g. 5 s, the experimental probability of the data timely delivery was 0.89 for the case of data delivery to node 1, 1.0 for the case of data delivery to node 2, 0.95 and 0.96 for the cases of data delivery to nodes 4 and 5, respectively. More detailed data on the experimental probability of data delivery within various time intervals, e.g. acceptable data delivery delays, is presented in Figure 10.7.

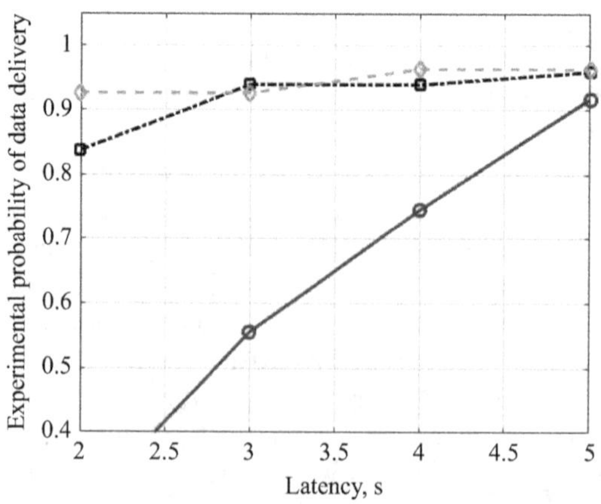

Figure 10.7 Experimental probability of data delivery within various time intervals

Data delivery latency per hop

Figure 10.6(b) shows intervals of average data delivery latency packet per network hop between the source node and each of the addressee nodes (indexes 1, 2, 4, 5, respectively). Although all the network hops have different lengths (0.5–2 km), data reaches the recipient via a multitude of hops in their different combinations.

In this connection, by dividing the entire data packet delivery latency by the number of hops corresponding to the route, the data delivery latency per network hop can be estimated. (Information on the data latency per network hop may be useful for subsequent comparison of the experimental operation results with differently organized networks covering approximately the same water area, i.e. about 4 sq. km.)

For the case of data delivery to node 1, the estimated data packet delivery latency per hop calculated in this manner was within 1.10–3.25 s. This time included the delay of the acoustic signal propagation in the channel, the pseudo-random back-off time (within the 0.3-s interval) of the signal retransmission by the network intermediate node, as well as the exponentially incrementing signal retransmission delay (2^n *s*, where definition of *n* is provided above). The ring-shaped marker indicates the distribution median which was 1.37 s.

In the cases of data delivery to node 2, packet delivery latency per one network hop was within a narrower range of 1.18–1.40 s (with a median of 1.19 s), and in the cases of data delivery to nodes 4 and 5 such latencies were within the limits close to those for the node 1 (but with medians even nearer to the lower edge of the relevant observed values interval: 0.95 and 0.88 s, respectively).

The median position closer to the lower edge is due to the fact that in the case of data packet simultaneous delivery along many short routes (2 hops), on one of the routes the packet retransmission pseudo-random delay (back-off time) was most likely positioned much closer to the lower edge of the delay interval than on other routes. Also, it can be seen by comparing the per-hop delivery latencies in different experiments, that with the route becoming longer because of a larger number of network hops and with the decreasing number of alternative routes, the per-hop data delivery latency increases. In an extreme case, if the random values are distributed uniformly within the set interval, with the number of spans growing infinitely and the number of alternative routes decreasing to one, the probability distribution of the sum of random values tends to a logarithmic one with the statistical expectation in the centre of the set uniform distribution. The per-hop data delivery latency median will be close to the statistical expectation of this logarithmic distribution. The data in Figure 10.6(b) and Table 10.2 confirm that with the route growing due to the number of network hops and with decreasing the number of alternative routes, the per-hop data delivery time, indeed, raises: from 0.88 s in the case of communication with node 5 (with the largest number of short routes) to 1.37 s in the case of communication with node 1 (with the largest number of long routes).

10.4.1.8 Case study conclusion

In general, this experiments confirmed the successful fulfilment of the remote (beyond the direct acoustic "visibility") data exchange among underwater communication nodes (sensors, vehicles, observatories), each capable of acting as a network relay. When there was no direct communication between the source and the recipient, data delivery was effected via such relay nodes.

It follows from the figures that the network demonstrated a possibility of transferring data from the source to the recipient along many different routes. In most cases, data were simultaneously delivered via a multiplicity of routes rather than along a single one. The network layer (SNCF) protocol's feature enabling the network relay nodes to redirect packets to other nodes brought about frequent duplication of data delivered to the end node (recipient).

On the one hand, this SNCF protocol's feature is an advantage, since it ensures high probability and the shortest time of data delivery to the end user (often along the route with the least number of hops). On the other hand, the repeated data delivery is related to increased power consumption of the network nodes. As such, the use of the SNCF protocol as is can be recommended for user applications in which saving the energy for the acoustic communication among the network nodes is not a vital requirement. In particular, the use of the SNCF protocol can be easily recommended for networking data among AUVs, or between an AUV and a surface support vessel, i.e. in applications where power consumed by the acoustic communication is significantly smaller than the power consumed by other devices/machinery on an AUV or ship. However, the use of the SNCF protocol can hardly be recommended for use in autonomous network systems, in which energy storage is normally low but long-term network operation is required (high endurance). Combinations of protocols permitting power consumed for the data delivery to be minimized (on the account of reduced redundancy and a probability of the first-attempt delivery) will be the subject of future research.

The most frequently observed (on the median) latencies of data delivery in a network correspond to the sum of propagation delays on a route (range function) and the sum of delays obtained from statistic characteristics of pseudo-random back-off time intervals (signal re-transmission by the relay node) on each/sum of the network hops. In the experiments conducted, the practically observed per-hop delivery latencies were most often 0.88–1.37 s. The observed latencies in the case of a message transmission along the entire route (from the source to the recipient) were most often 0.88–2.64 s. The latencies obtained in the experiment (at least in a network of similar configuration, size and UWA channel dynamics) provide a possibility to plan algorithms for operational coordination of mobile UWA network nodes, e.g. for the cases of their underwater group navigation in a large water area.

Due to the ability to operate in a UWA medium with rapidly varying impulse response (i.e. in communication conditions comprising large dynamics of the UWA medium), as well as due to small data delivery latencies in the network (at least in the operation mode without acknowledgement), the combination of the protocols (of physical, MAC, and routing layers) used in this work can be recommended in the tasks of network communications among underwater vehicles between each other, or between underwater vehicles and a surface support vessel in motion.

10.4.2 Case study – operation of UWA modems in a bi-modal network with a centralized topology

10.4.2.1 Operation conditions

In accordance with the technical specifications set forth for the WiMUST project [27], the following types of data should be delivered over the UWA communication channel: (1) fragments of seismic data, (2) control signals, navigation strings, and AUV status and control data, (3) service data, for example, for clock synchronization.

Navigation strings, control and monitoring signals of the AUV must be delivered in real time to all AUVs for arranging them into a formation, for joint manoeuvring, and for keeping a given formation of AUVs during their motion along a seismic surveying path. Service data should be delivered operatively for mission start or stop, synchronization of AUV measurements, mission modification, etc. Also, fragments of seismic data should be delivered as quickly as possible from the AUVs to the support ship at the surface. These fragments are important to assess the quality of the seismic data collected.

The architecture of the seismic survey system assumes the possibility of scaling according to the number of vehicles participating in the mission. Therefore, in the main operating modes, the delivery of navigation data from several accompanying (spatially separated) geo-referenced signal sources must be made unidirectional: to the AUVs without a reverse acknowledgement (bi-directional data exchange would result in a reduction in the positioning frequency, inversely proportional to the AUVs number). Provided that navigation data during the motion should be transmitted as often as possible (for example, with an interval of the order of a second), practically all available capacity of the UWA channel in the frequency band allocated for this task is completely reserved. Transmission of any additional data (apart of navigation data) in the reserved frequency range is practically impossible. At the same time, according to the WiMUST specification, seismic data should be delivered to the support vessel at the surface operatively (within less than 3 s for data exchange with each AUV); the amount of such data is relatively large: 10 kbps in each exchange cycle.

For simultaneous transmission of data, both for navigation and for monitoring the seismic mission, it was proposed to use simultaneously two different models of UWA modems operating in adjacent frequency bands. In particular, the UWA modems of Evologics, particularly S2CR42/65 [1] operating in the high-frequency (HF) range, 42–65 kHz, and S2CR18/34 [1] modems operating in the medium frequencies (MF), 18–34 kHz.

The modems S2CR42/65 (HF modems) have a relatively high data rate (in good conditions up to 30 kbit/s). In the tasks of the WiMUST project, it was decided to use these modems for transmitting seismic data from the AUVs to the support vessel. The UWA modems S2CR18/34 (MF-modems) are characterized by a moderate data transfer rate (in good conditions up to 13.8 kbps). In the tasks of the WiMUST project, it was decided to use them for transmission of relatively small data packets – control signals, navigation data, device status monitoring and service data.

HF-modems were to ensure the operability in the network of a simple centralized topology. MF-modems were to provide high efficiency on the network of decentralized topology, similarly to the task demonstrated in the previous subsection. For example, if the devices move beyond the operation range of given UWA modem, the decentralized topology network had to provide a connectivity sufficient to exchange data of any modem with any other one of the deployed network (group of AUVs).

Due to the proximity of frequency ranges and consequently induced inter-channel interference, simultaneous/asynchronous data transmission by means of the different modems can be a non-trivial task. In this regard, at the first step of the WiMUST project, the main task was to investigate the degree of mutual influence of the different modems models, and to evaluate their performance (in comparison to the project requirements) during their simultaneous operation.

10.4.2.2 Experimental conditions

An experiment was conducted in a fresh water lake (Werbellinsee near Berlin, Germany). Two HF modems were installed 2 m deep and in different tests, 25–150 m apart, in the lake with 4.5 m water depth. The two MF modems were installed at the same depth, close to the high-frequency modems (approx. 1 m apart), as shown in Figure 10.8.

The transducer of the HF modem is weakly resonant at 54 kHz, its frequency response drops by up to 6 dB at the band edges (at 42 and 65 kHz), and the directivity pattern is conical with 90° opening angle. During the experiments, the transmitted signal levels varied between 176 and 191 dB re 1 uPa/m. Outside of the nominal conical beam, up to an opening angle of 180°, the transducer's frequency response is almost linear across the band and the sensitivity drop is moderate. Relative to the signal level on the main axis of the directivity pattern, the sensitivity at 90° angle decreases by 14 dB (this allowed to use the modem for applications which require a hemispherical transducer directivity pattern for shortdistance transmissions, e.g. over several hundred metres). The MF modems have a weak resonance at 26 kHz, a frequency response with up to 6 dB drop at the band edges (at 18 and 34 kHz), a horizontally omnidirectional directivity pattern, and a weakly directional pattern in the vertical plane (opening over 120°). During the experiments, the transmitted signal levels varied between 178 and 188 dB re 1 uPa/m.

Each pair of MF and HF modems exchanged 10 kbit-data fragments in Burst Data mode (Burst Data is data transmission mode used by EvoLogics S2C technology [5] modems as part of the D-MAC protocol [28,29]). Each data record was split into 10 packets with short (20 ms) intervals between subsequent packet transmissions. Each packet contained two synchronization pulses, a header and payload data. Each packet train concluded with a service packet. The experiment lasted 2.5 h. The wind speed was strong, up to 21 m/s and the waves reached 40 cm high, causing a large variation between the signal impulse responses. During the experiments, the modem pairs were used simultaneously/asynchronously. Data exchange sessions lasted 30 min each. The tests allowed to experimentally evaluate the communication

MF <------1 m-------> HF <--------------25–150 m--------------> HF <----1 m---->MF

Figure 10.8 Modem placement during acoustic compatibility tests

performance for each UWA channel when different types of modems were simultaneously/asynchronously operated to deliver different data types: portions of seismic data, service data, positioning data, and ranging data for AUV position estimation and cooperative manoeuvring

10.4.2.3 Results

Figure 10.9(a) presents the spectrogram, obtained during operation of one pair of MF modems, and Figure 10.9(b) – the spectrogram of two pairs of MF and HF modems operating at the same time.

The spectrogram in Figure 10.9(a) was obtained by calculating the short-time Fourier transform with a 20 ms window (Hanning with a 50% overlap) over a 5-s interval of acoustic activity of the MF modems. The similarly obtained spectrogram in Figure 10.9(b) represents 5 s of acoustic activity of the MF and HF modems that simultaneously exchanged data.

The out-of-band power density of MF modems was moderately low (Figure 10.9(a)). Besides the relation between power spectral densities, and Figure 10.9(b) illustrates the out-of-band spectral densities of the MF and HF modems signals. With equal levels of the transmitted signals, at the corresponding modem's central frequency the power spectral density of the adjacent channel interference (coming from the neighbouring frequency band) reached about −18–−20 dB. Figure 10.10 shows the multipath intensity profiles, taken at the receiving side of the acoustic communication channel. As these examples demonstrate, the profiles were characterized by two groups of multipath arrivals. The first group consisted of the dominant arrival and the second arrival, weaker by 1–5 dB and delayed by 50 μs. The second group consisted of three multipaths, weaker by 7–10 dB and delayed by 640, 385, and 420 μs, respectively (relative to the dominant one). Varying (highly dynamic) multipath arrivals with variable levels and propagation delays made the channel a very challenging environment for high-speed communication. However, multipath propagation and dynamic multipath intensity profiles are quite common properties of UWA channels in shallow waters.

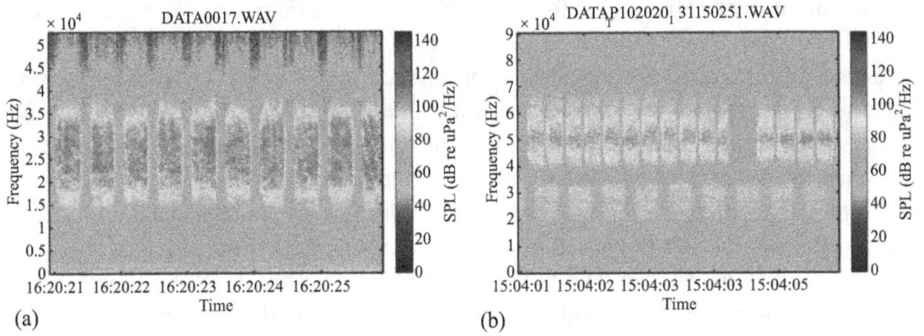

Figure 10.9 Acoustic activity spectrograms: (a) one pair of MF modems, and (b) two pairs of different modems (HF and MF) operating simultaneously

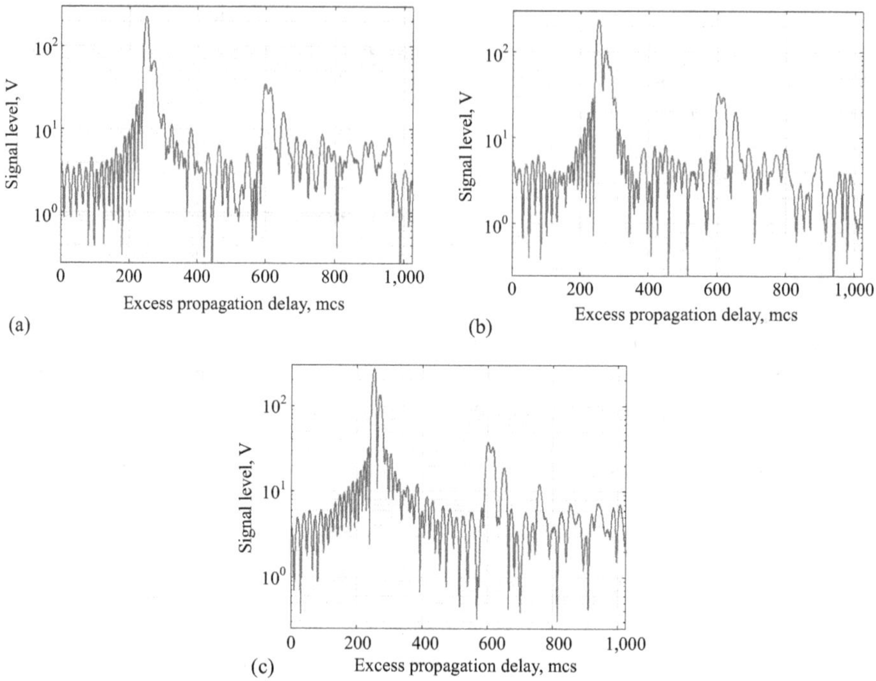

Figure 10.10 Examples of multipath intensity profiles, registered while conducting the experiments at Werbellinsee

Figure 10.11 illustrates the performance of the HF modems, operating in shallow water simultaneously with the MF modem. When working in parallel, the estimated data symbols had largely fluctuating power (measured here as RSSI – received signal strength indicator). The magnitude of the fluctuations depended on the HF modem's bitrate. In Figure 10.11(a) and (b), the fragments demonstrate the effect of out-of-band interference onto the HF modem from the simultaneously operating MF modem. Maximum fluctuations can be seen in Figure 10.11(a). These were observed at 11,904 bit/s transmission bitrate. Less fluctuations can be seen in Figure 10.11(b) at 9,259 bit/s. Minimum fluctuations are demonstrated in Figure 10.11(c) at bitrate 976 bit/s. For all these cases, the bitrate of the MF modems operating simultaneously with the HF modem was constant at 976 bit/s. Varying fluctuation RSSIs can be explained by different signal energy amounts contained within one modulated data symbol. As the transmission bitrate increased, the duration of one data symbol (signal modulation period) became shorter, and the energy per data symbol became smaller. While the out-of-band noise remained the same (transmission in the neighbouring band is carried out at constant bitrate), the increasing bitrate led to a reduction in the signal-to-noise ratio. Conversely, lowering the transmission bitrate led to an increase in the signal-to-noise ratio.

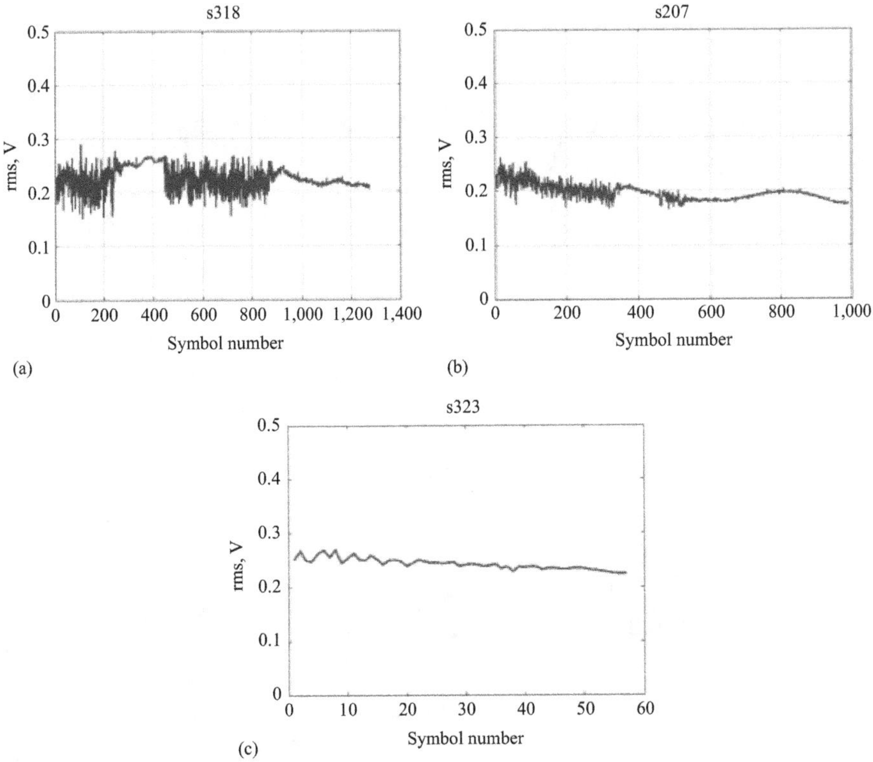

Figure 10.11 RSSI of data symbols, received by HF modems when operating in parallel with MF modems

Figure 10.12 demonstrates the fluctuations in data symbols estimates, received by the HF modems working in parallel with the MF modems. Figure 10.12(a) presents phase and RSSI spread of data symbol estimates at a bit rate of 11,904 bit/s (corresponding to the data exchange fragment in Figure 10.12(a). Figure 10.12(b) and c shows the phase and RSSI spread of data symbols at 9,256 and 976 bit/s, respectively (corresponding to the data exchange fragments in Figure 10.11(b) and (c).

Comparing the plots in Figure 10.12(a) and (c), one can notice that the phase spread is also growing with increasing bitrates. At the slowest transmission bitrate of 976 bit/s, the effects of out-of-band noise were least noticeable. At the maximum speed of 11,904 bit/s, the out-of-band noise was most prominent. The standard deviation of the signal phase from the expected discrete value (according to the differential QPSK) was 5.43° at highest transmission bitrate and 1.06° at the lowest one (3.3° at intermediate value). When expressing phase and RSSI spread of data symbols as signal-to-noise ratios, the ratio 12 dB corresponded to the maximum, 15 dB to the intermediate, and the ratio 24 dB – to the lowest transmission bitrate. Overall, transmission bitrates successfully achieved in tests with simultaneously operating modems remained within 8–14 kbit/s. Besides the experiments involving the simultaneous operation of HF and

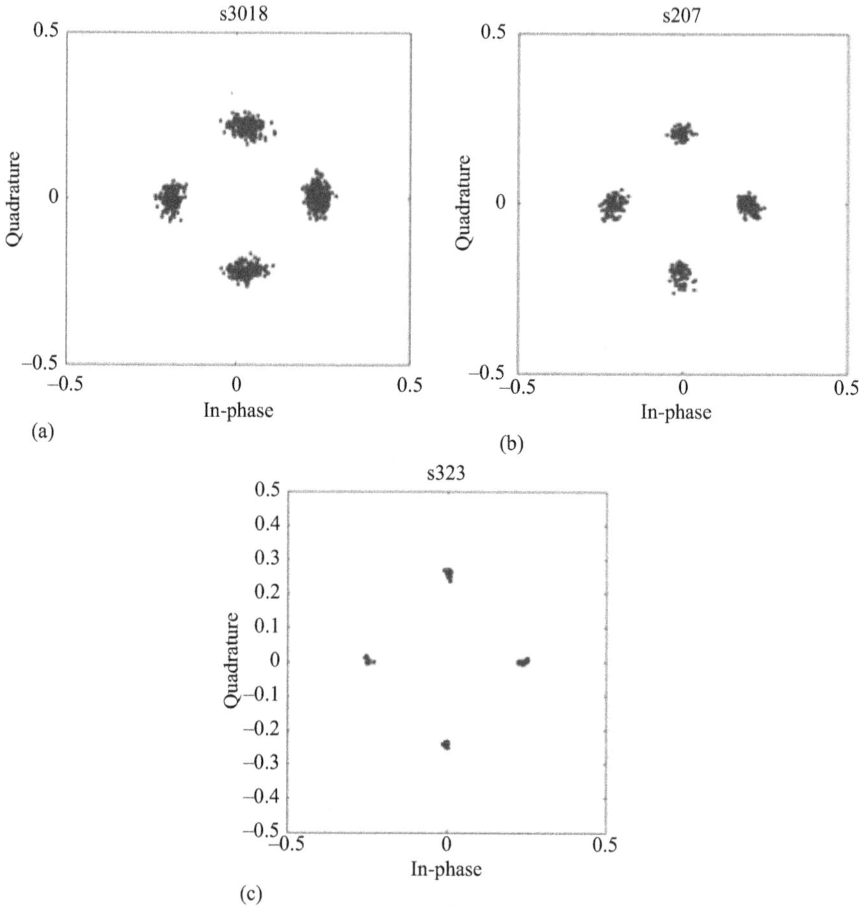

Figure 10.12 Spread of the measured data symbol values (RSSI and phases), received by HF modems when operating in parallel with MF modems

MF modems, the performance of the HF modems operating in isolation was also evaluated for the same conditions of the acoustic channel, i.e. without the parallel operation of MF modems. The transmission bitrates that were successfully achieved in the experiments were significantly higher, i.e. 16–25 kbit/s.

One should note that in both cases the mentioned transmission bitrates refer to the so-called nominal bitrate, i.e. the transmission bitrate, measured via the rate of the phase-shift keying of the signal). The so-called effective bitrate (data rate on the user interface, which is always lower than the nominal one due to data overhead, redundancy of error correction algorithms, etc.) comprised 2.3–5.2 kbit/s during simultaneous operation of HF and MF modems, and 4.2–8.1 kbit/s for HF modem-only transmissions. In general, the effects of out-of-band noise manifested with 1.5–1.7 times lower effective bitrates.

On average, the HF modem performance allowed to deliver 10 kbits of payload data (examples of seismic data) to the receiver, i.e. the operator on the support vessel, within 3 s. Therefore, when up to 10 AUVs operate in a group formation (as defined by the WiMUST project's specification), the data exchange cycle between each AUV and the support vessel comprises no more than 30 s. This interrogation rate conforms to the requirements of real-time seismic data acquisition and, therefore, was recommended for WiMUST applications.

In turn, performance evaluation for the MF modem that operates in parallel with the HF modem was based on the probability of a successful first-attempt data packet delivery. Transmission bitrate was constant, and the modems were used for D-MAC "instant message" exchange. The influence of out-of-band noise from the active HF modem manifested with an increased number of "damaged" packets, delivered to the recipient. The magnitude of noise influence was mostly defined by the difference between signal levels, emitted by both types of modems in their respective operating bands. In general, the probability of successful first-attempt data packet delivery was high, comprising 0.9–1.0 when signal levels, emitted by corresponding modems in neighbouring bands, were equal. Once signal levels, emitted by HF modems, were significantly higher, the probability of successful delivery of data packets from the first attempt could decrease dramatically. For example, if the HF modem's signal level was 12 dB higher, the probability of successful first-attempt data packet delivery by the MF modem dropped to 0.7.

The results of the experiments were used by WiMUST project participants to select the signal levels for the modems operating in various AUV formations.

The selected modems have been used in all lake and sea trials during the development of the WiMUST technology, including the final experiments performed on the Portuguese coast, near the town of Sines, at the end of January 2018.

10.4.2.4 Case study conclusion

A series of experiments in the lake and then the final experiment (final demonstration) of WiMUST performed in Sines, Portugal, in early 2018, revealed several important quantitative characteristics of a bimodal digital UWA communication link based on simultaneous operation of MF and HF modems.

In the operation mode planned for future WiMUST applications, the achieved performance benchmarks are sufficiently high for (1) real-time exchange of navigation and service data for mutual vehicle positioning within an AUV formation (using MF modems) and (2) delivering payload data to the support vessel using HF modems (sample data from seismic streamers towed by the AUVs).

The demonstrated values of effective HF modem bitrate were between 2.3 and 5.2 kbit/s (with an average value above 3.5 kbit/s). Taking the specifics of WiMUST project into account, this performance does meet the required delivery times of up to 10 kbits of control data from each AUV (the towed seismic streamer) within an interval that on average does not exceed 3 s.

The criterion for evaluating the performance of MF modems was the probability of successful first-attempt delivery of up to 512 bits of data. In conditions of practical interest this probability reached 0.9–1.0. Nevertheless, when the emitted signal levels were not balanced (the HF signal levels were 12 dB higher than MF signal levels), the probability of successful first-attempt data delivery dropped to 0.7. This led to a conclusion that transmission signal levels from both types of modems should be balanced to reach a compromise between successful data delivery for MF modems and sufficient effective bitrates of the HF modems.

10.4.3 Case study – UWA modems with integrated atomic clocks for positioning of a group of AUVs

Time synchronization of the nodes of a UWA network, particularly when the nodes are AUVs operating collectively in formation, is a necessary prerequisite for the concerted operation of the vehicles and the effective use of collected sensor measurements in diverse marine application scenarios. To meet this goal, one possible solution is to synchronize thee AUV clocks at the surface (typically by means of GPS) and then use low drift clocks to maintain the required precision when the vehicles are submerged. This chapter presents experimental results on the integration of chip-scale atomic clocks (CSAC) into the processing electronics of UWA modems that are used as underwater digital communication devices on board the "Medusa" and "Folaga" AUVs. Since one of the networking modes (particularly, medium access control modes) of the modems, built upon S2C technology, is based on transmission/reception of Synchronous Instant Messages, the modems with integrated atomic clocks become capable of precise measurements of the one-way signal propagation delay between the signal source and signal receiver units, and thus of precise estimation of distance between these units. This section describes experimental results on the CSAC accuracy achieved during multiple field tests, includes practical recommendations for CSAC disciplining and phase synchronization with the source of common time reference (GPS time), and provides experimental estimation on the positioning accuracy of AUVs that is achievable with S2C acoustic modems for communication and positioning purposes.

10.4.3.1 The task

When low-drift clocks are available, AUV positioning can be performed by measuring the duration of the one-way-propagation of the UWA signal from several geo-referenced sources to a signal receiver unit, and thus estimating the distances between the sources and the receiver. As modern AUVs are usually equipped with UWA communication units, the task of estimating the propagation delay can be assigned to the UWA modem.

Since there is no need for bidirectional data exchange, the solution for synchronous AUV positioning offers the extra benefits of shorter interrogation cycles and the opportunity to scale the AUV formation up to an arbitrary size without increasing the duration of the time window required to estimate the position of each individual vehicle.

In striking contrast to the solution based on bidirectional data exchange between geo-referenced sources and AUVs, the unidirectional approach adopted herein

Figure 10.13 AUVs and USVs used in the experiments

avoids overloading the channel with high-volumes of acoustic activity from multiple modems. It is also more energy-efficient, reduced the acoustic impact on the subsea environment (by reducing acoustic pollution) and facilitates a more discreet operation of the AUVs.

In what follows, using the terminology of underwater networking, underwater modems that provide the functionality of medium access and data routing will be referred to as "nodes". Geo-referenced signal sources (unmanned surface vehicles – USVs) and signal receivers on the targets to be positioned (the AUVs) will be referred to as "antenna nodes" and "positioning nodes", respectively.

Two types of AUVs were used during the different trials: the "Medusa" vehicles of IST, Portugal (Figure 10.13(a)), and the "Folaga" vehicles of ISME and Graaltech, Italy (Fig. 10.13b). Also, two types of autonomous surface vehicles (ASVs) were used in the trials: the DELFIM IST/ISR-Lisboa, Portugal (Figure 10.13(c)), and the ULISSE, ISME/UNIGE, Italy (Figure 10.13(d)).

Since an AUV is usually equipped with a pressure (depth) sensor, using just two antenna nodes is sufficient to solve the task of AUV positioning (the ambiguity of the position estimate can be eliminated with a priori information about the hemisphere of AUV location relative to the line connecting the antenna nodes, or by examining the Doppler shifts of the received signal).

10.4.3.2 UWA modem with integrated atomic clock

The results presented in this case study were obtained in the course of R&D work pursued under the EU funded WiMUST. In the scenarios adopted, two ASVs play the role of antenna nodes, while a group of AUVs towing streamers across the water column play the role of positioning nodes. In the objectives set forth in the project, high-accuracy AUV positioning is an extremely important requirement. Accurate positioning must be performed in so-called acoustic silence mode, and during an 8-h mission the clock drift must not exceed 50 μs. Furthermore, the requirements dictate that the atomic clocks be of small size and low weight, have low power consumption, and be commercially available.

To minimize size, the atomic clock was integrated into existing UWA modem electronics and provided an interface to distribute the timekeeping pulses to other modules of the AUV.

Currently, the market for compact atomic clocks offers only one commercial model that meets the technical requirements of the WiMUST project: the Microsemi SA.45s Chip Scale Atomic Clock (CSAC) [26,30]. According to its specifications,

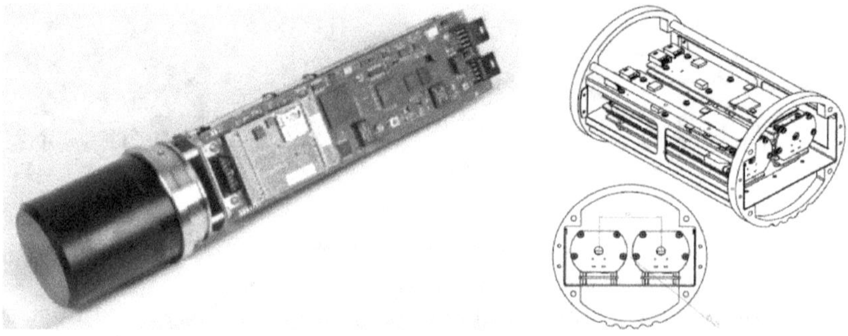

(a) (b)

Figure 10.14 Electronics of the S2CM18/34OEM modem: (a) modem before CSAC integration, (b) a model of an AUV section hosting two modems: S2CM18/34 OEM and S2CM42/65 OEM

the device is three orders of magnitude more accurate than OCXOs oscillators [31] and two orders of magnitude more accurate than TCXOs [32,33], consumes less than 120 mW, weighs 35 g, and takes up less than 16.5 cm^3. The CSAC provides access to 10 MHz meander and 1PPS on standard CMOS levels with short-time stability (the Allan deviation) around $3| \times |10^{-10}$ ppm over 1 s. The long-term aging drift grows by no more than $9| \times |10^{-9}$ ppm per month, the temperature range is wide, 10–70°C, and the maximum frequency deviation is $\pm 5| \times |10^{-10}$ ppm. The CSAC offers an input for the 1PPS reference signal (for example, the timekeeping signal from a GNSS receiver) that can be used to synchronize the 1PPS output of the atomic clock to ±10 ns accuracy.

Figure 10.14(a) shows the modem electronics of the standard S2CM18/34 OEM model manufactured by Evologics. The geometric constraints of the AUV section were taken into consideration during atomic clock integration (a section of Medusa AUV is presented in Figure 10.14(b). Figure 10.15 shows the S2CM18/34 OEM modem electronics with the integrated atomic clock. After integration, the accuracy of the CSAC was evaluated in the lab and during a series of open-sea trials (using the AUV Medusa). The main results demonstrated that after disciplining and phase synchronization with a 30 ns threshold and 600 s constant the phase offset of the CSAC's 1PPS output relative to the 1PPS from the GPS receiver did not exceed 20 ms after 3–5 h of autonomous operation. The projected accuracy of AUV position estimations during long-term autonomous deployment (several hours) was on the order of centimetres.

The modem contained the EviNS framework, which in this case, apart from networking protocols, included also a ROS driver for integration of the modem with the Medusa Electronics (Medusa AUV is based on ROS – widely spread Robot Operating System).

The experimental evaluation of the positioning accuracy achievable was also performed during extended sea trials with several AUVs (Medusa and Folaga) performing

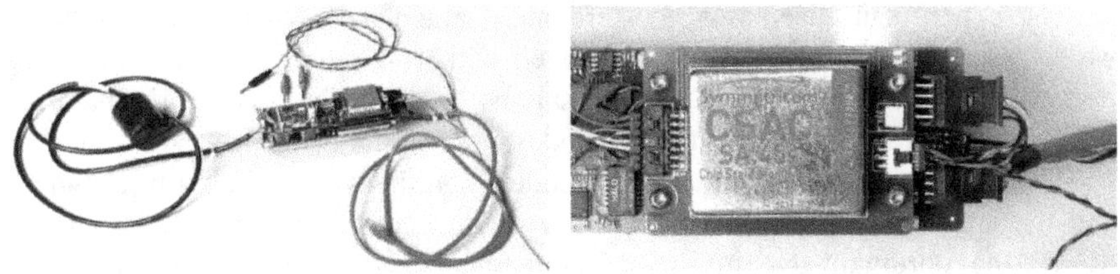

(a) (b)

Figure 10.15 S2CM18/34OEM modem electronics: (a) a photograph of the modem with all components, including the SCAC SA.4s and (b) the atomic clock module on the modem board

Figure 10.16 A Medusa AUV in a half-submerged state (the modem transducer is in the lower part). The picture shows also the encapsulated electronics for acoustic streamer data acquisition between the two cylindrical bodies

manoeuvres similar to those required in the WiMUST cooperative missions. The results of the sea trials are presented below.

10.4.3.3 Experimental results of AUV positioning using UWA modems with integrated atomic clocks

Sea trials aimed at assessing the performance of AUV positioning using underwater modems with built-in atomic clocks were performed in December 2016 near Sines, Portugal.

The experiments involved the double-hulled "Medusa" AUVs of IST, Portugal (Figure 10.16). The AUVs were half-submerged – the lower hull with the acoustic modem was underwater (modem at approx. 0.5 m depth), the upper hull was at the surface. In this arrangement, a GPS antenna was installed on a short mast that was always at least 25 cm above the surface, providing an opportunity to compare the data, obtained using UWA positioning, with data from GPS RTK positioning.

The operating conditions were the following: calm seas with 1 degree swell, wind at 2–3 m/s. The geo-referenced nodes and the AUVs moved no faster than 1.0 m/s.

The experiments used two geo-referenced antenna nodes installed on-board two ASVs (small-size catamarans) equipped with UWA modems that took turns transmitting their coordinates and time stamps to the AUVs that represented the underwater moving targets (positioning nodes).

The positioning nodes (underwater modems on the AUVs), being time-synced with the antenna nodes (underwater modems on geo-referenced signal sources), received the coordinates and timestamps and estimated their own geographic locations based on this data. To evaluate the accuracy of acoustic positioning, these coordinates were compared with coordinates obtained by RTK GPS at the moments of signal reception.

Figure 10.17 illustrates the sequence of signal transmissions during the motion of one of the AUVs along a straight line. For simplification, the subsequent time intervals are represented with integer non-dimensional numbers. The geo-referenced antenna nodes are labelled as ASV_1 and ASV_2 (ASV – Autonomous Surface Vehicle). The Positioning node (the AUV acting as a positioning target) is labelled AUV. Since all vehicles had access to GPS RTK, ASV_1 and ASV_2 transmitted data packets with their coordinates towards the AUV using their acoustic modems. The AUV received these packets and evaluated signal propagation delays (and, correspondingly, the distances) towards ASV_1 and ASV_2, used data from its depth sensor and with each data packet reception evaluated its own geographic location.

In particular, at the moment $t = 1$ the AUV received $p'_{1,1}$ data about ASV_1 location and evaluated distance r_1 towards ASV_1. Shortly after, at $t = 2$ moment, the AUV moved over by d_1 distance, received $p'_{2,2}$ data about ASV_2 location and evaluated distance r_2 towards ASV_2. Therefore, during its movement the AUV obtained a sequence of n measurements with distance evaluations r_1, r_2, r_n, received at moments t_1, t_2, t_n at points $p_{1,1}, p_{1,2}, p_{1,n}$ correspondingly.

Distance estimations $r_{[1:n]}$ were compared with distances between coordinates $p'_{[1:n]}$ and $p_{[1:n]}$, as well as between coordinates $p'_{[2:n]}$ and $p_{[2:n]}$, obtained by the GPS

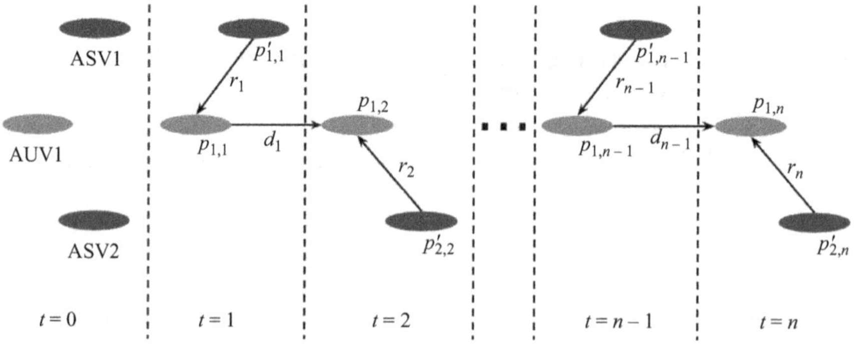

Figure 10.17 The concept of AUV positioning with underwater acoustic modems equipped with built-in atomic clocks (positioning in acoustic silence mode)

receiver on the AUV at timestamps t_1, t_2, t_n and from GPS receivers on the ASV_1 and ASV_2 at moments when they transmitted data about their respective locations.

Exact time measurements with the atomic clock enabled not solely the accurate estimation of the acoustic signal propagation delays. In fact, the underwater modem was equipped with a physical interface that allowed to provide timekeeping signals to other units of the AUV, in particular, to its on-board PC and the sensor data acquisition system. Data sets from inertial sensors and the micro-mechanical AHRS were used to model the AUV motions (for so-called odometry) that could in turn be used to increase the accuracy of AUV positioning/navigation even further. However, current tests focused on the possibility of odometry-independent positioning and navigation of the AUV using the underwater modem.

ASV_1 and ASV_2 had access to an accurate time source directly from the GPS receiver and used the 1PPS output to receive timekeeping signals (the scenario with 2–3 geo-referenced signal sources on board surface vehicles is foreseen for the WiMUST system). Accurate estimation of the ASV_1 and ASV_2 locations was performed directly with the GPS in RTK mode.

Figure 10.18 shows the actual trajectory of the AUV during the sea trials obtained with the GPS in red. The position estimates obtained using the set-up adopted and a maximum likelihood estimator are shown in green. The results of using the atomic clock for AUV positioning with underwater modems are presented in Figure 10.19 as well as the difference between distance estimations from the AUV to signal sources on the surface – between the estimation based on AUV modem data and the estimation based on GPS receiver data. Since positioning with GPS RTK can be considered accurate (the error is within a few centimetres), the results in Figure 10.19 illustrate the error of AUV positioning using modem data.

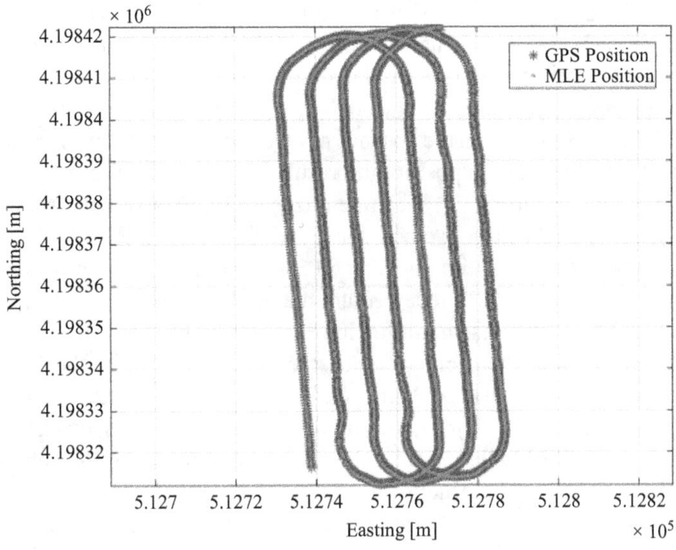

Figure 10.18 AUV motion trajectory during the modem positioning test

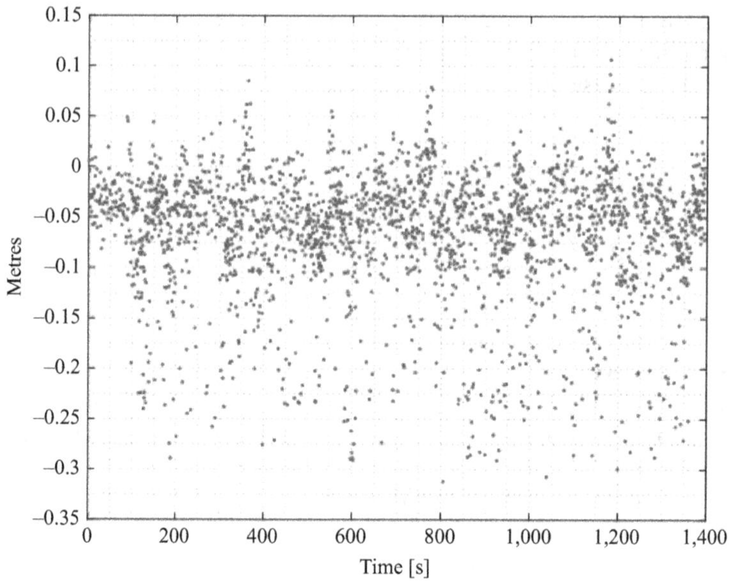

Figure 10.19 The difference between distance estimations towards a surface source using the modem and the GPS

As shown in Figure 10.19, the average difference is 4 cm. Non-zero difference can be explained by inaccurate accounting for sound speed variations in water (that varies with time and location). The standard deviation does not exceed 9 cm. In fact, the standard deviation represents the error estimate of AUV positioning with underwater modem data. It should be noted, that in reality the positioning error is less than the mentioned 9 cm. The AUV was half-submerged: the lower hull of Medusa AUV outfitted with the acoustic modem was underwater, whereas the upper hull with the GPS receiver was at the water surface. The AUV swell caused the lever arm between the modem and the GPS to vary. In the absence of lever arm compensation (by accounting for pitch and roll) the modem positioning results contained a certain error. Thus, the measured standard deviation (9 cm) characterizes not only the actual error of positioning by a modem with the atomic clock, but a sum of several errors: AUV positioning with modem data, AUV positioning with GPS and the error caused by variations of the lever arm between the GPS receiver and the modem, attached to the AUV.

Clearly, subsequent tests should include the possibility to account for the pitch and roll of the AUV at the moment when the modem transmits the GPS position data. Nevertheless, the measured 9 cm standard deviation indicates the potential for highly accurate positioning with UWA modem data (the centimetre-scale accuracy is comparable with GPS RTK), at least for operating distances of the vehicles within the WiMUST system, on the order of hundreds of metres).

The atomic clock drift over the whole measurement interval was around 2.5 μs per hour, conforming well with the results achieved in the lab and meeting the WiMUST system requirements.

10.4.3.4 Case study conclusion

The CSAC SA.45 miniature atomic clock was integrated with the electronics of a UWA modem, replacing the commonly used quartz clock and providing additional modem functionality. In particular, it enabled using the modem for highly accurate positioning/navigation of AUVs in acoustic silence mode, and as a source of timekeeping 1PPS signals for other AUV circuits.

Laboratory experiments allowed to develop practical recommendations for protocol maintenance and preparing the atomic clock for autonomous missions. In particular, when the atomic clock is disciplined every 3 weeks (without intermittent powering on and off), a short-term disciplining for 300–600 s is enough to refine the aging clock parameters. Afterwards, during an autonomous mission that lasts 8 h, the clock drift does not exceed 40–50 μs.

Experimental results regarding AUV positioning using the S2CR18/34OEM modem with integrated CSAC SA.45 atomic clock demonstrated the possibility of AUV positioning/navigation in acoustic silence mode with centimetre-grade accuracy (the RMS value) during a half-hour mission with an approximate clock drift of 2.5 μs per hour. This result meets the requirements of the WiMUST system and opens the way to solve multiple practical tasks that demand the deployment of large (scalable) AUV formations for long-term autonomous missions.

Acknowledgements

This work was conducted in the scope of the European Union's WiMUST project, Horizon 2020-ICT-2014-1 research and innovation programme, grant agreement No. 645141. The research was a collaborative effort of the WiMUST project partners:

- the Interuniversity Centre on Integrated Systems for the Marine Environment – ISME (Project coordinator) – Italy,
- Instituto Superior Tecnico, IST – Univ. Lisbon – Portugal,
- Centre for Technological Research of the Algarve – CINTAL – Portugal,
- University of Hertfordshire – United Kingdom.
- EvoLogics GmbH – Germany,
- Graal Tech S.r.l. – Italy,
- CGG – France,
- Geo Marine Survey Systems B.V. – the Netherlands, GeoSurveys – Consultores em Geofısica, Lda. – Portugal.

References

[1] EvoLogics GmbH. 2018. http://www.evologics.de.

[2] Kebkal K., and Bannasch, R., "Sweep-spread carrier for underwater communication over acoustic channels with strong multipath propagation," Journal of Acoustic Society of America, 2002, vol. 112(5), pp. 2043–2052.

[3] Kebkal O.G., Kebkal K.G. and Komar M. "Development of upper-layer protocols with S2CR acoustic modems emulator," Proceedings of the Conference on Underwater Communications: Channel Modelling and Validation, UCOMMS, 12–14 September 2012, Sestri Levante, Italy.

[4] Petrioli C., Petroccia R. and Potter J. "Performance evaluation of underwater MAC protocols: from simulation to at-sea testing," Proc. of OES/IEEE Oceans, Santander, Spain, June 2011.

[5] Masiero R., Azad S., Favaro F., *et al.*, "DESERT Underwater: an NSMiraclebased framework to DEsign, Simulate, Emulate and Realize Testbeds for Underwater network protocols," Proc. of IEEE/OES Oceans, Yeosu, Korea, June 2012.

[6] EviNS Framework, Internet resource: https://github.com/okebkal/evins.git.

[7] Kebkal O., Kebkal V. and Kebkal K., "EviNS: framework for development of underwater acoustic sensor networks and positioning systems," in OCEANS, 2015 IEEE – Italy, May 2015, pp. 1–6.

[8] Abreu P., Antonelli G., Arrichiello F., *et al.*, "Widely scalable mobile underwater sonar technology: an overview of the H2020 WiMUST project," Marine Technology Society Journal, Volume 50, Number 4, July/August 2016, pp. 42–53(12).

[9] Antonelli G., Cafaz A., Casalino G., *et al.*, "The Widely scalable Mobile Underwater Sonar Technology (WiMUST) H2020 project: first year status," Proc. OCEANS'16, Shanghai, China, April 10–13, 2016. DOI:10.1109/OCEANSAP.2016.7485587.

[10] Indiveri G., Antonelli G., Arrichiello F., *et al.*. "Overview and first year progress of the Widely Scalable Mobile Underwater Sonar Technology H2020 project, " Proc. 10th IFAC Conference on Control Applications in Marine Systems (CAMS'2016), Trondheim, Norway, September 2016. IFAC – Papers OnLine, Vol. 49, Issue 23, 2016, Pages 430–433.

[11] Pacini F., Paoli G., Kebkal O., *et al.*, "Integrated communication network for underwater applications: the SWARMs approach," 4th Underwater Communications and Networking Conference (UComms 2018), Lerici, Italy, 28–30 August 2018.

[12] Rahman A., Olesinski W., and Gburzynski, P., "Controlled flooding in wireless ad-hoc networks," Wireless Ad-Hoc Networks, 2004 International Workshop on. IEEE, 2004, pp. 73–78.

[13] Guerra F., Casari P., and Zorzi M., "World ocean simulation system (WOSS): A simulation tool for underwater networks with realistic propagation modeling," in Proceedings of the Fourth ACM International Workshop on UnderWater Networks, ser. WUWNet'09. New York, NY, USA: ACM, 2009, pp. 4:1–4:8. Pa3Me: http://doi.acm.org/10.1145/1654130.1654134.

[14] van Dam T. and Langendoen K. "An adaptive energy efficient mac protocol for wireless sensor networks," Proceedings of the first international conference on Embedded networked sensor systems, pages 171–180. ACM Press, 2003.

[15] Ye W., Heidemann J., and Estrin, D., "An energy-efficient MAC protocol for wireless sensor networks," Proceedings of IEEE INFOCOM, 2002.

[16] Chitre M., Shahabudeen S., and Stojanovic M., "Underwater acoustic communications and networking: recent advances and future challenges," Marine Technology Society Journal, vol. 42, no.1, pp. 103–116, Spring 2008.

[17] Jiejun, K., Jun-hong C., Dapeng W., and Gerla, M. Building underwater ad-hoc networks and sensor networks for large scale real-time aquatic applications. IEEE Military Communications Conference, 2005. MILCOM2005.

[18] Heidemann, J., Wei Y., Wills J., Syed A., and Yuan L. "Research challenges and applications for underwater sensor networking," IEEE Wireless Communications and Networking Conference, 2006. WCNC 2006.

[19] Azad S., Casari P., Petrioli C., Petroccia R., and Zorzi M., "On the impact of the environment on MAC and routing in shallow water scenarios," IEEE Proc. of MTS/OCEANS, Satnatder, Spain, June 2011.

[20] Otnes R., Asterjadhi A., Casari P., *et al.*, Underwater acoustic networking techniques, Springer Briefs in Electrical and Computer Engineering, DOI: 10.1007/978/3642252242.1, 2012.

[21] Syed A.A., Ye W., and Heidemann J., "Comparison and evaluation of the T-Lohi MAC for underwater acoustic sensor networks," IEEE Journal on Selected Areas in Communications, vol. 26, No. 9, 2008, pp. 1731–1743. DOI: http://www.isi.edu/~johnh/PAPERS/Syed08b.pdf.

[22] Peleato B., and Stojanovic M., "Distance aware collision avoidance protocol for ad-hoc underwater acoustic sensor networks," IEEE Communication Letters, 2007, pp. 1025–1027.

[23] Cavin D., and Schiper A., "Probabilistic broadcast for flooding in wireless mobile ad hoc networks," Proc. Conf. Wireless Communications and Networking, 2003. WCNC 2003. 2003 IEEE, vol. 2.

[24] Liang W., Yu H., Liu L., Li B., and Che C., "Information-carrying based routing protocol for underwater acoustic sensor network," International Conference on Mechatronics and Automation, ICMA 2007, pp.729–734, 2007.

[25] Wahid A., and Dongkyun K., "Analyzing routing protocols for underwater wireless sensor networks," International Journal of Communication Networks and Information Security (IJCNIS) Vol. 2, No. 3, December 2010, pp. 253–261.

[26] Lutwak, R., Emmons, D., Riley, W., and Garvey, R.M. "The Chip-scale atomic clock – coherent population trapping vs. conventional interrogation," Proc. 34th Annual Precise Time and Time Interval Systems and Applications Meeting, 3–5, December, 2002, Reston, VA, pp. 539–550.

[27] Abreu P., Antonelli G., Arrichiello F., *et al.*, "Widely scalable mobile underwater sonar technology: An overview of the H2020 WiMUST project," Marine Technology Society Journal 01/2016; 50(4):42–53.

[28] Kebkal O., Komar M., and Kebkal K., "D-MAC: Hybrid media access control for underwater acoustic sensor networks," Proceedings of IEEE Communications Workshops (ICC), IEEE International Conference, Cape Town, South Africa, 2010. ISBN: 978-1-4244-6824-9.

[29] Kebkal O., Komar M., Kebkal K., and Bannasch R., "D-MAC: media access control architecture for underwater acoustic sensor networks," In: OCEANS

2011 IEEE, 6–9 June 2011, Santander, Spain. IEEE Oceanic Engineering Society. ISBN 978-1-4577-0087-3.
[30] Gardner, A.T., and Collins, J.A., "A second look at Chip Scale Atomic Clocks for long term precision timing," OCEANS 2016 MTS/IEEE Monterey, Monterey, CA, 2016, pp. 1–9.
[31] Poddar, A.K., and Rohde, U.L., "Techniques minimize the phase noise in crystal oscillator circuits," IEEE International Symposium on Frequency Control (FCS), 2012.
[32] Rohde, U.L., Poddar A.K., and Apte A.M., "Getting its measure: oscillator phase noise measurement techniques and limitations," IEEE Microwave Magazine, vol.14, issue 6, 2013.
[33] Stofanik V., Sam I.A.E., and Minarik B.M., "Direct digital synthesizer clock frequency versus temperature dependency compensation using two look-up tables," Proceedings of the 2004 IEEE Joint 50th Anniversary Conference UFFC, pp. 770–774.

Chapter 11
Commercial applications of ASVs

Dan Hook[1]

Abstract

Autonomous surface vessels (ASVs) have seen a rapid rise in their technical progress and capabilities, use cases and associated business models. This chapter discusses the current and emerging applications in three key sectors; i.e., defence, commercial and scientific.

It includes a technology overview covering platforms, structures, energy systems, propulsion systems, internal sensors, external sensors, communications, processing, launch and recovery, auxiliary systems, remote operation centres and software.

An overview of ASV types is provided covering different sizes and configurations and the chapter concludes with a look ahead to the future and how ASV technology may be applied in large sectors such as transport.

11.1 Introduction

Historically, the development and adoption of autonomous surface vessels (ASVs) has not drawn the same attention as air, land or even sub-surface systems. However, in the last ten years it has been one of the fastest and most exciting areas of robotics and autonomy to observe and partake in. The topic is made even more exciting by the fact that there is still so much to do for providing benefits to safety, efficiency and productivity. The Maritime Environment and associated industries and operations are vast and there is always a new opportunity for robotic systems to make their impact. It is with this in mind that I would like to advise the reader that this chapter does not cover every application of the technology. The pace of development is so fast that I would predict that a new application is discussed and trialled within the timeframe that most people will read this book.

This chapter will focus on the commercial applications of Autonomous Surface Vessel (ASV) technology. These are wide in terms of complexity, value, water

[1]L3 ASV, Portchester, UK

depth, distance from shore and frequency of operation. A summary list of commercial industries currently utilising ASV technology is provided below:

- Hydrographic survey/charting,
- Offshore oil and gas,
- Offshore renewables (tide, wave and wind),
- Dredging and coastal construction,
- Maritime shipping,
- Water supply and monitoring,
- Security.

Although this chapter focuses on industrial applications, for completeness in discussing the topic I wanted to frame the commercial applications with the related defence and scientific applications. Many of the companies and research organisations involved with ASV development work in all of these sectors, and indeed, there is very significant overlaps between technologies, equipment and techniques.

11.2 Defence applications

To outline the importance of maritime autonomous systems (MAS) to the defence world I asked the current Royal Navy Fleet Robotics Officer to share his thoughts on this disruptive technology.

MARITIME AUTONOMOUS SYSTEMS IN THE ROYAL NAVY
Commander Peter J. Pipkin BEng(hons) MSc PgDip CEng MIET Royal Navy Fleet Robotics Officer, Maritime Capability

It would seem almost certain that a future Royal Navy will deliver significant areas of capability through unmanned and autonomous systems. These technologies are already proliferating in our everyday lives, in the workplace and the home. They offer disruptive and transformational opportunities to a multitude of tasks, from self-navigating vehicles, be they cars, aircraft or boats, to autonomous algorithms processing big data at speeds and efficiencies that people could not hope to replicate. They enable better access to dangerous environments and they can work continuously for hours, days, weeks or even months or years in an increasing number of systems. All of these attributes have the potential to deliver varying degrees of efficiency and effectiveness into Royal Navy operations. The challenging questions are, which operations, to what breadth and how soon?

To try and answer some of these questions the Royal Navy hosted the Unmanned Warrior demonstrations in October 2016. Unmanned Warrior was a collaborative event conceived, planned and delivered by Defence, Industry and Academia that saw over 40 participating organisations operating approximately 50 systems in the realistic and demanding environment of the Joint Warrior exercise programme. MAS were deployed in the air, on the surface and under the water across a spectrum of operational tasking in order to rapidly develop a better understanding of where MAS offers credible capability choices for the Royal Navy.

The breadth of demonstrations at Unmanned Warrior spanned a number of maturity levels. At the higher technology readiness levels the aim was to investigate, understand and develop where and how systems could support current operations. Intelligence, Surveillance and Reconnaissance is a good example of where multiple configurations of platforms and sensors were widely tested. At the conceptual level the ambition was to challenge dogma and doctrine by going back to the first principles and examining how MAS could enable a completely different approach. Such transformation has the potential to change the paradigm of like-for-like, evolutionary capability replacement that currently exists and arguably that hampers Defence's ability to exploit rapidly evolving technologies more easily and quickly. Importantly, this was all done in the collaborative environment that had existed since day one, with all parts of the 'enterprise' contributing and recognising the considerable value in doing so.

A significant part of the Royal Navy's approach to MAS is increasingly focussed on the human element too. Unmanned, by definition, removes the human further from the point of operation or delivery. That does not mean people are not involved; for example, it is UK policy that the operation of weapon systems will be under human control. In a dangerous environment, such as mine countermeasures, this is clearly a good thing and will ensure that our people are less exposed to activities that might result in casualties. Unmanned systems can be designed better for their intended tasks without the costs and constraints required of operator protection. They can be operated in more hostile and extreme environments without the risk to personnel and they can be left to operate without getting distracted or fatigued. But the adoption of MAS is not about replacing the officers and sailors that are still the Royal Navy's greatest asset; people will still be needed. Some of them will be tasking, operating or exploiting unmanned vehicles, more and more of them will be assisted by autonomous systems, particularly where autonomy can enhance and support data processing, understanding and decision making in a world where the information workload is increasing at a rate that humans cannot keep pace with. Defence is not alone in this regard and the Royal Navy plays an active part in contributing to the wider debate on the conceptual, ethical, legal and regulatory aspects of responsible MAS operation.

Unmanned Warrior was one step on the path to the Royal Navy's adoption of MAS. Of its many successes, the most important was the clear evidence that MAS is going to be a fundamental component of the future Royal Navy.

The applications which are currently in service, being developed, demonstrated or designed are discussed in Sections 11.2.1–11.2.6.

11.2.1 Mine counter measures (MCM)

The MCM community was quick to see the benefits of ASVs and in particular, the advantage of keeping 'the man out of the minefield'. A significant number of ASVs have been developed and proven in a range of MCM roles including sweeping (towing an influence sweep system to trigger the mine), hunting (towing a sonar system to find the mine) and disposal by launching a one-shot or re-usable mine disposal remotely operated vehicle (ROV) to destroy the mine. An additional ASV role has been to act as a surface communications and positioning gateway for Autonomous Underwater Vehicle (AUV) mine hunting operations.

11.2.2 Submarine warfare (ASW)

At the time of writing, this application is still very much in the research, development and initial testing phase. ASVs have been deployed with active dipping sonars and passive thin line towed array hydrophones (TLTA). A project of particular note is the USA's Defense Advanced Research Projects Agency (DARPA) funded ASW Continuous Trail Unmanned Vessel (ACTUV) project, which consists of a large (132' long) unmanned surface vehicle (USV) equipped with a wide range of ASW sonars.

11.2.3 Command, Control, Communications, Computers, Information/Intelligence, Surveillance, Targeting Acquisition and Reconnaissance (C4ISTAR)

A large number of ASVs have been deployed in the C4ISTAR role. A wide range of payloads including Electro-optics (day and night), radio surveillance and communications have been fitted to typically high speed ASVs but also some low speed, low signature environmentally powered systems. A limited number of these vessels have been weaponised.

11.2.4 Targets and training systems

Live fire targets is one of the oldest applications for unmanned boats with early systems dating back to pre-World War II. Unmanned boats ranging in size from approximately 3 to 15 m have been used in live fire weapon trials and training exercises. The targets can carry augmented signature equipment and operate either individually or as swarms. Many of the unmanned air vehicle (UAV) systems have historic links to target applications and the same is partly true in the ASV community.

11.2.5 Met-Ocean (METOC)

ASVs have been operated to support Met-Ocean activities by carrying Acoustic Doppler Current Profilers (ADCPS), Conductivity Temperature and Depth (CTD), dissolved oxygen, coloured dissolved organic matter (CDOM), chlorophyll and a host of other sensors. There is a significant overlap between the scientific, commercial and defence applications in this role.

11.2.6 Rapid environmental assessment (REA)

In addition to the METOC sensors discussed above there have been several ASVs deployed with sonar and camera systems for conducting REA. In this application the ASV can enter an area, conduct a survey and transmit real-time results back to the

Military Commander. A good example of this is an ASV system that was used for measuring the gradient of beaches, wave heights and bottom type ahead of amphibious landings.

11.3 Scientific applications overview

To outline the importance of ASVs/USVs to the scientific community I asked Professor Russell Wynn of the National Oceanography Centre in the UK to give an overview. He has been very active in the use of marine robotics in leading edge scientific research.

USVs and SCIENCE APPLICATIONS
Professor Russell B Wynn
National Oceanography Centre

The global scientific community is increasingly looking to USVs as a potential solution to key challenges in oceanographic research, including long-term measurement of ocean circulation and surface weather conditions (and their responses to climate change), as well as more targeted applications, such as water quality and marine protected area monitoring. This focus on unmanned systems is partly due to the increasingly high cost of operating traditional research vessels, and also the urgent need for more spatially and temporally extensive data in the under-sampled ocean environment.

Used in isolation, USVs can provide real-time data from the critical interface between ocean and atmosphere, allowing scientists to better constrain fluxes of heat and CO_2 between these two media. By combining USVs with underwater and aerial unmanned vehicles, it will be possible in the future to look at these and other fluxes from the seabed to the lower atmosphere, potentially providing a step-change in our ability to accurately model and predict global heat and carbon budgets.

When used in combination with satellite remote sensing and ocean model data, USVs can be tasked to target areas of high biodiversity that are of interest to conservation scientists, such as oceanic fronts that divide water masses of different properties. As well as being able to make direct measurements of parameters that validate frontal location (e.g., sea surface temperature and chlorophyll fluorescence), USVs can directly image marine biota such as seabirds and marine mammals, as well as marine macro-litter and manned surface vessels. They are also able to carry passive acoustic monitors that are effective at detecting presence or absence of echo-locating marine mammals, as well as levels of background oceanic noise that include both natural noise (sediment transport, wave action) and anthropogenic noise (e.g., shipping traffic and seismic surveys).

The above capability means that USVs are a useful tool for monitoring of offshore Marine Protected Areas (MPAs). In addition to biodiversity monitoring, recent proof-of-concept studies have demonstrated their utility for long-range surveillance missions to monitor illegal and unregulated fishing, providing validation of satellite remote sensing data.

Current USV developments that are of value to the science community include increased capability to undertake direct sampling while at sea, for example, through samplers mounted on the vehicle or on a winch system to enable vertical profiling, and increased inter-operability with other unmanned platforms, e.g., the ability to launch small submarine or aerial vehicles from USVs to investigate under-ice or over adjacent land.

ASVs are being utilised in a varied range of scientific research applications. These include everything from open ocean transits to inland waterway and riverine research. As scientists gain confidence in unmanned systems they are considering them for ever more challenging tasks and integrating more complex payloads. They are also increasingly utilising unmanned systems to make up for a short fall in available ship time.

11.3.1 Examples of scientific applications

Ocean profiling: In this application the USV is configured with either a hull or winch mounted CTD probe. It is then sent to transit along a route measuring and storing data onboard and often transmitting it to shore via a communication link. This capability has been proven from a range of different types of ASV including both diesel and electric powered and wave/wind powered systems.

Ocean current monitoring: ADCPS are available in a wide range of frequencies to suit for measuring the water current (speed and direction) in different depths of water. An appropriate ADCP is integrated onto an ASV with a GPS system and usually able to log data onboard and transmit results or data packages to shore.

River current monitoring: This application is usually undertaken by smaller, electric powered USVs with hull mounted ADCP systems. It allows for the volumetric flow rate of the river to be accurately measured and often in areas where it would have been too dangerous or logistically challenging to attempt in a manned vessel.

Fish, mammal and birdlife baseline studies: ASVs can provide a quiet, unobtrusive and long endurance platform for the deployment of sensors to assess the populations of fish, marine mammals or birdlife. ASVs have been deployed with acoustic monitoring equipment to detect marine mammals, cameras to detect birdlife and sonars for fish stock assessment. The USV can be deployed to ‘hold station’ in one location or patrol on a pre-planned route dependent upon the requirements of the scientist. This type of deployment is often conducted before, during and after marine construction works.

Fish tracking: ASVs have been utilised as tracking vessels in fish behavioural studies where fish have been caught, tagged and released with acoustic transmitters. The relatively (compared to ships) low noise ASVs are able to hold station, cover search grids or even follow individual fish.

Tracking and communication with underwater vehicles: Scientists have seen the value in deploying unmanned ocean buoyancy gliders and self-propelled AUVs for a wide range of tasks. By pairing the underwater vehicle with a surface vehicle they can significantly enhance their positioning and data transfer capability. This is a particularly interesting area of development and one that I am confident will deliver some disruptive capability over the coming years.

There exists a strong 'eco-system' between scientists, sensor suppliers and USV developers. This relationship is generating a rapid pace of application development and the number of regular deployments is likely to increase significantly over the coming years.

11.4 Technology overview

A modern ASV is not just the physical vessel but a tightly integrated engineering system comprising of the physical platform, communication link(s), external and internal passive and active sensors, processors, energy supply and management, launch and recovery, remote operation consoles/computers, auxiliary support systems and importantly a structured, safe and effective software solution.

An ASV system is only as strong or reliable as the weakest link in this chain of systems and components so each needs to have been developed, tested and proven in the harsh marine environment. As ASVs are put in to ever more challenging environments and further from intervention support, the skills and knowledge of the engineers and scientists within the industry have to grow fast.

The systems are developed with rigorous design standards, sub-component/component and system level testing and in many cases full system redundancy.

The systems currently available on the market are from a wide range of suppliers and each have their own detailed system arrangement. However, it is possible to simplify the system diagram to one that is valid for all systems and this is shown in Figure 11.1.

In the following sub-sections I have provided a brief overview of the systems and technology currently used in the USV industry.

Platforms: A very large number of hull forms, arrangements and sizes of USV have been constructed. These have included systems from approximately 1 m in length up to 90 m at the time of writing. The majority of systems have used a mono-hull configuration but systems have also been built on catamaran, trimaran, semi-submersible and other more niche hull forms. Many of the systems are constructed

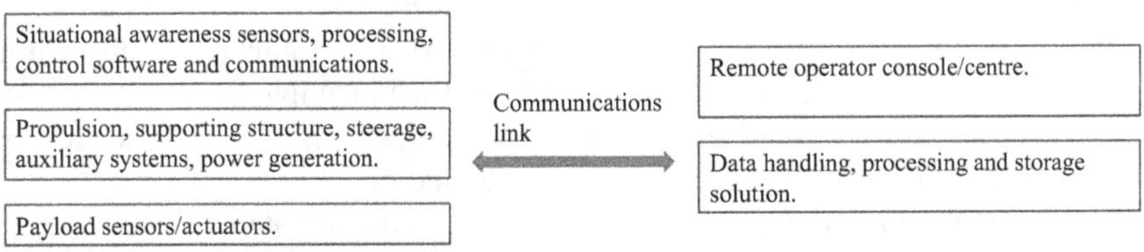

Figure 11.1 High level, common system overview

as dedicated unmanned platforms but some are based upon conversions or upgrades of existing craft designs.

Structures: All of the ASVs are built around a buoyant structure and the materials have included aluminium, steel, poly-ethylene, glass re-enforced plastics, advanced composites and inflatable materials. The structural material is generally selected to suit the budget, construction volume/number, the required ruggedness, the performance and weight constraints.

Energy systems: A large number of the systems to date have relied on diesel engines as their primary source of power. These typically drive the propulsion system directly and have alternators for electrical power generation. In some cases the diesel engine acts as a dedicated electrical generator which then powers electric motors for propulsion. Some systems have been powered by outboard petrol engines and others by electrical batteries powering electric motors. There are a significant number of energy harvesting craft and these use either photovoltaic solar panels, wind turbines and batteries as their source of electrical power and/or propulsion. It is becoming increasingly common to find hybrid arrangements of two or more of these energy sources installed on a USV.

Propulsion systems: Many ASVs have a high speed requirement and others a high tow force requirement, whilst some need to stay at sea for months at a time and make low noise emissions. The range of propulsion systems used varies to meet these requirements and has included submerged propellers, surface piercing propellers, waterjets, sails and rigid aero-foils, wave propelled foil arrangements and even air-propellers.

Internal sensors: For a remote operator to evaluate system performance, make decisions and find fault it is necessary on most ASVs to fit a range of internal sensors. These sensors are connected to the vessels command and control system and typically used to trigger alarms or effect the vessels actions. Typical examples include voltmeters, current meters, temperature sensors, pressure sensors, leak sensors, flow meters and strain gauges. It is expected that onboard health monitoring, predictive maintenance and fault tolerant design will be a large area of development in ASVs over the coming years.

External sensors: For safe and effective navigation the system needs to know where it is, which direction it is facing, what speed it is travelling at (over ground and sometimes through the water) and what static and dynamic objects are surrounding it. It may also need to know about its environment (temperature, humidity, water depth, wave height and wind speed) to make decisions onboard or feedback information to a remote operator to use in decision making. These external sensors vary significantly between the types of platform but typically include a Global Positioning System (GPS), magnetic compass, motion reference unit (MRU), Automatic Identification System (AIS), radar(s), daylight camera(s) and thermal imaging camera(s). Systems have also been built with multi-parameter meteorological sensors, light detection and ranging (LIDAR) and an ever-increasing range of additional sensor types.

Communication systems: The type of communication systems fitted to a USV depend very much upon its intended operation. The system installed will need to provide the required range, bandwidth, security and backup or redundant features.

Table 11.1 Typical communication systems

	Line of Sight	**Beyond Line of Sight (BLOS)/ Over the Horizon (OTH)**
Lower bandwidth	Serial Modem (UHF/VHF) General Packet Radio Service (GPRS)	HF Radio Iridium Short Burst Data Modem (SBDM) Iridium Dial-up/Router-Based Unrestricted Digital Internetworking Connectivity Solutions (RUDICS)
Higher bandwidth	L-Band IP Radios 4G Long Term Evolution Radio (LTE)* S-band/Wi-Fi	L-Band Satcom Ku-Band Satcom Ka-band Satcom

*Note: This network is well established in some locations and allows over the horizon communications from operator to vessel.

This often requires the installation of at least two communication systems that could be used for different parts of a mission. Communication systems can be discussed in terms of those which provide 'line of sight' range and those which provide 'over the horizon' connectivity, such as satellite communications and very low frequency radios. Within these two groups there are then variations in bandwidth ranging from systems capable of only a few bytes per second to systems capable of relaying multiple channels of high definition video and payload data. Table 11.1 provides an overview of some of the options currently seen on ASVs.

Processors: All commercial ASV's rely on onboard processing to undertake control of critical functions such as propulsion, navigation and communications. There are a range of approaches to onboard processing that are all coupled to the software architecture and required function and intelligence the ASV needs. These different approaches can be loosely grouped into three types, low processing capacity embedded systems, high processing capacity application based systems and hybrid systems using both low and high capacity processing.

For the vast majority of ASV's the hybrid approach is used, often with an low capacity embedded processor being responsible for the basic control and interfacing of sub-systems and the high capacity processor such as an ARM Cortex A or X86 being used for sensor processing, mission management or higher level self-navigating autonomous systems. The rapid development of ever more powerful and energy efficient processors has been instrumental in enabling unmanned systems to proliferate in recent years (Figure 11.2).

Launch and recovery: Nearly all of the USV systems up to approximately 20 m in length will at some stage during their regular deployments be required to be launched and then recovered from the water. This is a critical part of the system and any operation, and has impacts on operational efficiency, safety and overall effectiveness. Some very good ASV systems have failed to be adopted because of a weak link in the Launch and recovery part of the system.

Figure 11.2 L3 ASV ASView-Base, an example of a base station containing radios, processors and power supplies

It would be possible to write a whole book on the different types of Launch and Recovery (LAR) systems that are in use or development. Most of the successful systems include a means of safely lowering/sliding the ASV in to the water and then remotely releasing it without the need for a manned support boat or divers. Likewise the best systems are then able to re-connect the ASV to a recovery system without the need for a support boat. All systems need to be designed and tested to appropriate design standards/regulations and effective training provided to the operators.

Auxiliary systems: This is a very wide ranging and 'catch all' title to cover the important sub-systems of any ASV that carry and distribute fuel/energy, provide cooling, bilge pumping, electrical distribution, hydraulic actuation and so forth. It is particularly important on unmanned platforms that these auxiliary systems are designed and built to a high standard and capable of remote monitoring, diagnosis and control.

Remote operation consoles/computers: Even highly autonomous systems will require human interaction at some point. The control interface for this will involve a hardware element and software element with some form of Graphical User Interface (GUI).

The hardware elements could include any of the following:

- A tablet or smartphone,
- A 'belly-pack' for manual operations (with or without a screen for a GUI) (Figure 11.3),
- A rugged laptop (with or without additional screens),
- An industrial desk or rack mounted computer with operator consoles and screens,
- A control room with multiple configurable screens and operator stations.

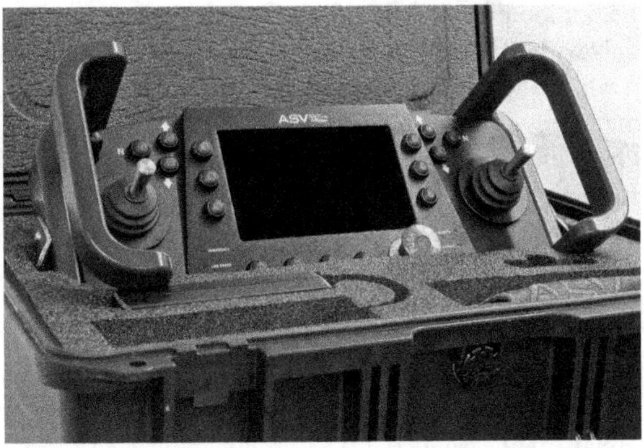

Figure 11.3 L3 ASV ASView-Helm, an example of a remote operator control interface or commonly called 'BellyPack'

Control software: This is a critical component of any system and is certainly where much of the important and valuable intellectual property (IP) is being generated within the industry.

Other chapters within this book cover some of the detailed algorithms, architectures and developments being produced within this very active domain.

The complexity of the software within ASV systems varies considerably from small, remotely operated systems up to autonomous systems with many sensor inputs and control function outputs. In general most organisations involved with ASV development have adopted a modular and scalable approach to allow them to continue adding functionality into the future.

Functional payloads: The payload is the term commonly used within the unmanned industry to describe the sensors/actuators/mechanism/cargo that the unmanned system is carrying and which actually performs the task or function of the system.

I will discuss some of the specific payloads in the applications sections of this chapter but in general they all need to be integrated with the ASV and consideration made for their effect upon stability, trim, drag, structural loads, electrical requirements, data bandwidth and storage requirements and control.

11.5 ASV types

In the previous section I discussed some of the technologies and sub-systems used within ASV platforms. In this section I will show some particular examples of systems that either exist and are operational or are likely to be in the near future.

The term Autonomous Surface Vessel (ASV) or alternatively USV has been used to describe a very wide range of platforms. These range in size from 1 m long or less up to approximately 90 m at the time of writing and this is likely to increase to over 200 m in the not too distant future. They vary in complexity from simple direct remote control only with a basic payload sensor to systems capable of full autonomous operation with a wide range of complex payloads onboard.

There are many possible ways to group them but for simplicity I have chosen overall length. I have included some photographs of example systems and I thank the manufacturers for their support. These systems have been selected as examples, and there are many similar systems available from other suppliers which are not included here.

11.5.1 Small systems

There are a very large number of small systems available on the market, some of them as commercial off the shelf (COTS) products but also those constructed by organisations bespoke for their own use. By small I refer to systems in the 0.5–3 m long range. These are systems that can generally be man-handled, transported in pick-up trucks and vans and very unlikely to increase risk to other seagoing vessels.

They tend to be either electric/propeller driven or making innovative use of the environment for wave or wind propulsion.

They are generally too small to carry large and high bandwidth satellite communications equipment so primarily rely on radio communications (Line of Sight), 4G LTE (coastal) or low bandwidth Iridium type satellite communications.

Their payload size, power and mass is limited but generally capable of carrying a range of shallow water sonar systems, water sensors, smaller high frequency ADCP units, cameras and basic meteorological sensors.

Some example systems are described in Figures 11.4–11.7.

Figure 11.4 Ocean Science Z-boat (electrically propelled, multi-role monohull)

Figure 11.5 Autonaut USV

Figure 11.6 Liquid Robotics Waveglider (wave powered surface float with sub-surface tethered unit)

Figure 11.7 L3 ASV C-Cat 2 (electric waterjet powered multi-role catamaran)

Figure 11.8 L3 ASV C-Worker 4 (diesel electric powered monohull with large moonpool for payload modules)

Figure 11.9 L3 ASV C-Worker 7

11.5.2 Medium systems

The 'medium' size class of vessels covers a very wide range of systems, applications, complexities and price points. These are generally systems that can be moved around the world by trailer/cradle and launched/operated from large ships or the dockside but take cranes and experienced personnel to do so. They would range in length from approximately 3 m+ up to 24 m in length.

They are capable of carrying a wide range of sensors and communication equipment allowing for not only direct manual control but also highly automated and even autonomous open ocean operation.

Figure 11.10 Atlas Elektronik Ltd ARCIMS (diesel waterjet powered catamaran for a range of defence applications)

Figure 11.11 L3 ASV C-Sweep 'Halcyon' variant designed and built for Thales UK (twin diesel engine propelled monohull developed for a range of MCM and ASW applications)

Many of the systems in this category are provided as multi-role or re-configurable platforms capable of carrying a very wide range of different payloads for different tasks.

Some example systems are described in Figures 11.8–11.12.

11.5.3 Large systems

Large systems include anything above approximately 24 m in length and have been one of the most active discussion and conference topics in marine autonomy over the

Figure 11.12 *Sail drone (wind powered trimaran)*

Figure 11.13 *Kongsberg and Automated Ships Ltd 'Hronn' – (Unmanned Offshore Supply Vessel Catamaran)*

Figure 11.14 *Rolls Royce – concept picture*

Figure 11.15 Space X – Autonomous Spaceport Drone Ship

last three years. At the time of writing very few systems have been built but proposals include custom designed and special purpose systems and conversions or addition of capability to existing vessels.

These large systems will be able to transit many thousands of miles, carry large payloads and an almost unbounded range of sensors and communication equipment.

Some example systems are provided in Figures 11.13–11.15.

Rolls Royce – (Unmanned tugs, container ships, cargo vessels and ferries. The concept picture shown here is of collaborative fleet of container vessels where possibly only the front one is crewed)

11.6 Industrial applications

Autonomous Surface Vessels (ASVs) have been undertaking a wide range of commercial applications since the early 2000s. Many of the deployments have been in a trials role but increasingly these are becoming regular operations with the commercial and contractual pressures that entails. This has driven reliability upwards and encouraged developers to introduce systems that allow tasks to be completed quickly, efficiently and profitably.

There are many different ways of categorising the industrial applications. I have divided them into proven applications, those in development and future applications. This is a fluid list and many readers may have knowledge that upgrades or downgrades some applications but to my knowledge in 2017 this is where the industry stands.

11.6.1 Proven

Applications in this category are those that have been completed by USVs, more than once and in a commercial environment. They are applications that will be repeated over and over and even more successfully as the systems continue to develop.

11.6.1.1 Inshore survey

This is a shallow water task and usually entails a shore launched USV conducting bathymetric survey with either a Single Beam Echo Sounder (SBES) or Multi-Beam Echo Sounder (MBES) and occasionally Side Scan Sonar (SSS). The vessel will be equipped with a decimetre accurate DGPS or DGNSS system and operated usually in a mixed mode between manual control and pre-programmed lines. Dependent upon the system the data are either logged onboard, transmitted and stored on the shore in near real time or both. Smaller USVs (<3 m) are generally used in this application.

One benefit of USVs in this application is that they allow areas to be surveyed, which may be difficult or dangerous to get conventional manned survey boats to.

11.6.1.2 Hydrographic coastal mapping

This application is to collect bathymetric data for a wide range of uses including navigation charts, pre-construction surveys and seabed monitoring. By coastal it typically includes waters within a countries Exclusive Economic Zone (EEZ) which can reach out as far as 200 nautical miles from shore. This means a wide range of water depths and sea conditions will be experienced. There have been some very successful survey operations using USVs in this role and they have either run in parallel with a manned survey vessel (force multiplier), from a shore control station or from another ship engaged in a different task (e.g., geotechnical survey). A slightly larger USV is used in this application (typically 4–8 m) and is typically equipped with an MBES, DGPS, sound velocity (SV) sensor, survey grade MRU and gyro. Dependant on the client's requirement, additional sensors such as a SSS, magnetometer and sub-bottom profiler (SBP) can be integrated to build a more comprehensive geophysical dataset.

11.6.1.3 Sediment/ plume tracking

During many marine construction or dredging tasks there is a requirement to monitor the density and distribution of disturbed sediment in the water column. This can sometimes be undertaken with fixed buoys or conventional boats but USVs offer an easily reconfigured and low impact option. The ASV would be equipped with an appropriate turbidity sensor interfaced to a logging, monitoring and alarm system. The size and type of USV required is primarily driven by the endurance requirement coupled with the local tidal/current conditions.

11.6.1.4 Current monitoring

An ADCP can be deployed on a USV with a DGPS system and used to measure water current throughout the water column. This is used in a variety of industries including siting tidal turbines, monitoring current during construction and de-commissioning tasks and assessing outflows from rivers and dams.

The size of ADCP is linked directly to its operating frequency and water depth capability. In turn this will determine the size of USV required along with the usual endurance, speed and sea state factors.

A particularly important example of this application has been monitoring the strong eddy currents that can occur seasonally in places such as the Gulf of Mexico. These violent and sometimes unpredictable currents can cause dangerous conditions for tasks, such as drilling and pipeline installation. The conditions can become so hazardous that it is sometimes necessary to uncouple Floating Production Storage and Offloading (FPSO) vessels at significant cost. A USV can be a cost effective method of monitoring these currents and providing early warnings or equally importantly avoiding false alarms and wasting money.

11.6.1.5 Water quality monitoring/sampling

A USV can be equipped with wide range of sensors, either hull mounted, pole mounted or on a deployable winch. The systems can be commanded to hold station, follow pre-programmed routes or even in some cases undertake advanced search patterns to find sources and concentration gradients. These capabilities along with their own minimal impact on the environment make them ideal for monitoring water quality and properties. This type of work is completed in a commercial sense in a wide variety of locations including river outlets, factory and power station outlets, lakes and reservoirs and increasingly within commercial ports where pollution and ballast water processes are a hot topic.

11.6.1.6 ROV tracking

An Ultra Short Baseline (USBL) acoustic positioning system can be used to position an ROV accurately to within just a few centimetres dependant on water depth. The USBL head is typically installed on a purpose built, hull mounted, retractable pole on a large construction support vessel. In some applications the vessel is not the one from which the ROV is deployed but actually a secondary vessel dedicated to the positioning/tracking task. USVs have been proven as able to deploy a USBL system and track ROVs. The smaller size and lower background noise (self-noise) of the USV ensures more stable acoustic performance and a significant cost saving over a secondary ship. A typical application is touchdown monitoring in deepwater pipe laying operations.

11.6.1.7 AUV tracking and communication

This application is very similar to ROV tracking as the USBL system integrated on a USV can be equally effective in tracking untethered AUVs. This particular application can offer significant commercial advantages as it can 'free' the AUV from the support ship and allow the AUV to go and perform geophysical survey or inspection work, typically on pipelines, whilst the support ship undertakes other tasks. An integral part of this system is the communications link in the form of an acoustic modem, often integrated in the USBL head, which can reliably transfer telemetry and selected payload data up from the AUV and send commands down to the AUV in real time. It should be noted that USV and AUV control stations are located on the support vessel and must be in communication range of the USV making this set up ideal for infield operations.

11.6.1.8 Compatt/transponder calibration

An alternative to the USBL positioning solution is the deployment of a Long Base Line (LBL) transponder array which subsea assets can use for accurate positioning in deeper waters. When these transponders are first installed to the seabed they themselves need accurate positioning. This is completed from the surface using a USBL system in an exercise commonly referred to as 'boxing-in'. The surface USV is equipped with a USBL system, high accuracy DGPS and survey grade MRU and gyro and then operated in a circle of a radius from the transponder typically 2 to 3 times water depth. The exercise is repeated in the opposite direction and can also incorporate position fixes at 4 cardinal points around the transponder to derive an accurate real world position. The USV is very efficient at this task as it can be programmed to accurately complete the positioning circles, has a low self-noise allowing for more stable acoustic performance and higher accuracy measurements. This frees up the ship to lay more transponders or complete other heavy tasks rather than having to complete positioning work.

11.6.1.9 Acoustic modem/data transfer

Not all subsea or seabed structures are connected to the shore with data cables. Some systems are standalone and have a regular data transfer/download requirement. Several different types of USV have been proven to be very effective in this role. They are able to accurately station keep above the subsea structures data modem and 'harvest' the data. As in other applications this data transfer rate can be higher than on a ship due to the low self-noise and positioning accuracy. The data flow can be bi-directional, which allows for settings/controls to be adjusted on the seabed and this has been demonstrated in near real time via a satellite link on the USV.

11.6.1.10 Open ocean profiling

USVs, particularly the wave propelled type, have been proven in conducting long distance open ocean profiling missions. They can be equipped with surface hull mounted, towed or winching sensors and the data both logged onboard and transmitted back to shore. This application is conducted frequently in the oceanography and scientific research domains but also in some commercial applications where organisations want to understand the physical oceanography in a certain area.

11.6.1.11 Passive acoustic monitoring (seismic, construction, decommissioning)

The impact upon marine mammal life from sound in the ocean is slowly becoming better understood. This is leading to requirements for limitations on the sound emitted and also for monitoring of the sound propagation and monitoring for the presence of marine mammals. In some circumstances to achieve the required detection sensitivity and warning radius the hydrophones need to be located at some distance from the source. USVs have proved to be an ideal way of achieving this, they can tow a passive acoustic monitoring (PAM) array and act as escorts to seismic vessels or perimeter patrollers for offshore construction tasks such as wind farm turbine piling.

11.6.1.12 Fish stock assessment and tracking, fish stock management

Aquaculture is becoming an increasingly important solution to feeding the worlds growing population. Of equal importance is optimising current fishing practices and research to ensure that areas are not over fished and that correct quota volumes can be set. USVs have been proven as ideal platforms to deploy a range of fish stock assessment and fish tracking equipment. In general, the USV is smaller in displacement and installed power than the equivalent fishery research vessel and therefore, has less impact on fish behaviour and less impact on measured results.

11.6.1.13 LIDAR scanning

LIDAR works on similar principles to Radar and sonar but uses a laser. The modern survey grade LIDAR systems when coupled with a GPS and MRU can build very detailed 3D point clouds of structures above water. This capability has been installed on USVs, which have then been able to navigate around very accurate paths and build high resolution models for engineers to evaluate. This has been used on structures such as dock walls, pontoons, ships and bridges and is likely to grow in to a commonly used tool for offshore structure inspection. A useful addition to this sensor is to add a multi-beam sonar to the USV and by careful planning of tides it is then possible to build a seamless 3D dataset of the structure above and below water.

11.6.1.14 Surface hydrocarbon monitoring

An industrial application is to monitor for the spill or leakage of hydrocarbons. One method for this is to deploy sensitive hydrocarbon 'sniffing' sensors on the surface and patrol along pipelines and around platforms. These sensors can be easily and falsely triggered by normal emissions from ships, so are much more suited to deployment on very low power or even passive USVs, floats and buoys. The additional advantage of a USV in this role is that the lower day rates of operation significantly reduce project costs where long-term monitoring is required over a large area.

11.6.2 In development

There are many USV research and development projects underway to offer new industrial applications. Many of these are likely to generate in service functional systems within a few years and some of these includes the following.

ROV deployment, operation and recovery: A large-scale USV capable of transporting, launching, operating, recovering and then re-launching a commercial ROV. There will be several different manufacturers supplying solutions like this with some variety in size range covering everything from small inspection or 'eyeball' ROVs in relatively shallow water through to large, work class deep water systems. There are several technical challenges presented by not having the human ROV operations team onboard the launch vessel but these will soon be overcome by the multi-disciplinary engineering teams working on them.

This capability will provide an attractive solution to operators wanting to complete subsea work without the additional cost of a support ship. The real economic advantage will be achieved when the complete system can operate over the horizon on a satellite communications link.

AUV deployment, operation and recovery: In a similar way to the ROV system above there will be commercial USV applications where single or multiple AUVs will be operated from an ASV system. The commercial AUVs range significantly in size from low power, almost disposable systems of only a few kg displacement right up to deep water inspection systems weighing several tons. A particular challenge is for the USV to be able to recover not only a live, functioning AUV but also one that has run out of battery or suffered a technical fault.

11.6.3 Future

Anybody attending unmanned systems conferences, reading industry magazines or technology blogs will be in-undated with a large number of future applications for unmanned systems. USVs will form a large part of the future marine industry. Just some of the exciting commercial applications they will be undertaking are described below.

Transport: There have been several studies in to using unmanned surface systems for transport. This has included smaller scale systems on lakes/rivers/canals right up to full size, unmanned container and bulk ships from organisations such as Rolls Royce, DnV and Wartsila. The utilisation of unmanned technology in transport will allow for a reduction in at sea labour costs and create high technology shore-based jobs. This will have the advantage of making some very low speed transits, a commercial opportunity for some cargoes and routes. It will allow for the removal of some weight, volume and equipment costs from the ship by removing accommodation requirements and this in turn could lead to alternative and more efficient ship designs.

Tug boat: There are some tug boat operations which are inherently high risk that could benefit from remote operations and unmanned control. There are also some complex manoeuvres where it is possible that an unmanned system will be more accurate and efficient leading to fuel and time savings.

Ferries: There is a Rolls Royce program underway to develop an in-shore ferry that is able to navigate between pick-up and drop-off points automatically and crew on board will be for passenger service and safety.

The list of future applications goes on and could include open ocean deepwater search and survey, firefighting, oil spill response, life-saving, mineral reclamation, dredging, ocean clean-up of litter and more. These are exciting times within the industry and I would encourage the reader to consider how their technology, area of work or problem could be impacted by USVs.

11.7 Conclusion

Represented within this chapter are just some of the applications being considered for ASVs. As system developers and users continue to gain confidence and add further functionality this range of applications will continue to grow.

A particular positive trend that I have noticed is the increasing number of payload/sensor and equipment suppliers offering components which have been specifically designed for unmanned operations. This new breed of lower power, compact and more open architecture components will also increase the number of new systems hitting the water and completing real-world work.

I would encourage readers of this book to consider how unmanned marine systems could affect their business, research area or future plans. The technology is moving very fast and offers a disruptive opportunity to reduce costs, improve safety, improve data quality and volume.

Index

www.ingramcontent.com/pod-product-compliance
Lightning Source LLC
LaVergne TN
LVHW011842200926
848484LV00002B/391
* 9 7 8 1 8 3 9 5 3 0 2 9 6 *